SET-VALUED MAPPINGS AND
ENLARGEMENTS OF MONOTONE OPERATORS

Optimization and Its Applications

VOLUME 8

Aims and Scope
Optimization has been expanding in all directions at an astonishing rate during the last few decades. New algorithmic and theoretical techniques have been developed, the diffusion into other disciplines has proceeded at a rapid pace, and our knowledge of all aspects of the field has grown even more profound. At the same time, one of the most striking trends in optimization is the constantly increasing emphasis on the interdisciplinary nature of the field. Optimization has been a basic tool in all areas of applied mathematics, engineering, medicine, economics and other sciences.

The series *Optimization and Its Applications* publishes undergraduate and graduate textbooks, monographs and state-of-the-art expository works that focus on algorithms for solving optimization problems and also study applications involving such problems. Some of the topics covered include nonlinear optimization (convex and nonconvex), network flow problems, stochastic optimization, optimal control, discrete optimization, multi-objective programming, description of software packages, approximation techniques and heuristic approaches.

SET-VALUED MAPPINGS AND ENLARGEMENTS OF MONOTONE OPERATORS

By

REGINA S. BURACHIK
University of South Australia, Mawson Lakes, Australia

ALFREDO N. IUSEM
IMPA, Rio de Janiero, Brazil

 Springer

Regina S. Burachik
University of South Australia
School of Mathematics and Statistics
Mawson Lakes
Australia

Alfredo N. Iusem
IMPA
Instituto de Matematica Pura e Aplicada
Rio de Janeiro
Brazil

ISBN-13: 978-1-4419-4346-0   e-ISBN-13: 978-0-387-69757-4

To Yalçin and Ethel

Acknowledgments

We are indebted to the Instituto de Matemática Pura e Aplicada, in Rio de Janeiro, to the Departamento de Sistemas of the Coordenação de Programas de Pós-Graduação em Engenharia of the Universidade Federal de Rio de Janeiro and to the Department of Mathematics and Statistics of the University of South Australia, which provided us with the intellectual environment that made the writing of this book possible.

We are grateful to Juan Enrique Martínez-Legaz and to Maycol Falla Luza who read parts of the manuscript of this book, and made valuable suggestions for improvements.

Contents

List of Figures

Preface

Point-to-set analysis studies mappings F from a given space X to another space Y, such that $F(x)$ is a subset of Y. Because we can see $F(x)$ as a set that changes with x, we are concerned first with sequences (or more generally nets), of sets. Such nets can also be seen as point-to-set mappings, where X is a directed set. Through appropriate generalizations, several basic continuity notions and results, and therefore quite a bit of general topology, can be extended both to nets of sets and to point-to-set mappings, taking into account the topological properties of X and Y; that is, without considering a topology for the family of subsets of Y. A significant part of this theory, turning around fixed points of point-to-set mappings and culminating with Kakutani's theorem, was built during the first half of the twentieth century.

We include in this book a self-contained review of the basics of set-valued analysis, together with the most significant fixed point and related results (e.g., Ekeland's variational principle, and Kakutani's and Caristi's theorems).

A particular class of point-to-set mappings, namely the class of maximal monotone operators, was introduced in the 1960s. This class of mappings is a natural extension of derivatives of convex functions to the realm of set-valued mappings, because its prototypical example is the gradient of a convex and differentiable real-valued function. We give in Chapter 4 a detailed review of some key topics of the theory of maximal monotone operators, including Rockafellar's fundamental results on domains and ranges of maximal monotone operators and Minty's theorems.

Maximal monotone operators enjoy without further assumptions some continuity properties, but in general they fail to be continuous, in the point-to-set sense. However, in the 1960s, Brøndsted and Rockafellar showed that a particular subset of maximal monotone operators, namely the subdifferentials of convex functions, can be approximated by a continuous point-to-set mapping. In other words, a subdifferential of a convex function can be embedded in a family of point-to-set mappings parameterized by a positive number. This positive parameter measures the proximity of these approximations to the original subdifferential mapping. Such a family received the name of ε-subdifferential, where ε is the parameter. This embedding process, called "enlargement", can be extended to arbitrary maximal monotone operators. This is the core of the second part of this book. Such enlargements are defined and analyzed in Chapter 5. In Chapter 6 they are used in the formulation of several algorithms for problems of finding zeroes of point-to-set operators, and more generally, to solving variational inequalities. The latter problem is a natural extension of nonsmooth convex optimization problems. Among the algorithms treated in Chapter 6, we mention bundle methods for maximal monotone operators, extragradient methods for nonsmooth variational inequalities, and robust versions of proximal point schemes in Banach spaces.

This book is addressed to mathematicians, engineers, and economists interested in acquiring a solid mathematical foundation in topics such as point-to-set operators, variational inequalities, general equilibrium theory, and nonsmooth optimization, among others. It is also addressed to researchers interested in the new and rapidly growing area of enlargements of monotone operators and their applications.

The first four chapters of the book can also be used for teaching a one-quarter course in set-valued analysis and maximal monotone operators at the MSc or PhD level. The only requisites, besides a level of mathematical maturity corresponding to a starting graduate student in mathematics, are some basic results of general topology and functional analysis.

Chapter 1

Introduction

1.1 Set-valued analysis

Point-to-set mappings appear naturally in many branches of science, such as control theory [215], game theory [226], economics [7], biomathematics [96], physics [118], and so on. The following section describes in some detail several problems involving point-to-set mappings, drawn from different areas within mathematics itself.

Any ill-conditioned problem forces us to consider point-to-set applications. For instance, they may represent the solution set when our problem has more than one solution. It is clear that the use of these mappings will be satisfactory only if we are able to extend to them, in some way, the main tools of point-to-point (i.e., classical) analysis.

The need for some sort of set-valued analysis was foreseen at the beginning of the twentieth century by the founders of the functional calculus: Painlevé, Hausdorff, Bouligand, and Kuratowski. The latter considered them in his important book on topology. The writers of Bourbaki's volume on topology decided to avoid the set-valued approach by considering these mappings as point-to-point mappings from X to $\mathcal{P}(Y)$, the set of subsets of Y. This point-to-point approach has some drawbacks, among which:

- The original nature of the problem is lost, and the (possible) structure of Y cannot be exploited.

- Continuity in this framework is a too-restrictive condition, because of the complicated topology of $\mathcal{P}(Y)$, thus satisfied by few mappings.

- The extra complexity provoked by this approach gives an erroneous impression of the difficulty of point-to-set analysis, which can indeed be studied in a more direct way.

In contrast with this point-to-point approach, point-to-set analysis follows the theory developed by the above-mentioned authors, who studied a set-valued map by means of its graph. This takes us back to the analytical geometry theory of

1

the seventeenth century, developed by Fermat, Descartes, and Viète, among others, who also studied functions via their graphs.

1.2   Examples of point-to-set mappings

Why should we study point-to-set mappings? We describe below some examples in which these objects appear naturally.

Example 1.2.1. Inverse Problems

Let X and Y be topological spaces, $G : X \to Y$ a mapping, and consider the problem

$$\text{Find } x \in X \text{ such that } G(x) = y. \tag{1.1}$$

The set of solutions of the problem above is given by

$$F(y) := \{z \in X \ : G(z) = y\}.$$

Recall that (1.1) is said to be *well posed* (in Hadamard's sense) when

- It has a unique solution, and

- This solution depends continuously on the data.

This amounts to requiring F to be single-valued and continuous. For *ill-posed problems* (i.e., those that are not well posed), which are a very common occurrence in applications, F may be empty, or multivalued, that is, point-to-set in general.

Example 1.2.2. Parameterized Optimization Problems

Take X and Y as in Example 1.2.1 and $W : X \times Y \to \mathbb{R}$. Consider the problem of determining $V : Y \to \mathbb{R}$ given by:

$$V(y) := \inf_{x \in X} W(x, y). \tag{1.2}$$

The function V is called the *marginal function* or *value function* associated with W. For a given $y \in Y$, the set

$$G(y) := \{x \in X \ : W(x, y) = V(y)\}$$

is a point-to-set mapping, which is also called *the marginal function* associated with (1.2). It is important to determine conditions that ensure that $G(y)$ is not empty and to study its behavior as a function of y. We discuss more about marginal functions in Chapter 3, Section 3.8.

Example 1.2.3. Optimality Conditions for Nondifferentiable Functions

Fermat's optimality condition ensures that any extreme point x^* of a differentiable function $f : \mathbb{R}^n \to \mathbb{R}$ must verify $\nabla f(x^*) = 0$. When dealing with convex nondifferentiable functions, this condition can still be used, but the concept of derivative must be extended to this "nonsmooth" case. The *subdifferential* of f at x is defined as

$$\partial f(x) := \{w \in \mathbb{R}^n : f(y) \geq f(x) + \langle w, y - x \rangle \; \forall y \in \mathbb{R}^n\}.$$

This defines a point-to-set mapping $\partial f : \mathbb{R}^n \to \mathcal{P}(\mathbb{R}^n)$. It can be seen that $\partial f(x) = \{\nabla f(x)\}$ if and only if f is differentiable at x (see e.g., [187, Theorem 25.1]). We show in Chapter 4 that the optimality condition can now be expressed as $0 \in \partial f(x^*)$, thus recovering the classical Fermat result. The concept of subdifferential was introduced in the 1960s by Moreau, Rockafellar, and others (see [160, 187]).

Example 1.2.4. Economic Theory and Game Theory

In the 1930s, von Neumann suggested the extension of Brouwer's fixed point theorem to point-to-set mappings, which Kakutani achieved in 1941 [112]. This result was used by Arrow and Debreu in the 1950s, [6, 71], for proving the existence of Walrasian equilibria, which was conjectured by Léon Walras himself at the end of the nineteenth century, and provided the first formalized support for Adam Smith's *invisible hand*. In general, point-to-set mappings arise naturally in economics when dealing with problems with nonunique solutions, and similar situations lead to the appearance of these mappings in game theory; see [226].

Example 1.2.5. Variational Inequalities

Variational inequalities have been a classical subject in mathematical physics, [118], and have recently been considered within optimization theory as a natural extension of minimization problems [89]. Let X be a Banach space, X^* its dual, $T : X \to \mathcal{P}(X^*)$ a mapping, and C a closed subset of X. The problem consists of finding $\bar{x} \in C$ such that there exists $\bar{u} \in T(\bar{x})$ satisfying $\langle \bar{u}, x - \bar{x} \rangle \geq 0$ for all $x \in C$.

If C is convex and T is the subdifferential of a convex function $g : X \to \mathbb{R}$, then the solution set of the variational inequality coincides with the set of minimizers of g in C. The subdifferential is point-to-set when g is not smooth, therefore it is natural to consider variational inequalities with point-to-set mappings, rather than point-to-point. In the realm of point-to-set mappings, the notion that extends in a natural way the behavior of derivatives of real-valued functions is called *maximal monotonicity*. It is satisfied by subdifferentials of convex functions and it plays a central role in set-valued analysis. Chapter 4 of this book is devoted to this family of operators.

1.3 Description of the contents

The material in this book is organized as follows. Chapter 2 is dedicated to point-to-set mappings and their continuity properties. Chapter 3 contains the basic core of convex analysis, leading to several classical fixed point results for set-valued mappings, including the well-known theorems due to Caristi, Ekeland, Ky Fan, and Kakutani. This chapter also includes some results on derivatives of point-to-set mappings, and its last section describes certain duality principles for variational inclusions. Chapter 4 deals with one important class of point-to-set mappings:

maximal monotone operators, establishing their most important properties. The final two chapters contain recent results that have not yet been published in books. Chapter 5 is devoted to enlargements of maximal monotone applications. The most important example of such an enlargement is the ε-subdifferential of a convex function. An enlargement can be defined for an arbitrary maximal monotone operator, in such a way that it inherits many of the good properties of the ε-subdifferential. We establish the properties of a whole family of such enlargements, and apply them to develop extragradient type methods for nonsmooth monotone variational inequalities, bundle methods, and a variational sum of two maximal monotone operators.

Chapter 6 applies the theory of the preceding chapters to the proximal point algorithm in infinite-dimensional spaces, reporting recently obtained results, including inexact versions in Banach spaces with constant relative errors and a convergence analysis in Hilbert spaces with similar error criteria for nonmonotone operators.

In Chapters 2 and 3 we have drawn extensively from the excellent books [18] and [194]. The contents of Chapters 2-4 can be considered classical and it can be found in many well-known books, besides those just mentioned (e.g., [206, 172]). However, most of this material has been organized so as to fit precisely the needs of the last two chapters, which are basically new. We have made the book as close as possible to being self-contained, including proofs of almost all results above the level of classical functional analysis.

Chapter 2

Set Convergence and Point-to-Set Mappings

2.1 Convergence of sets

Sequences of sets appear naturally in many branches of applied mathematics. In Examples 1.2.1 and 1.2.2, we could have a sequence of problems of the kind (1.1) or (1.2), defined by a sequence $\{y_n\}_{n\in\mathbb{N}}$. This generates sequences of sets $\{F(y_n)\}_{n\in\mathbb{N}}$ and $\{G(y_n)\}_{n\in\mathbb{N}}$. We would like to know how these sets change when the sequence $\{y_n\}_{n\in\mathbb{N}}$ approaches some fixed y. In Example 1.2.5, we may want to approximate the original problem by a sequence of related problems, thus producing a sequence of corresponding feasible sets and a sequence of corresponding solution sets. How do these sequences approximate the original feasible set and the original solution set? Convergence of sets is the appropriate tool for analyzing these situations. Convergence of a family of sets to a given set C is expressed in terms of convergence, in the underlying space, of elements in each set of the family to a given point in C. Therefore the properties of the underlying space are reflected in the convergence properties of the family of sets. When the underlying space satisfies the first countability axiom N_1, (i.e., when every point has a countable basis of neighborhoods), then the convergence of sequences of sets can be expressed in terms of convergence of sequences of point in these sets. This simplification is not possible when the space is not N_1. For the latter spaces we need to use nets, a generalization of the concept of sequences, in order to study convergence of families of sets.

Spaces that are not N_1 are of interest in future chapters. An important example, which motivates us to consider these spaces, is a Banach space endowed with the weak topology. Throughout this chapter, we use nets in most of the basic concepts and derive the corresponding sequential version for the case in which the space is N_1. Because some readers may be unfamiliar with the concept of nets, we give here a brief introduction to this concept.

2.1.1 Nets and subnets

A sequence in a topological space X is a function $\xi : \mathbb{N} \to X$. The fact that the domain of this function is the set $\mathbb{N}$ plays a central role in the analysis of sequences. Indeed, the definitions of "tail" and "subsequence" strongly rely on the order-related properties of $\mathbb{N}$. The definition of net requires the use of a domain I with less restrictive order properties than those enjoyed by $\mathbb{N}$. We now give the relevant definitions. A *partial order* relation $\succeq$ defined in a set I is defined as a binary relation that is *reflexive*, (i.e., $i \succeq i$ for all $i \in I$), *antisymmetric*, (i.e., if $i \succeq j$ and $j \succeq i$ then $i = j$), and *transitive* (i.e., if $i \succeq j$ and $j \succeq k$ then $i \succeq k$). A partial order is said to be *total* when every pair $i, j \in I$ verifies either $i \succeq j$ or $j \succeq i$. For instance, the usual order in the sets $\mathbb{N}, \mathbb{Z}, \mathbb{R}$ is total. The set $\mathcal{P}(X)$ of parts of X with the order relation $A \succeq B$ defined by $A \subset B$ for all $A, B \in \mathcal{P}(X)$, is a partial order that is not total. A set I with a partial order $\succeq$ is said to be *directed* if for all $i, j \in I$ there exists $k \in I$ such that $k \succeq i$ and $k \succeq j$. In other words, a set is directed when every pair of elements has an upper bound. Every total order is directed, because in this case the upper bound of any pair of indices is the largest index of the pair. So we see that the set $\mathbb{N}$ with the usual order is a very special case of a directed set. A less trivial example of a directed set is the collection of all closed subsets of X, with the partial order of the inclusion. Indeed, given a pair of closed subsets A, B of X, the subset $A \cup B$ belongs to the collection and is a successor of both A and B. Given a topological space X, a set $U \subset X$ is a *neighborhood* of a given point x whenever there exists an open set W such that $x \in W \subset U$. The family of neighborhoods of a point provides another important example of a directed set with the partial order defined by the *reverse inclusion*, that is, $A \succeq B$ if and only if $B \supset A$. More generally, any collection of sets $\mathcal{A} \subset \mathcal{P}(X)$ closed under finite intersections is a directed set with the reverse inclusion.

Definition 2.1.1. *Given a directed set I and a topological space X, a net is a function $\xi : I \to X$, where the set I is a partially ordered set that is directed. We denote such a net $\{\xi_i\}_{i \in I}$.*

Because $I = \mathbb{N}$ is a directed set, every sequence is a net. We mentioned above two important concepts related to sequences: "tails" and "subsequences." Both concepts rely on choosing particular subsets of $\mathbb{N}$. More precisely, the family of subsets

$$\mathcal{N}_\infty := \{J \subset \mathbb{N} : \mathbb{N} \setminus J \text{ is finite}\},$$

represents all the tails of $\mathbb{N}$, and they are used for defining the tails of the sequence. The family

$$\mathcal{N}_\sharp := \{J \subset \mathbb{N} : J \text{ is infinite}\},$$

represents all the subsequences of $\mathbb{N}$ and they are used for defining all the subsequences of the sequence. Clearly, $\mathcal{N}_\infty \subset \mathcal{N}_\sharp$. We define next the subsets of an arbitrary directed set I that play an analogous role to that of $\mathcal{N}_\infty$ and $\mathcal{N}_\sharp$ in $\mathbb{N}$. The subsets that represent the "tails" of the net are called *terminal sets*. A subset J of a directed set I is *terminal* if there exists $j_0 \in I$ such that $k \in J$ for all $k \succeq j_0$.

A subset K of a directed set I is said to be *cofinal* when for all $i \in I$ there exists $k \in K$ such that $k \succeq i$. It is clear that $\mathcal{N}_\infty$ is the family of all the terminal subsets of $\mathbb{N}$, and $\mathcal{N}_\sharp$ is the family of all cofinal sets of $\mathbb{N}$. The fact that $\mathcal{N}_\infty \subset \mathcal{N}_\sharp$ can be extended to the general case. We prove this fact below, together with some other useful properties regarding cofinal and terminal sets.

Proposition 2.1.2. *Let I be a directed set.*

(i) *If $K \subset I$ is terminal, then K is cofinal.*

(ii) *An arbitrary union of cofinal subsets is cofinal.*

(iii) *A finite intersection of terminal subsets of I is terminal.*

(iv) *If $J \subset I$ is terminal and $K \subset I$ is cofinal, then $J \cap K$ is cofinal.*

(v) *If $J \subset I$ is terminal and $K \supset J$, then K is terminal.*

(vi) *$J \subset I$ is not terminal if and only if $I \setminus J$ is cofinal.*

(vii) *$J \subset I$ is not cofinal if and only if $I \setminus J$ is terminal.*

Proof. For proving (i), take any $i \in I$ and $k \in K$ a threshold for K. Because I is directed there exist $j \succeq i$ and $j \succeq k$. By definition of threshold, we must have $j \in K$. Using now the facts that $j \succeq i$ and $j \in K$, we conclude that K is cofinal. The proof of (ii) is direct and is left to the reader. For proving (iii), it is enough to check it for two sets. If J and K are terminal subsets of I, then there exist $\bar{j} \in J$ and $\bar{k} \in K$ such that $j \in J$ for all $j \succeq \bar{j}$ and $k \in K$ for all $k \succeq \bar{k}$. Because the set I is directed, there exists $i_0 \in I$ such that $i_0 \succeq \bar{j}$ and $i_0 \succeq \bar{k}$. Then $i_0 \in J \cap K$ and every $i \succeq i_0$ verifies $i \in J \cap K$. So $J \cap K$ is terminal. Let us prove (iv). Let j_0 be a threshold for J and take any $i \in I$. Because I is directed there exists $i_1 \in I$ such that $i_1 \succeq i$ and $i_1 \succeq j_0$. Because K is confinal there exists $k_0 \in K$ such that $k_0 \succeq i_1$. Using now the transitivity of the partial order, and the fact that $i_1 \succeq j_0$, we conclude that $k_0 \in J$. Therefore $k_0 \in K \cap J$. We also have $k_0 \succeq i_1 \succeq i$ and hence $J \cap K$ is cofinal. The proofs of (v)–(vii) are straightforward from the definitions. $\square$

Terminal sets are used for defining limits of nets, whereas cofinal sets are used for defining cluster points.

Definition 2.1.3. *Let $\{\xi_i\}_{i \in I}$ be a net in X.*

(i) *A point $\bar{\xi} \in X$ is a* limit *of a net $\{\xi_i\}_{i \in I} \subset X$ if for every neighborhood V of $\bar{\xi}$ there exists a terminal subset $J \subset I$ such that $\{\xi_j\}_{j \in J} \subset V$. This fact is denoted $\bar{\xi} = \lim_{i \in I} \xi_i$ or $\xi \to_{i \in I} \bar{\xi}$.*

(ii) *A point $\bar{\xi} \in X$ is a* cluster point *of a net $\{\xi_i\}_{i \in I} \subset X$ if for every neighborhood V of $\bar{\xi}$ there exists a cofinal subset $J \subset I$ such that $\{\xi_j\}_{j \in J} \subset V$.*

Uniqueness of a limit is a desirable property that holds under a mild assumption on X: whenever $x \neq y$ there exist neighborhoods U, V of x and y, respectively, such that $U \cap V = \emptyset$. Such spaces are called Hausdorff spaces. It is clear that, under this assumption, a limit of a net is unique. We assume from now on that the topological space X is a Hausdorff space.

Definition 2.1.4. *Given a net $\{\xi_i\}_{i \in I}$ in X and a set $S \subset X$, we say that the net is* eventually *in S when there exists a terminal subset $J \subset I$ such that $x_j \in S$ for all $j \in J$. We say that the net is* frequently *in S when there exists a cofinal subset $J \subset I$ such that $x_j \in S$ for all $j \in J$.*

Remark 2.1.5. *By Definition 2.1.4, we have that $\bar{\xi}$ is a cluster point of the net $\{\xi_i\}_{i \in I}$ if and only if for every neighborhood W of $\bar{\xi}$, the net is frequently in W. Similarly, $\bar{\xi}$ is a limit of the net $\{\xi_i\}_{i \in I}$ if and only if for every neighborhood W of $\bar{\xi}$, the net is eventually in W.*

Recall that a topological space satisfies the first countability axiom N_1 when every point has a countable basis of neighborhoods. In other words, for all $x \in X$ there exists a countable family of neighborhoods $\mathcal{U}_x = \{U_m\}_{m \in \mathbb{N}}$ of x such that each neighborhood of x contains some element of $\mathcal{U}_x$. The main result regarding limits of nets is the characterization of closed sets (see [115, Chapter 2, Theorems 2 and 8]).

Theorem 2.1.6. *A subset $A \subset X$ is closed if and only if $\bar{\xi} \in A$ for all $\bar{\xi}$ such that there exists a net $\{\xi_i\}_{i \in I} \subset A$ that converges to $\bar{\xi}$. If X is N_1, the previous statement is true with sequences instead of nets.*

Recall that every cluster point of a sequence is characterized as a limit of a subsequence. A similar situation occurs with nets, where subsequences are replaced by subnets. A naïve definition of subnets would be to restrict the function ξ to a cofinal subset of I. However, the cofinal subsets of I are not enough, because there are nets whose cluster points cannot be expressed as a limit of any subnet obtained as the restriction to a cofinal subset of I (see [115, Problem 2E]). In order to cover these pathological cases, the proper definition should allow the subnet to have an index set with a cardinality bigger than that of I.

Definition 2.1.7. *Let I, J be two directed sets and let $\{\xi_i\}_{i \in I} \subset X$ be a net in X. A net $\{\eta_j\}_{j \in J} \subset X$ is a* subnet *of $\{\xi_i\}_{i \in I}$ when there exists a function $\phi : J \to I$ such that*

(SN1) $\eta_j = \xi_{\phi(j)}$ for all $j \in J$. In other words, $\xi \circ \phi = \eta$.

(SN2) For all $i \in I$ there exists $j \in J$ such that $\phi(j') \succeq i$ for all $j' \succeq j$.

The use of an exogenous directed set J and a function ϕ between J and I allows the cardinality of J to be higher than that of I. When J is a cofinal subset of I, then taking ϕ as the inclusion we obtain a subnet of $\{\xi_i\}_{i\in I}$. With this definition of subnets, the announced characterization of cluster points holds (see [115, Chapter 2, Theorem 6]).

Theorem 2.1.8. *A point $\bar{\xi} \in X$ is a cluster point of the net $\{\xi_i\}_{i\in I}$ if and only if it is the limit of a subnet of the given net.*

Three other useful results involving nets are stated below.

Theorem 2.1.9.

(a) *X is compact if and only if each net in X has a cluster point. Consequently, X is compact if and only if each net in X has a subnet that converges to some point in X.*

(b) *Let X, Y be topological spaces and $f : X \to Y$. Then f is continuous if and only if for each net $\{\xi_i\}_{i\in I} \subset X$ converging to $\bar{\xi}$, the net $\{f(\xi_i)\}_{i\in I} \subset Y$ converges to $f(\bar{\xi})$.*

(c) *If X is a topological space and the net $\{\xi_i\}_{i\in I} \subset X$ converges to $\bar{\xi}$, then every subnet of $\{\xi_i\}_i$ converges to $\bar{\xi}$.*

Proof. For (a) and (b), see, for example, [115, Chapter 5, Theorem 2 and Chapter 3, Theorem 1(f)]. Let us prove (c). Let $\{\eta_j\}_{j\in J}$ be a subnet of $\{\xi_i\}_{i\in I}$ and fix $W \in \mathcal{U}_{\bar{\xi}}$. Because $\{\xi_i\}_{i\in I}$ converges to $\bar{\xi}$, there exists $i_W \in I$ such that

$$\xi_i \in W \quad \text{for all } i \geq i_W. \tag{2.1}$$

Using (SN2) in Definition 2.1.7 for $i := i_W$, we know that there exists $j_W \in J$ such that $\phi(j) \succeq i_W$ for all $j \succeq j_W$. The latter fact, together with (SN1) and (2.1), yields

$$\eta_j = \xi_{\phi(j)} \in W \quad \text{for all } j \geq j_W. \tag{2.2}$$

So $\{\eta_j\}_j$ is eventually in W, and hence it converges to $\bar{\xi}$. $\quad\square$

Let X be a topological space and take $x \in X$. A standard procedure to construct a net in X convergent to x is the following. Take as the index set the family $\mathcal{U}_x$ of neighborhoods of x. We have seen above that this index set is directed with the reverse inclusion as a partial order. For every $V \in \mathcal{U}_x$, choose $x_V \in V$. Hence the net $\{x_V\}_{V\in\mathcal{U}_x}$ converges to x. We refer to such a net as a *standard* net convergent to x.

2.2 Nets of sets

A net of sets is a natural extension of a net of points, where instead of having a point-to-point function from a directed set I to X, we have a point-to-set one.

Definition 2.2.1. *Let X be a topological space and let I be a directed set. A net of sets $\{C_i\}_{i \in I} \subset X$ is a function $C : I \rightrightarrows X$; that is, $C_i \subset X$ for all $i \in I$. In the particular case in which $I = \mathbb{N}$ we obtain a* sequence of sets $\{C_n\}_{n \in \mathbb{N}} \subset X$.

Definition 2.2.2. *Take a terminal subset J of I and choose $x_j \in C_j$ for all $j \in J$. The net $\{x_j\}_{j \in J}$ is called a* selection *of the net $\{C_i\}_{i \in I}$. We may simply say in this case that $x_i \in C_i$* eventually. *Take now a cofinal subset K of I. If $x_k \in C_k$ for all $k \in K$, then we say that $x_i \in C_i$* frequently. *If a selection $\{x_j\}_{j \in J}$ of $\{C_i\}_{i \in I}$ converges to a point x, then we write $\lim_i x_i = x$. When K is cofinal and $\{x_k\}_{k \in K}$ converges to a point x, we write $\lim_{i \in K} x_i = x$ or $x_i \to_K x$.*

Our next step consists of defining the interior limit and the exterior limit of a net of sets. Recall that $\bar{x}$ is a cluster point of a net $\{x_i\}_{i \in I}$ when $x_i \in W$ frequently for every neighborhood W of $\bar{x}$. If for every neighborhood of $\bar{x}$ we have that $x_i \in W$ eventually, then $\bar{x}$ is a limit of the net. We use these ideas to define the interior and exterior limits of a net of sets.

Definition 2.2.3. *Let X be a topological space and let $\{C_i\}_{i \in I}$ be a net of sets in X. We define the sets $\lim \text{ext}_{i \in I} C_i$ and $\lim \text{int}_{i \in I} C_i$ in the following way.*

(i) *$x \in \lim \text{ext}_{i \in I} C_i$ if and only if for each neighborhood W of x we have $W \cap C_i \neq \emptyset$ frequently (i.e., $W \cap C_i \neq \emptyset$ for i in a cofinal subset of I).*

(ii) *$x \in \lim \text{int}_{i \in I} C_i$ if and only if for each neighborhood W of x we have $W \cap C_i \neq \emptyset$ eventually (i.e., $W \cap C_i \neq \emptyset$ for i in a terminal subset of I).*

From Definition 2.2.3 and the fact that every terminal set is cofinal, we have that $\lim \text{int}_i C_i \subset \lim \text{ext}_i C_i$. When the converse inclusion holds too, the net of sets converges, as we state next.

Definition 2.2.4. *A net of sets $\{C_i\}_{i \in I}$* converges *to a set C when*

$$\lim_i \text{int } C_i = \lim_i \text{ext } C_i = C.$$

This situation is denoted $\lim_i C_i = C$ or $C_i \to C$.

When the space is N_1 and we have a sequence of sets, we can describe the interior and exterior limits in terms of sequences and subsequences. The fact that this simplification does not hold for spaces which are not N_1 is proved in [24, Exercise 5.2.19].

Proposition 2.2.5. *Let $\{C_n\}_{n \in \mathbb{N}} \subset X$ be a sequence of sets.*

(i) *If there exist $n_0 \in \mathbb{N}$ and $x_n \in C_n$ such that $\{x_n\}_{n \geq n_0}$ converges to x, then $x \in \lim \text{int}_{n \in \mathbb{N}} C_n$. When X is N_1, the converse of the previous statement holds.*

(ii) *Assume that there exists $J \in \mathcal{N}_\sharp$ such that for each $j \in J$ there exists $x_j \in C_j$ with $\{x_j\} \to_J x$. Then $x \in \lim \text{ext}_{n \in \mathbb{N}} C_n$. When X is N_1, the converse of the previous statement holds.*

Proof. The first statements of both (i) and (ii) follow directly from the definitions. Let us establish the second statement in (i). Assume that X is N_1 and let $\mathcal{U}_x = \{U_m\}_{m \in \mathbb{N}}$ be a countable family of neighborhoods of x. We must define a sequence $\{x_n\}$ and an index n_0 such that $x_n \in C_n$ for all $n \geq n_0$ and $x_n \to x$. Take a family of neighborhoods of x that is *nested* (i.e., $U_{m+1} \subset U_m$ for all m). Because $x \in \lim \text{int}_{n \in \mathbb{N}} C_n$, we know that for each $m \in \mathbb{N}$, there exists $N_m \in \mathbb{N}$ such that $U_m \cap C_n \neq \emptyset$ for all $n \geq N_m$. Because the family is nested, we can assume that $N_m < N_{m+1}$ for all $m \in \mathbb{N}$. Fix $n_0 \in \mathbb{N}$. The set $J := \{n \in \mathbb{N} \mid n \geq N_{n_0}\} \in \mathcal{N}_\infty$ and we can choose $y_{nm} \in U_m \cap C_n$ for all $n \in J$. For every $n \in J$, there exists N_m such that $N_m \leq n < N_{m+1}$. Define the sequence $x_n = y_{nm}$ for every $n \in J$ such that $N_m \leq n < N_{m+1}$. Fix a neighborhood U of x, and take k such that $U_k \subset U$. Then, from the definition of x_n and the fact that the family of neighborhoods is nested, it can be checked that $x_n \in U_k$ for all $n \geq N_k$. Hence $\{x_n\}_{n \geq N_{n_0}}$ converges to x. Let us prove now the second statement in (ii). Let $\mathcal{U}_x = \{U_m\}_{m \in \mathbb{N}}$ be a family of neighborhoods of x, as in (i). Because $x \in \lim \text{ext}_{n \in \mathbb{N}} C_n$ there exists $n_1 \in \mathbb{N}$ such that $U_1 \cap C_{n_1} \neq \emptyset$. The set $\{n \in \mathbb{N} \mid U_2 \cap C_n \neq \emptyset\}$ is infinite and hence we can choose some n_2 in the latter set such that $n_2 > n_1$. In this way we construct an infinite set $J := \{n_k\}_{k \in \mathbb{N}}$ such that $U_k \cap C_{n_k} \neq \emptyset$ for all k. Taking $x_k \in U_k \cap C_{n_k}$ for all k we obtain a sequence that converges to x. This completes the proof of (ii) $\square$

The external and internal limits $\lim \text{ext}_i C_i$ and $\lim \text{int}_i C_i$ always exist (although they can be empty). Because they are constructed out of limits of nets, it is natural for them to be closed sets. This fact is established in the following result, due to Choquet [62]. Given $A \subset X$, A^c denotes the complement of A (i.e., $X \setminus A$), $\overline{A}$ its closure, A^o its interior, and ∂A its boundary.

Theorem 2.2.6. *Let X be a topological space and let $\{C_i\}_{i \in I}$ be a net of sets in X. Then*

(i)
$$\lim_{i \in I} \text{int } C_i = \cap \left\{ \overline{\cup_{i \in K} C_i} \; : \; K \text{ is a cofinal subset of } I \right\}.$$

(ii)
$$\lim_{i \in I} \text{ext } C_i = \cap \left\{ \overline{\cup_{i \in K} C_i} \; : \; K \text{ is a terminal subset of } I \right\}.$$

As a consequence, the sets $\lim \text{ext}_{i \in I} C_i$ and $\lim \text{int}_{i \in I} C_i$ are closed sets.

Proof. (i) Note that $x \in \lim \text{int}_{i \in I} C_i$ if and only if for every neighborhood W of x we have that $W \cap C_i \neq \emptyset$ for all $i \in J$, where J is a terminal subset of I. Fix now a cofinal set $K \subset I$. Then by Proposition 2.1.2(iv) we have that $J \cap K$ is also cofinal and

$W \cap C_i \neq \emptyset$ for all $i \in J \cap K$. Hence $W \cap (\cup_{k \in K} C_k) \neq \emptyset$. Because W is an arbitrary neighborhood of x, we conclude that $x \in \overline{\cup_{i \in K} C_i}$. Using now that K is an arbitrary cofinal set of I, we conclude that $x \in \cap \{ \overline{\cup_{i \in K} C_i} : K$ is a cofinal subset of $I \}$. Conversely, assume that $x \notin \lim \mathrm{int}_{i \in I} C_i$. Then there exists W a neighborhood of x such that the set of indices $J := \{ i \in I : W \cap C_i \neq \emptyset \}$ is not terminal. Because $W \cap C_i = \emptyset$ for all $i \in J^c$ we have that $W \cap (\cup_{i \in J^c} C_i) = \emptyset$. Hence

$$x \notin \overline{\cup_{i \in J^c} C_i}. \tag{2.3}$$

By Proposition 2.1.2(vi) J^c is cofinal and hence (2.3) implies that

$$x \notin \cap \{ \overline{\cup_{i \in K} C_i} : K \text{is a cofinal subset of } I \}.$$

The proof of (ii) follows similar steps and is left as an exercise. $\square$

When X is a metric space, the interior and exterior limits can be expressed in terms of the (numerical) sequence $\{ d(x, C_n) \}_n$. Let us denote the distance in X by $d : X \times X \to \mathbb{R}$ and recall that for $K \subset X$,

$$d(x, K) := \inf_{y \in K} d(x, y).$$

In particular, $d(x, \emptyset) = +\infty$. Denote by $B(x, \rho)$ the closed ball centered at x with radius ρ; that is, $B(x, \rho) := \{ z \in X : d(z, x) \leq \rho \}$.

Proposition 2.2.7. *Assume that X is a metric space. Let $\{ C_n \}_{n \in \mathbb{N}}$ be a sequence of sets such that $C_n \subset X$ for all $n \in \mathbb{N}$. Then,*

(i) $\lim \mathrm{ext}_n C_n := \{ x \in X : \lim \inf_n d(x, C_n) = 0 \}.$ (2.4)

(ii) $\lim \mathrm{int}_n C_n := \{ x \in X : \lim \sup_n d(x, C_n) = 0 \}$

$\qquad\qquad\quad = \{ x \in X : \lim_n d(x, C_n) = 0 \}.$ (2.5)

Proof. (i) Take $x \in \lim \mathrm{ext}_n C_n$. By Proposition 2.2.5(ii) there exists a $J \in \mathcal{N}_\sharp$ and a corresponding subsequence $\{ x_j \}_{j \in J}$ such that $x_j \to_J x$. If for some $a > 0$ we have $\lim \inf_n d(x, C_n) > a > 0$ then by definition of $\lim \inf$ there exists $k_0 \in \mathbb{N}$ such that $\inf_{n \geq k_0} d(x, C_n) > a$. On the other hand, because $x_j \to_J x$ there exists $j_0 \in J$ such that $j_0 \geq k_0$ and $d(x, x_j) < a/2$ for all $j \in J$, $j \geq j_0$. We can write

$$a/2 > d(x, x_{j_0}) \geq d(x, C_{j_0}) \geq \inf_{j \geq k_0} d(x, C_j) > a,$$

which is a contradiction. For the converse inclusion, note that the equality $\lim \inf_n d(x, C_n) = 0$ allows us to find $J \in \mathcal{N}_\sharp$ such that $x_j \to_J x$. Therefore by Proposition 2.2.5(ii) we must have $x \in \lim \mathrm{ext}_n C_n$. The proof of (ii) follows steps similar to those in (i). $\square$

Example 2.2.8. Consider the following sequence of sets

$$C_n := \begin{cases} \{1/n\} \times \mathbb{R} & \text{if } n \text{ is odd,} \\ \{1/n\} \times \{1\} & \text{if } n \text{ is even.} \end{cases}$$

Then, $\lim \operatorname{int}_n C_n = \{(0,1)\}$ and $\lim \operatorname{ext}_n C_n = \{0\} \times \mathbb{R}$ (see Figure 2.1).

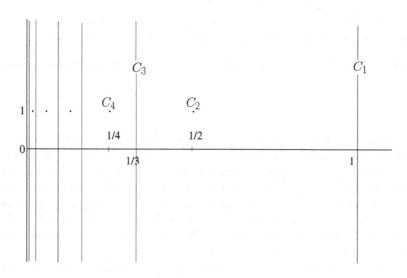

Figure 2.1. *Internal and external limits*

Because always $\lim \operatorname{int}_i C_i \subset \lim \operatorname{ext}_i C_i$, for proving convergence of the net $\{C_i\}_{i \in I}$ it can be useful to find an auxiliary set C such that $\lim \operatorname{ext}_i C_i \subset C \subset \lim \operatorname{int}_i C_i$. The result below contains criteria for checking whether a set C satisfies some of these inclusions. In this result, we consider the following property of a topological space X.

(P): If F is closed and $x \notin F$ then there exists a neighborhood U of x such that $\overline{U} \cap F = \emptyset$. *Regular* topological spaces (i.e., those satisfying the third separability condition (T3) stated below; see [115]), trivially satisfy (P). We recall next the definition of (T3): if F is closed and $x \notin F$ then there exist open sets $U, V \subset X$ such that $F \subset U$, $x \in V$, and $U \cap V = \emptyset$. Property (P) also holds under a weaker separability condition than (T3). This condition is called $(T_{2\frac{1}{2}})$: If x, y are two different points in X, then there exist neigborhoods U_x and U_y of x and y, respectively, such that $\overline{U_x} \cap \overline{U_y} = \emptyset$. Such topological spaces are called *completely Hausdorff* spaces.

When working with N_1 spaces, we always assume that the countable family of neighborhoods $\mathcal{U}_x$ around a given point x is nested; that is, $U_{m+1} \subset U_m$ for all m.

Proposition 2.2.9 (Criteria for checking convergence). *Assume that X is a topological space. Consider a net $\{C_i\}_{i \in I}$ of subsets of X, and a set $C \subset X$.*

(a) *$C \subset \lim \operatorname{int}_i C_i$ if and only if for every open set U such that $C \cap U \neq \emptyset$, there exists a terminal subset J of I such that $C_j \cap U \neq \emptyset$ for all $j \in J$.*

(b) *Assume (only in this item) that C is closed and that X satisfies property (P).*

If for every closed set B such that $C \cap B = \emptyset$, there exists a terminal subset J of I such that $C_j \cap B = \emptyset$ for all $j \in J$, then $\lim \mathrm{ext}_i\, C_i \subset C$.

(c) *If $\lim \mathrm{ext}_i\, C_i \subset C$, then for every compact set B such that $C \cap B = \emptyset$, there exists a terminal subset J of I such that $C_j \cap B = \emptyset$ for all $j \in J$.*

(d) *Assume that C is closed, X is N_1, and $I = \mathbb{N}$; that is, we have a sequence of sets $\{C_n\}$. Then the converse of (c) holds.*

Proof.

(a) Suppose that $C \subset \lim \mathrm{int}_i\, C_i$ and take an open set U such that $C \cap U \neq \emptyset$. Then there exists some $x \in C \cap U$. The assumption on C implies that $x \in \lim \mathrm{int}_i\, C_i$. Hence U is a neighborhood of x and by definition of interior limit we must have $U \cap C_i \neq \emptyset$ for a terminal subset of I. This proves the "only if" statement. For proving the "if" statement, consider C satisfying the assumption on open sets and take $x \in C$. For every neighborhood W of x we have that $W \cap C \neq \emptyset$, and by the assumption on C we get $W \cap C_i \neq \emptyset$ for i in a terminal subset of I, which implies that $x \in \lim \mathrm{int}_i\, C_i$.

(b) Take $z \in \lim \mathrm{ext}_i\, C_i$ and suppose that $z \notin C$. Because C is closed, property (P) implies that there exists $U \in \mathcal{U}_z$ with $\overline{U} \cap C = \emptyset$. Applying now the assumption to $B := \overline{U}$ we conclude that there exists a terminal subset J of I such that

$$C_j \cap \overline{U} = \emptyset \text{ for all } j \in J. \tag{2.6}$$

Because $z \in \lim \mathrm{ext}_i\, C_i$ we have that $U \cap C_i \neq \emptyset$ for all i in a cofinal subset of I. This contradicts (2.6), establishing the conclusion.

(c) Assume that the conclusion does not hold. Then there exists a compact set B_0 such that $C \cap B_0 = \emptyset$ and for every terminal subset J of I there exists $j \in J$ with $C_j \cap B_0 \neq \emptyset$. For every fixed $i \in I$, the set $J(i) := \{l \in I : l \succeq i\}$ is terminal. Then, by our assumption, there exists $l_i \in J(i)$ such that $C_{l_i} \cap B_0 \neq \emptyset$. For every $i \in I$, select $x_{l_i} \in C_{l_i} \cap B_0$. By Proposition 2.1.2(ii), the set $K := \cup_{i \in I} l_i$ is cofinal in I and by construction $\{x_k\}_{k \in K} \subset B_0$. Because B_0 is compact, by Theorem 2.1.9(a) there exists a cluster point $\bar{x}$ of $\{x_k\}_{k \in K}$ with $\bar{x} \in B_0$. Take a neighborhood W of $\bar{x}$. Because $\bar{x}$ is a cluster point of $\{x_k\}_{k \in K}$, we must have $\{x_k\}_{k \in K'} \subset W$ for a cofinal subset K' of K. Therefore $W \cap C_i \neq \emptyset$ for all $i \in K'$. Because $K' \subset K$ is cofinal in I, we conclude that $\bar{x} \in \lim \mathrm{ext}_i\, C_i$. Using the assumption that $\lim \mathrm{ext}_i\, C_i \subset C$, we get that $\bar{x} \in C$. So $\bar{x} \in C \cap B_0$, contradicting the assumption on B_0.

(d) Assume that C satisfies the conclusion of (c). Suppose that there exists $z \in \lim \mathrm{ext}_n\, C_n$ such that

$$z \notin C. \tag{2.7}$$

The space is N_1, therefore there exists a countable nested family of neighborhoods $\{U_m\}_m$ of z. By (2.7) we can assume that $U_m \cap C = \emptyset$ for all $m \in \mathbb{N}$. The fact that $z \in \lim \mathrm{ext}_n\, C_n$ implies that for all $m \in \mathbb{N}$ there exists a cofinal (i.e., an infinite) set $K_m \subset \mathbb{N}$ such that $U_m \cap C_i \neq \emptyset$ for every $i \in K_m$. For $m := 1$ take an index $i_1 \in K_1$. For this index i_1 take $x_1 \in U_1 \cap C_{i_1}$. The latter set is not empty by definition of K_1. For $n := 2$ take an index $i_2 \in K_2$ such that $i_2 \succsim i_1$. We can do this because K_2 is infinite. For this index i_2 take $x_2 \in U_2 \cap C_{i_2}$. In general, given x_m and an index $i_m \in K_m$, choose an index $i_{m+1} \in K_{m+1}$ such that $i_{m+1} \succsim i_m$ and take $x_{m+1} \in U_{m+1} \cap C_{i_{m+1}}$. By construction, the set of indices $\{i_m\}_m \subset \mathbb{N}$ is infinite, and hence cofinal. The sequence $\{x_m\}$ converges to z, because $x_m \in U_m$ for all m and the family of neighborhoods $\{U_m\}_m$ is nested. Note that only a finite number of elements of the sequence $\{x_m\}$ can belong to C, because otherwise, inasmuch as C is closed and the sequence converges to z, we would have $z \in C$, contradicting (2.7). So, there exists $m_0 \in \mathbb{N}$ such that $x_m \notin C$ for all $m \geq m_0$. Consider the compact set $B := \{x_m : m \geq m_0\} \cup \{z\}$. By construction, $B \cap C = \emptyset$. Applying the assumption to B, there exists $J \in \mathcal{N}_\infty$ such that

$$B \cap C_j = \emptyset, \qquad (2.8)$$

for all $j \in J$. Because J is terminal, there exists $\bar{j} \in \mathbb{N}$ such that $j \in J$ for all $j \geq \bar{j}$. Consider the set of indices $J_0 := \{i_m\}_{m \geq m_0}$. This set J_0 is cofinal and hence there exists some $i_{\bar{m}} \in J_0 \cap J$. By definition of $\{x_m\}$ we must have $x_{\bar{m}} \in C_{i_{\bar{m}}} \cap B$, contradicting (2.8). Therefore we must have $z \in C$. $\quad\square$

Remark 2.2.10. Proposition 2.2.9(d) is not true when C is not closed. Take $X = \mathbb{R}^2$, $C := \{x \in \mathbb{R}^2 : x_1^2 + x_2^2 > 1\}$, and $C_n := \{x \in \mathbb{R}^2 : x_1^2 + x_2^2 = 1 + 1/n\}$ for all n (see Figure 2.2). Then

$$\lim_n C_n = \lim \operatorname{ext}_n C_n = \{x \in \mathbb{R}^2 : x_1^2 + x_2^2 = 1\} \not\subset C.$$

It is easy to see that whenever a compact set B satisfies $B \cap C = \emptyset$, it holds that $B \cap C_n = \emptyset$ for all n. Also, when X is not N_1 we cannot guarantee the compactness of the set B constructed in the proof of Proposition 2.2.9(d).

When X is a metric space, we can use the distance for checking convergence of a sequence of sets.

Proposition 2.2.11 (Criteria for checking convergence in metric spaces).
Assume that X is a metric space. Consider a sequence $\{C_n\}$ of subsets of X, and a set $C \subset X$.

(a) $C \subset \lim \operatorname{int}_n C_n$ *if and only if* $d(x, C) \geq \lim \sup_n d(x, C_n)$ *for all* $x \in X$.

(b) *Assume that C is closed. If* $d(x, C) \leq \lim \inf_n d(x, C_n)$ *for all* $x \in X$ *then* $C \supset \lim \operatorname{ext}_n C_n$.

(c) *Assume that X is a finite-dimensional Banach space. If* $C \supset \lim \operatorname{ext}_n C_n$ *then* $d(x, C) \leq \lim \inf_n d(x, C_n)$ *for all* $x \in X$.

Proof. (a) Define $d := d(x, \overline{C}) = d(x, C) \geq 0$. Then $B(x, d + \varepsilon)^\circ \cap C \neq \emptyset$ for all $\varepsilon > 0$. By Proposition 2.2.9(a) there exists $J \in \mathcal{N}_\infty$ such that $B(x, d + \varepsilon)^\circ \cap C_j \neq \emptyset$

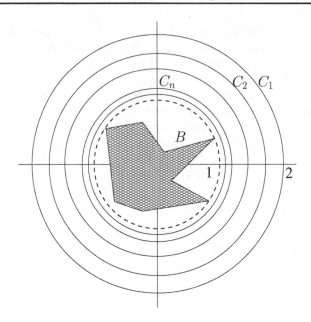

Figure 2.2. *Counterexample for convergence criteria*

for all $j \in J$. In other words, there exists $J \in \mathcal{N}_\infty$ such that $d(x, C_j) < d + \varepsilon$ for all $j \in J$, which implies that

$$\limsup_n d(x, C_n) \leq \sup_{j \in J} d(x, C_j) \leq d + \varepsilon.$$

Inasmuch as $\varepsilon > 0$ is arbitrary, we get the desired inequality.

For the converse result, observe that by the assumption every point $x \in C$ satisfies $\limsup_n d(x, C_n) = 0$ and hence $x \in \liminf_n C_n$ by (2.5).

(b) Take $x \in \limext_n C_n$. By (2.4) we have $\liminf_n d(x, C_n) = 0$, which together with the assumption yields $d(x, C) = 0$. Because C is closed, we conclude that $x \in C$.

(c) Suppose that $C \supset \limext_n C_n$ and let $d := d(x, \overline{C}) = d(x, C) \geq 0$. If $d = 0$ the conclusion obviously holds. Assume now that $d > 0$ and take $\varepsilon \in (0, d)$. Then $B(x, d - \varepsilon) \cap C = \emptyset$. Because X is finite-dimensional, $B(x, d - \varepsilon)$ is compact, and by Proposition 2.2.9(c) there exists $J \in \mathcal{N}_\infty$ such that $B(x, d - \varepsilon) \cap C_j = \emptyset$ for all $j \in J$. In other words, there exists $J \in \mathcal{N}_\infty$ such that $d(x, C_j) > d - \varepsilon$ for all $j \in J$, which implies that

$$\liminf_n d(x, C_n) \geq \inf_{j \in J} d(x, C_j) \geq d - \varepsilon.$$

Inasmuch as $\varepsilon \in (0, d)$ is arbitrary, we get the desired inequality.   □

Tychonoff's theorem states that the arbitrary product of compact spaces is compact. Mrowka [163] applied this result in order to obtain a compactness property of an arbitrary net of sets. Namely, he proved that every net of sets has a set' convergent subnet. Recall that the discrete topology in a space Y has for open sets all possible subsets of Y. Given two nonempty sets A, B, by A^B we denote the set of all functions from B to A.

Theorem 2.2.12. *Let X be a Hausdorff space and $\{C_i\}_{i \in I}$ a net in X. Then there exists a convergent subnet $\{\tilde{C}_j\}_{j \in J}$ of $\{C_i\}_{i \in I}$.*

Proof. Let $\mathcal{B}$ be the base of the topology of X and consider $Y := \{0, 1\}^{\mathcal{B}}$, where $\{0, 1\}$ is endowed with the discrete topology. Because $\{0, 1\}$ is compact, by Tychonoff's theorem we conclude that Y is compact as well. For every $i \in I$, define the net $\{f_i\}_{i \in I} \subset Y$ as

$$f_i(W) := \begin{cases} 1 & \text{if } W \cap C_i \neq \emptyset \\ 0 & \text{if } W \cap C_i = \emptyset. \end{cases}$$

Because Y is compact, by Theorem 2.1.9(a) there exists a convergent subnet $\{g_j\}_{j \in J}$ of $\{f_i\}_{i \in I}$. Let $g \in Y$ be the limit of the convergent subnet $\{g_j\}_{j \in J}$. By the definition of subnet, there exists a function $\varphi : J \to I$ such that $f_{\varphi(j)} = g_j$ for all $j \in J$, where φ and J verify condition (SN2) of the definition of subnet. This implies that the net $\{\tilde{C}_j\}_{j \in J}$ defined by $\tilde{C}_j := C_{\varphi(j)}$ is a subnet of $\{C_i\}_{i \in I}$. Let $W \in \mathcal{B}$ be such that $W \cap \tilde{C}_j \neq \emptyset$ frequently in J. We claim that $W \cap \tilde{C}_j \neq \emptyset$ eventually in J. Note that this claim yields $\lim \text{int}_j \tilde{C}_j = \lim \text{ext}_j \tilde{C}_j$. Let us prove the claim. If $W \cap \tilde{C}_j \neq \emptyset$ frequently in J then, by definition of $\tilde{C}_j$, we have that $W \cap C_{\varphi(j)} \neq \emptyset$ frequently in J. Using now the definition of f_i, we get $f_{\varphi(j)}(W) = 1$ frequently in J. In other words, $g_j(W) = 1$ frequently in J. Thus, the limit g of $\{g_j\}_{j \in J}$ must verify $g(W) = 1$, because otherwise g_j must be eventually equal to zero, a contradiction. So we have that $g(W) = 1$. Using again the fact that g is the limit of $\{g_j\}_{j \in J}$, we conclude that $g_j(W) = 1$ eventually. Backtracking the previous argument, we obtain that $W \cap \tilde{C}_j \neq \emptyset$ eventually in J. Therefore, $\lim \text{int}_j \tilde{C}_j = \lim \text{ext}_j \tilde{C}_j$ and hence the subnet $\{\tilde{C}_j\}_{j \in J}$ converges to this common set. $\square$

Recall that a topological space X satisfies the second countability axiom N_2 when there exists a countable family of open sets $\mathcal{B} := \{U_m\}_{m \in \mathbb{N}}$ such that any open set can be expressed as a union of elements in $\mathcal{B}$. We denote by $A \subsetneq B$ the fact that $A \subset B$ and $A \neq B$.

Remark 2.2.13. Assume that we have a sequence of sets $\{C_n\}_{n \in \mathbb{N}}$. Looking at this sequence as a net, we know by Theorem 2.2.12 that it has a convergent subnet. However, for general topological spaces, this subnet is not necessarily a subsequence of sets. In other words, the index set of the subnet is not necessarily a cofinal subset of $\mathbb{N}$. In [24, Proposition 5.2.13] it is shown that in metric spaces which are not N_2 there exist sequences of sets without a convergent subsequence. On the other hand, when the space is N_2, we can always extract a convergent subsequence of nets. We prove this fact next.

Proposition 2.2.14. *Assume that X satisfies N_2. If $\lim \operatorname{ext}_n C_n \neq \emptyset$, then there exist $J \in \mathcal{N}_\sharp$ and a nonempty set C such that $C_j \to_J C$.*

Proof. Fix an element $z \in \lim \operatorname{ext}_n C_n$. Then there exist $J_0 \in \mathcal{N}_\sharp$ and a sequence $\{x_j\}_{j \in J_0}$ such that $x_j \in C_j$ for all $j \in J_0$ and $x_j \to_{J_0} z$. Let $j_0 := \min J_0$. Consider the index set

$$I_1 := \{j \in J_0 : C_j \cap U_1 \neq \emptyset\},$$

where U_1 is the first element of the countable basis $\mathcal{B}$. Define now an index set $J_1 \subsetneq J_0$ in the following way.

$$J_1 := \begin{cases} \{j \in I_1 : j > j_0\}, & \text{if } I_1 \in \mathcal{N}_\sharp, \\ \{j \in J_0 \setminus I_1 : j > j_0\}, & \text{otherwise.} \end{cases}$$

By construction, $J_1 \subsetneq J_0$ and $J_1 \in \mathcal{N}_\sharp$. Continuing this process, given J_{k-1} and $j_{k-1} := \min J_{k-1}$, the kth index set $J_k \subsetneq J_{k-1}$ is defined by considering the set

$$I_k := \{j \in J_{k-1} : C_j \cap U_k \neq \emptyset\},$$

where U_k is the kth element of $\mathcal{B}$. Now define

$$J_k := \begin{cases} \{j \in I_k : j > j_{k-1}\}, & \text{if } I_k \in \mathcal{N}_\sharp, \\ \{j \in J_{k-1} \setminus I_k : j > j_{k-1}\}, & \text{otherwise.} \end{cases}$$

Thus this sequence of index sets satisfies

$$J_0 \supsetneq J_1 \supsetneq J_2 \supsetneq \cdots \supsetneq J_{k-1} \supsetneq J_k \supsetneq \cdots,$$

and $J_k \in \mathcal{N}_\sharp$ for all k. Also, the sequence $\{j_k\}_{k \in \mathbb{N}}$ of the first elements of each J_k is strictly increasing. Consider now the index set J consisting of all these minimum elements; that is,

$$J := \{j_0, j_1, j_2, \ldots, j_{k-1}, j_k, \ldots\}.$$

Then $J \in \mathcal{N}_\sharp$ because all these elements are different. We claim that for any fixed $k \in \mathbb{N}$, only one of the two possibilities below can occur for the basis element U_k.

(I) There exists $m_k \in \mathbb{N}$ such that $U_k \cap C_j \neq \emptyset$ for all $j \in J$ such that $j \geq m_k$. Equivalently,

$$\{j \in J : U_k \cap C_j \neq \emptyset\} \supset \{j \in J : j \geq m_k\}. \tag{2.9}$$

(II) There exists $m_k \in \mathbb{N}$ such that $U_k \cap C_j = \emptyset$ for all $j \in J$ such that $j \geq m_k$. Equivalently,

$$\{j \in J : U_k \cap C_j = \emptyset\} \supset \{j \in J : j \geq m_k\}. \tag{2.10}$$

For proving this claim, take $k \in \mathbb{N}$ (or equivalently, $U_k \in \mathcal{B}$). Either $I_k \in \mathcal{N}_\sharp$ or $I_k \notin \mathcal{N}_\sharp$. Consider first the case in which $I_k \in \mathcal{N}_\sharp$. In this case we have $J_k \subset I_k$.

By construction, $J_\ell \subset J_k$, for all $\ell \geq k$. In particular, the first elements of all these sets are in J_k. In other words,

$$\{j \in J : j \geq j_k\} = \{j_k, j_{k+1}, \ldots\} \subset J_k \subset I_k = \{j \in J_{k-1} : C_j \cap U_k \neq \emptyset\}.$$

Note that the leftmost set is contained in J, so by intersecting both sides with J we have that case (I) holds with $m_k := j_k$ in (2.9). Consider now the case in which $I_k \notin \mathcal{N}_\sharp$. This means that the set $I_k = \{j \in J_{k-1} : C_j \cap U_k \neq \emptyset\}$ is finite. Then there exists $j_\ell \in J$ such that $j_\ell > j$ for all $j \in I_k$. Because $I_k \subset J_{k-1}$, this implies that j_ℓ will be strictly bigger than the first element of J_{k-1} (i.e., $j_\ell > j_{k-1}$). By construction of J, if $j \in J$ is such that $j \geq j_\ell$, then there exists $p \in \mathbb{N}$ ($p \geq \ell > k-1$) such that $j = j_p = \min J_p$. Therefore, $j = j_p \in J_p \subset J_{k-1}$. So, if $j \in J$ is such that $j \geq j_\ell$, then $j \in J_{k-1}$ and $j \notin I_k$, by definition of j_ℓ. It follows that

$$\{j \in J : j \geq j_\ell\} \subset \{j \in J_{k-1} : j \notin I_k\} = \{j \in J_{k-1} : C_j \cap U_k = \emptyset\}.$$

Note again that the leftmost set is contained in J, so by intersecting both sides with J we have that case (II) holds with $m_k := j_\ell$ in (2.10). The claim holds. Now define $C := \lim \mathrm{ext}_J C_j$. Because $J \subset J_0$, we have that $x_j \to_J z \in C$. So, $C \neq \emptyset$. Our aim is to prove that $C \subset \lim \mathrm{int}_J C_j$, in which case $C_j \to_J C$, by definition of C. In order to prove the claim, we use Proposition 2.2.9(a). It is enough to show that for every basis element $U_k \in \mathcal{B}$ such that

$$C \cap U_k \neq \emptyset, \tag{2.11}$$

there exists $L_0 \in \mathcal{N}_\infty$ such that $C_j \cap U_k \neq \emptyset$ for all $j \in L_0 \cap J$. So, assume that $U_k \in \mathcal{B}$ verifies (2.11) and take $x \in C \cap U_k$. By definition of C, there exist an infinite set $L \subset J$ and a sequence $\{y_j\}_{j \in L}$ such that $y_j \in C_j$ for all $j \in L$ and $y_j \to_L x$. Because $x \in U_k$, there exists $L_0 \in \mathcal{N}_\infty$ such that $y_j \in U_k$ for all $j \in L_0 \cap L$. Therefore, for this U_k there exists an infinite set of indices $L_0 \cap L$ such that $y_j \in C_j \cap U_k$. It follows that this U_k must satisfy condition (I), which implies the existence of an m_k such that $C_j \cap U_k \neq \emptyset$ for all $j \in J$, $j \geq m_k$. By Proposition 2.2.9(a), we conclude that $C \subset \lim \mathrm{int}_J C_j$ and hence $C_j \to_J C$. $\square$

For a sequence $\{\alpha_n\}$ of real numbers, it is well-known that its smallest cluster point is given by $\liminf_n \alpha_n$ and its largest cluster point is given by $\limsup_n \alpha_n$. A remarkable fact is that Proposition 2.2.14 allows us to establish an analogous situation for sequences of sets. Of course, cluster points must be understood in the sense of set convergence, and the smallest and largest cluster points are replaced by the intersection and the union of all possible cluster points, respectively. More precisely, given a sequence $\{C_n\}$ of subsets of X, define the set

$$\mathcal{CP} := \{D : \text{ there exists } J \in \mathcal{N}_\sharp \text{ such that } C_j \to_J D\}$$

$$= \{D : D \text{ is a cluster point in the sense of set convergence}\}. \tag{2.12}$$

Corollary 2.2.15. *Let X be N_2 and let $\{C_n\}$ be a sequence of subsets of X. Consider the set $\mathcal{CP}$ defined as in (2.12). Then*

(i) $\liminf_n C_n = \bigcap_{D \in \mathcal{CP}} D$.

(ii) $\limext_n C_n = \bigcup_{D \in \mathcal{CP}} D$.

Proof.
    (i) Take $x \in \liminf_n C_n$. By the definition of $\liminf$ there exist an index set $J_x \in \mathcal{N}_\infty$ and a corresponding sequence $\{x_j\}_{j \in J_x}$ such that $x_j \in C_j$ for all $j \in J_x$ with $x_j \to x$. We must prove that $x \in D$ for all $D \in \mathcal{CP}$. In order to do this, take $D \in \mathcal{CP}$. By the definition of $\mathcal{CP}$ there exists $\hat{J} \in \mathcal{N}_\sharp$ such that $C_j \to_{\hat{J}} D$. Note that $\hat{J} \cap J_x \in \mathcal{N}_\sharp$. It is clear that the sequence $\{x_j\}_{j \in \hat{J} \cap J_x}$ satisfies $x_j \in C_j$ for all $j \in \hat{J} \cap J_x$ with $x_j \to_{\hat{J} \cap J_x} x$. Because $J_x \in \mathcal{N}_\infty$, this means that $x \in \liminf_{\hat{J}} C_j = D$, as required. For the converse inclusion, take $x \in \bigcap_{D \in \mathcal{CP}} D$ and suppose that $x \notin \liminf_n C_n$. Then there exist a neighborhood V of x and an index set $J_x \in \mathcal{N}_\sharp$ such that for all $j \in J_x$ it holds that

$$C_j \cap V = \emptyset. \tag{2.13}$$

We have two alternatives for the subsequence $\{C_j\}_{j \in J_x}$:

(I) $\limext_{j \in J_x} C_j = \emptyset$, or

(II) $\limext_{j \in J_x} C_j \neq \emptyset$.

In case (I), we have $C_j \to_{J_x} \emptyset$, so $\emptyset \in \mathcal{CP}$ and hence $\bigcap_{D \in \mathcal{CP}} D = \emptyset$, contradicting the fact that $x \in \bigcap_{D \in \mathcal{CP}} D$. So, alternative (II) holds and we can use Proposition 2.2.14 in order to conclude that there exist a nonempty set C and an index set $\hat{J} \in \mathcal{N}_\sharp$, $\hat{J} \subset J_x$ such that

$$C_j \to_{\hat{J}} C. \tag{2.14}$$

This implies that $C \in \mathcal{CP}$, and by the assumption on x, we have $x \in C$. This fact, together with (2.14), implies that there exist an infinite index set $J' \subset \hat{J} \subset J_x$ and a sequence $\{y_j\}_{j \in J'}$, with $y_j \in C_j$ for all $j \in J'$, such that $y_j \to_{J'} x$. Then, for large enough $j \in J'$ we have $C_j \cap V \neq \emptyset$, which contradicts (2.13). Thus, $x \in \liminf_n C_n$.
    (ii) If $D \in \mathcal{CP}$ and $x \in D$, then by definition of $\mathcal{CP}$ it holds that $x \in \limext_n C_n$. This proves that $\bigcup_{D \in \mathcal{CP}} D \subset \limext_n C_n$. For the converse inclusion, take $x \in \limext_n C_n$. By the definition of $\limext$ there exist an index set $J_x \in \mathcal{N}_\sharp$ and a corresponding sequence $\{x_j\}_{j \in J_x}$ such that $x_j \in C_j$ for all $j \in J_x$ with $x_j \to_{J_x} x$. Because $\limext_{J_x} C_j \neq \emptyset$, by Proposition 2.2.14 there exist a nonempty set C and an index set $\hat{J} \in \mathcal{N}_\sharp$, $\hat{J} \subset J_x$ such that $C_j \to_{\hat{J}} C$. Because $x_j \in C_j$ for all $j \in \hat{J} \subset J_x$, we also have $x_j \to_{\hat{J}} x$ and hence $x \in \liminf_{\hat{J}} C_j = C$. Then there exists $C \in \mathcal{CP}$ such that $x \in C$ and we conclude that $x \in \bigcup_{D \in \mathcal{CP}} D$.
$\square$

Exercises

2.1. Prove that

(i) $\lim \operatorname{int}_i C_i \subset \lim \operatorname{ext}_i C_i$.

(ii) $\lim_i C_i = \emptyset$ if and only if $\lim \operatorname{ext}_i C_i = \emptyset$.

(iii) $\lim \operatorname{int}_i C_i = \lim \operatorname{int}_i \overline{C_i}$ (idem lim ext).

(iv) If $B_i \subset C_i$ for all $i \in I$ then $\lim \operatorname{int}_i B_i \subset \lim \operatorname{int}_i C_i$ (idem lim ext).

(v) If $C_i \subset B_i \subset D_i$ for all $i \in I$ and $\lim_i C_i = \lim_i D_i = C$, then $\lim \operatorname{int}_i B_i = C$.

2.2. Let $\{C_n\}$ be a sequence of sets.

(i) If $C_{n+1} \subset C_n$ for all $n \in \mathbb{N}$, then $\lim_n C_n$ exists and $\lim_n C_n = \cap_n \overline{C_n}$.

(ii) If $C_{n+1} \supset C_n$ for all $n \in \mathbb{N}$, then $\lim_n C_n$ exists and $\lim_n C_n = \overline{\bigcup_n C_n}$.

2.3. Let $\{C_n\}$ be a sequence of sets. Assume that $\lim_n C_n = C$ with $C_n \subset C$ and C_n closed for all $n \in \mathbb{N}$. Define $D_n := \bigcup_{i=1}^n C_i$. Prove that $\lim_n D_n = C$. State and prove an analogous result for $\lim_n C_n = C$ with $C_n \supset C$.

2.4. In items (i) and (ii) below, assume that X is a metric space.

(i) Take $\{x_n\} \subset X$, $\{\rho_n\} \subset \mathbb{R}_+$, x and $\rho > 0$ such that $\lim_n x_n = x$ and $\lim_n \rho_n = \rho$. Define $C_n := B(x_n, \rho_n)$. Then $\lim_n C_n = B(x, \rho)$.

(ii) With the notation of (i), assume that $\lim_n \rho_n = \infty$. Then $\lim_n C_n = X$ and $\lim_n X \setminus C_n = \emptyset$.

(iii) Take $D \subset X$ such that $\overline{D} = X$. Define $C_n := D$ for all $n \in \mathbb{N}$. Then $\lim_n C_n = X$.

(iv) Find $\{C_n\}$ such that every C_n is closed, $\lim_n C_n = X$ but $\lim_n X \setminus C_n \neq \emptyset$.

(v) If $C_n = D_1$ when n is even and $C_n = D_2$ when n is odd, then $\lim \operatorname{ext}_n C_n = \overline{D_1} \bigcup \overline{D_2}$ and $\lim \operatorname{int}_n C_n = \overline{D_1} \cap \overline{D_2}$.

2.5. Prove the following facts:

(i) $\lim \operatorname{ext}_i (A_i \cap B_i) \subset \lim \operatorname{ext}_i A_i \cap \lim \operatorname{ext}_i B_i$.

(ii) $\lim \operatorname{int}_i (A_i \cap B_i) \subset \lim \operatorname{int}_i A_i \cap \lim \operatorname{int}_i B_i$.

(iii) $\lim \operatorname{ext}_i (A_i \cup B_i) = \lim \operatorname{ext}_i A_i \cup \lim \operatorname{ext}_i B_i$.

(iv) $\lim \operatorname{int}_i (A_i \cup B_i) \supset \lim \operatorname{int}_i A_i \cup \lim \operatorname{int}_i B_i$.

(v) Prove that all the inclusions above can be strict.

2.6. Assume that $f : X \to Y$ is continuous. Then

(i) $f(\lim \operatorname{ext}_i C_i) \subset \lim \operatorname{ext}_i f(C_i)$.

(ii) $f(\lim \operatorname{int}_i C_i) \subset \lim \operatorname{int}_i f(C_i)$.

(iii) $\lim \operatorname{ext}_i f^{-1}(C_i) \subset f^{-1}(\lim \operatorname{ext}_i C_i)$.

(iv) $\lim \operatorname{int}_i f^{-1}(C_i) \subset f^{-1}(\lim \operatorname{int}_i C_i)$.

(v) Prove that all the inclusions above can be strict. Hint: for (i)–(ii) try $f : \mathbb{R} \to \mathbb{R}$ defined by $f(t) := e^{-t}$ and $C_n := [n, n+1]$; for (iii)–(iv) try $f : \mathbb{R} \to \mathbb{R}$ defined by $f(t) := \cos t$ and $C_n := [1 + 1/n, +\infty)$.

(vi) Assume that f is also topologically proper; that is, if $\{f(x_n)\}$ converges in Y, then $\{x_n\}$ has cluster points. Then equality holds in (i).

(vii) Assume that f is topologically proper and bijective. Then equality holds in (iii).

(viii) Find a topologically proper and surjective f for which the inclusion in (iii) is strict. Hint: try

$$f(t) = \begin{cases} t + (1 - 2\pi) & \text{if } t > 2\pi, \\ \cos t & \text{if } t \in [-\pi/2, 2\pi], \\ t + \pi/2 & \text{if } t < -\pi/2, \end{cases}$$

and $C_n := [1 + 1/n, +\infty)$.

(ix) Show a topologically proper and injective f for which the inclusion in (iii) is strict. Hint: try $f : [-1, 1] \to \mathbb{R}$ defined by $f(t) := t$ and $C_n := [1 + 1/n, +\infty)$.

2.7. Prove that the converse of Proposition 2.2.9(b) is not true. Hint: take $C := \{0\}$ and $C_n := [n, +\infty)$ for all n. Then $C \supset \lim \text{ext}_n C_n = \emptyset$, but the closed set $B := [1, +\infty)$ does not satisfy the conclusion.

2.8. Prove that Proposition 1.2.8(b) does not hold if C is not closed (see Remark 2.2.10).

2.9. Prove that Proposition 1.2.8(c) does not hold if X is not finite-dimensional. Hint: take X a Hilbert space, $C = \{z : \|z\| = 2\}$, and $C_n = \{e_n\}$ for all n, where $\{e_n\}$ is an orthonormal basis. Then clearly $C \supset \lim \text{ext}_n C_n = \emptyset$, but the inequality involving distances is not true for $x = 0$.

2.3   Point-to-set mappings

Let X and Y be arbitrary Hausdorff topological spaces and let $F : X \rightrightarrows Y$ be a mapping defined on X and taking values in the family of subsets of Y; that is, such that $F(x) \subset Y$ for all $x \in X$. The possibility that $F(x) = \emptyset \subset Y$ for some $x \in X$ is admitted.

Definition 2.3.1. *Consider X, Y, and F as above. Then F is characterized by its graph, denoted $\text{Gph}(F)$ and defined as*

$$\text{Gph}(F) := \{(x, v) \in X \times Y : v \in F(x)\}.$$

The projection of $\text{Gph}(F)$ onto its first argument is the domain of F, denoted $D(F)$ and given by:

$$D(F) := \{x \in X : F(x) \neq \emptyset\}.$$

The projection of $\mathrm{Gph}(F)$ *onto its second argument is the range of* F, *denoted* $R(F)$ *and given by:*

$$R(F) := \{y \in Y \ : \exists x \in D(F) \ \text{such that} \ y \in F(x)\}.$$

Definition 2.3.2. *Given a point-to-set mapping* $F : X \rightrightarrows Y$, *its inverse* F^{-1} : $Y \rightrightarrows X$ *is defined by its graph* $\mathrm{Gph}(F^{-1})$ *as*

$$\mathrm{Gph}(F^{-1}) := \{(v, x) \in Y \times X \ : v \in F(x)\}.$$

In other words, F^{-1} is obtained by interchanging the arguments in $\mathrm{Gph}(F)$.

2.4 Operating with point-to-set mappings

Let X be a topological space and Y a topological vector space. We follow Minkowski [148] in the definition of addition and scalar multiplication of point-to-set mappings.

Definition 2.4.1. *Consider* $F_1, F_2 : X \rightrightarrows Y$ *and define* $F_1 + F_2 : X \rightrightarrows Y$ *as*

$$(F_1 + F_2)(x) := \{v_1 + v_2 \ : v_1 \in F_1(x) \ \text{and} \ v_2 \in F_2(x)\}.$$

Definition 2.4.2. *Consider* $F : X \rightrightarrows Y$. *For any* $\lambda \in \mathbb{R}$, *define* $\lambda F : X \rightrightarrows Y$ *as*

$$\lambda F(x) := \{\lambda v \ : v \in F(x)\}.$$

Composition of point-to-set mappings is also standard.

Definition 2.4.3. *Consider* $F : X \rightrightarrows Y$ *and* $G : Z \rightrightarrows X$ *and define* $F \circ G : Z \rightrightarrows Y$ *as*

$$(F \circ G)(x) := \cup_{z \in G(x)} F(z).$$

Remark 2.4.4. Observe that when F is point-to-point and $M \subset Y$, the set $F^{-1}(M)$ can be seen as $\{x \in X \ : F(x) \cap M \neq \emptyset\}$, or as the set $\{x \in X \ : \{F(x)\} \subset M\}$.

These two interpretations allow us to define two different notions of the inverse image of a set $A \subset Y$ through a point-to-set mapping F.

Definition 2.4.5.

(i) *The* inverse image *of* A *is defined as*

$$F^{-1}(A) := \{x \in X \ : F(x) \cap A \neq \emptyset\}.$$

(ii) *The* core *of* A *is defined as*

$$F^{+1}(A) := \{x \in X \ : F(x) \subset A\}.$$

2.5   Semicontinuity of point-to-set mappings

Assume that X and Y are Hausdorff topological spaces and $F : X \rightrightarrows Y$ is a point-to-set mapping. We consider three kinds of semicontinuity for such an F.

Definition 2.5.1.

(a) F is said to be outer-semicontinuous(OSC) at $x \in D(F)$ if whenever a net $\{x_i\}_{i \in I} \subset X$ converges to x then $\lim \text{ext}_i F(x_i) \subset F(x)$.

(b) F is said to be inner-semicontinuous(ISC) at $x \in D(F)$ if for any net $\{x_i\}_{i \in I} \subset X$ convergent to x we have $F(x) \subset \lim \text{int}_i F(x_i)$.

(c) F is said to be upper-semicontinuous(USC) at $x \in D(F)$ if for all open $W \subset Y$ such that $W \supset F(x)$ there exists a neighborhood U of x such that $F(x') \subset W$ for all $x' \in U$.

We say that F is OSC *(respectively,* ISC, USC*) if (a) (respectively, (b), (c)) holds for all $x \in D(F)$.*

Remark 2.5.2. If F is OSC or ISC at x, then we must have $F(x)$ a closed set. Indeed, assume $y \in \overline{F(x)}$ and define the net $\{x_i\}$ as $x_i = x$ for all $i \in I$, which trivially converges to x. Because $y \in \overline{F(x)}$ for every W neighborhood of y we have that $W \cap F(x) = W \cap F(x_i) \neq \emptyset$ for all $i \in I$. This gives $y \in \lim \text{int}_i F(x_i) \subset \lim \text{ext}_i F(x_i) = F(x)$, where we are using in the equality the fact that $F(x_i) = F(x)$ for all i.

Two kinds of continuity are of interest for point-to-set mappings. We are concerned mostly with the first one.

Definition 2.5.3.

(i) $F : X \rightrightarrows Y$ is continuous at $x \in D(F)$ if it is OSC and ISC at x.

(ii) $F : X \rightrightarrows Y$ is K-continuous at $x \in D(F)$ if it is USC and ISC at x (K for Kuratowski, [127]).

By the definition of interior and exterior limits, we have $\lim \text{int}_i F(x_i) \subset \lim \text{ext}_i F(x_i)$. When F is continuous and the net $\{x_i\}$ converges to x, we must have $F(x) \subset \lim \text{int}_i F(x_i) \subset \lim \text{ext}_i F(x_i) \subset F(x)$. In other words,

$$\lim_i F(x_i) = F(x),$$

which resembles point-to-point continuity.

We show next that outer-semicontinuity at every point is equivalent to closedness of the graph.

Theorem 2.5.4. *Assume that X and Y are Hausdorff topological spaces and take a point-to-set mapping $F : X \rightrightarrows Y$. The following statements are equivalent.*

(a) $\mathrm{Gph}(F)$ *is closed.*

(b) *For each* $y \notin F(x)$, *there exist neighborhoods* $W \in \mathcal{U}_x$ *and* $V \in \mathcal{U}_y$ *such that* $F(W) \cap V = \emptyset$.

(c) F *is* OSC.

Proof. Let us prove that (a) implies (b). Assume that $\mathrm{Gph}(F)$ is closed and fix $y \notin F(x)$. Closedness of the graph implies that the set $F(x)$ is closed. Hence there exists a neighborhood $V \in \mathcal{U}_y$ such that $F(x) \cap V = \emptyset$. We claim that there exist neighborhoods W_0 of x and $V_0 \in \mathcal{U}_y$ with $V_0 \subset V$ such that $F(W_0) \cap V_0 = \emptyset$. Suppose that for every pair of neighborhoods W of x and $U \in \mathcal{U}_y$ with $U \subset V$ we have $F(W) \cap U \neq \emptyset$. Take the directed set $I = \mathcal{U}_x \times \mathcal{U}_y$ with the partial order of the reverse inclusion in both coordinates. For every $(W, U) \in I$ we can choose $x_{WU} \in W$ and $y_{WU} \in F(x_{WU}) \cap U$. We claim that the nets $\{x_{WU}\}$ and $\{y_{WU}\}$ converge to x and y, respectively. Indeed, take $W_1 \in \mathcal{U}_x$. Consider the subset of I given by $J := \{(W, U) \mid W \subset W_1, \ U \in \mathcal{U}_y\}$. The set J is clearly terminal and for all $(W, U) \in J$ we have $x_{WU} \in W \subset W_1$. Hence the net $\{x_{WU}\}$ converges to x. Fix now $U_1 \subset V$ with $U_1 \in \mathcal{U}_y$. Consider the subset of I given by $\hat{J} := \{(W, U) \mid U \subset U_1, \ W \in \mathcal{U}_x\}$. Again the set $\hat{J}$ is terminal in I and for all $(W, U) \in \hat{J}$ we have $y_{WU} \in U \subset U_1$. Therefore the claim on the nets $\{x_{WU}\}$ and $\{y_{WU}\}$ is true. Altogether, the net $\{(x_{WU}, y_{WU})\} \subset \mathrm{Gph}(F)$ converges to (x, y). Assumption (a) and Theorem 2.1.6 yield $(x, y) \in \mathrm{Gph}(F)$, so that $y \in F(x)$, which contradicts the assumption on y. Therefore there exist neighborhoods W_0 of x and $V_0 \in \mathcal{U}_y$ with $V_0 \subset V$ such that $F(W_0) \cap V_0 = \emptyset$, which implies (b). For the converse, take a net $\{(x_i, y_i)\}_{i \in I} \subset \mathrm{Gph}(F)$ converging to (x, y) and assume that $y \notin F(x)$. By (b), there exist neighborhoods $W \in \mathcal{U}_x$ and $V \in \mathcal{U}_y$ such that $F(W) \cap V = \emptyset$. Because $\{(x_i, y_i)\}$ converges to (x, y), there exists a terminal subset J of I such that $x_i \in W$ and $y_i \in V$ for all $i \in J$. Because $(x_i, y_i) \in \mathrm{Gph}(F)$, we have that $y_i \in F(x_i) \cap V \subset F(W) \cap V = \emptyset$ for all $i \in J$, which is a contradiction, and therefore $y \in F(x)$. Hence, $\mathrm{Gph}(F)$ is closed by Theorem 2.1.6. Let us prove now that (b) implies (c). Take a net $\{x_i\}_{i \in I}$ converging to x and suppose that there exists $y \in \lim \mathrm{ext}_i \, F(x_i)$ such that $y \notin F(x)$. Take neighborhoods $W \in \mathcal{U}_x$ and $V \in \mathcal{U}_y$ as in (b). Because $y \in \lim \mathrm{ext}_i \, F(x_i)$ there exists a set J cofinal in I such that $F(x_i) \cap V \neq \emptyset$ for all $i \in J$. For a terminal subset K of I, we have that $x_i \in W$ for all $i \in K$. Hence for all i in the cofinal set $K \cap J$ we have $\emptyset \neq F(x_i) \cap V \subset F(W) \cap V = \emptyset$. This contradiction entails that $y \in F(x)$, which yields that F is OSC. We complete the proof by establishing that (c) implies (b). Assume that there exists $y \notin F(x)$ for which the conclusion in (b) is false. As in the argument above, we construct a net $\{(x_{WU}, y_{WU})\}$ converging to (x, y). Using this fact and (c), we get that $y \in \lim \mathrm{int}_I \, F(x_{WU}) \subset \lim \mathrm{ext}_I \, F(x_{WU}) \subset F(x)$, which contradicts the assumption on y. $\square$

A direct consequence of the previous result and Theorem 2.1.6 is the following sequential characterization of outer-semicontinuous mappings for N_1 spaces.

Theorem 2.5.5. *Assume that X, Y are N_1 and take $F : X \rightrightarrows Y$. The following statements are equivalent.*

(a) *Whenever a sequence $\{(x_n, y_n)\} \subset \mathrm{Gph}(F)$ converges to $(x, y) \in X \times Y$, it holds that $y \in F(x)$.*

(b) *F is OSC.*

Inner-semicontinuity of point-to-set mappings can also be expressed in terms of sequences when the spaces are N_1.

Theorem 2.5.6. *Assume that X, Y are N_1 and take $F : X \rightrightarrows Y$. Fix $x \in D(F)$. The following statements are equivalent.*

(a) *For any $y \in F(x)$ and for any sequence $\{x_n\} \subset D(F)$ such that $x_n \to x$ there exists a sequence $\{y_n\}$ such that $y_n \in F(x_n)$ for all $n \in \mathbb{N}$ and $y_n \to y$.*

(b) *F is ISC at x.*

Proof. Assume that (a) holds for F and take a net $\{z_i\} \subset X$ converging to x. We must prove that $F(x) \subset \liminf_i F(z_i)$. Assume that there exists $y \in F(x)$ such that $y \notin \liminf_i F(z_i)$. This means that there exists a neighborhood U_k of the countable basis $\mathcal{U}_y$ and a cofinal subset J of I such that $U_k \cap F(z_i) = \emptyset$ for all $i \in J$. Using the fact that the whole net $\{z_i\}$ converges to x, we have that the subnet $\{z_j\}_{j \in J}$ also converges to x. Because X is N_1, we can construct a countable set of indices $\{i_n \in J \mid n \in \mathbb{N}\}$ such that the sequence $\{z_{i_n}\}_n$ converges to x. Assumption (a) yields the existence of a sequence $y_n \in F(z_{i_n})$ with $y_n \to y$. Take now n_0 such that $y_n \in U_k$ for all $n \geq n_0$. For all $n \geq n_0$ we can write

$$y_n \in U_k \cap F(z_{i_n}).$$

By construction, $i_n \in J$ which yields $U_k \cap F(z_{i_n}) = \emptyset$, a contradiction. Hence $y \in \liminf_i F(z_i)$, which entails that F is ISC at x.

Assume now that F is ISC at x. Fix $y \in F(x)$ and take a sequence $\{x_n\} \subset D(F)$ such that $x_n \to x$. Every sequence is a net, thus we have that $F(x) \subset \liminf_n F(x_n)$, by inner-semicontinuity of F. Hence $y \in \liminf_n F(x_n)$. By Proposition 2.2.5(i) with $C_n = F(x_n)$, there exists $y_n \in F(x_n)$ with $y_n \to y$. Therefore (a) holds. $\square$

Remark 2.5.7. The definition of USC can be seen as a natural extension of the continuity of functions. Indeed, for point-to-point mappings upper-semicontinuity and continuity become equivalent. However, this concept cannot be used for expressing continuity of point-to-set mappings in which $F(x)$ is unbounded and varies "continuously" with a parameter, as the following example shows.

Example 2.5.8. Define $F : [0, 2\pi] \rightrightarrows \mathbb{R}^2$ as

$$F(\alpha) := \{\lambda(\cos\alpha, \sin\alpha) : \lambda \geq 0\}.$$

F is continuous, but it is not USC at any $\alpha \in [0, 2\pi]$. As it is shown in Figure 2.3, there exists an open set $W \supset F(\alpha)$ such that for $\beta \to \alpha$ the sets $F(\beta)$ are not contained in W.

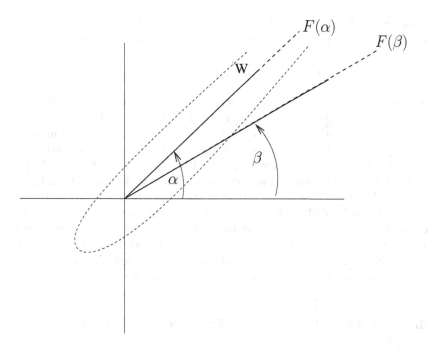

Figure 2.3. *Continuous but not upper-semicontinuous mapping*

In resemblance to the classical definition of continuity, the following proposition expresses semicontinuity of point-to-set mappings in terms of inverse images. The result below can be found in [26, Ch. VI, Theorems 1 and 2].

Proposition 2.5.9 (Criteria for semicontinuity). *Assume that $x \in D(F)$.*

(a) F is USC at x if and only if $F^{+1}(W)$ is a neighborhood of x for all open $W \subset Y$ such that $W \supset F(x)$. Consequently, F is USC if and only if $F^{+1}(W)$ is open for every open $W \subset Y$.

(b) F is ISC at x if and only if $F^{-1}(W)$ is a neighborhood of x for all open $W \subset Y$ such that $W \cap F(x) \neq \emptyset$. Consequently, F is ISC if and only if $F^{-1}(W)$ is open in Y for all open $W \subset Y$.

Proof. (a) The result follows immediately from the definitions.

(b) Assume that F is ISC at x and let $W \subset Y$ be an open set such that $W \cap F(x) \neq \emptyset$. We clearly have that $x \in F^{-1}(W)$. Assume that $F^{-1}(W)$ is not a neighborhood of x. Denote as $\mathcal{U}_x$ the family of all neighborhoods of x. We know that the set $\mathcal{U}_x$ is directed with the reverse inclusion. If the conclusion is not true, then for all $V \in \mathcal{U}_x$, $V \not\subset F^{-1}(W)$. In other words, for all $V \in \mathcal{U}_x$ there exists $x_V \in V$ such that

$$F(x_V) \cap W = \emptyset. \tag{2.15}$$

Because $x_V \in V$ for all $V \in \mathcal{U}_x$, we have that $\{x_V\}$ is a standard net, and hence it tends to x. By inner-semicontinuity of F we conclude that

$$F(x) \subset \liminf_{V \in \mathcal{U}_x} F(x_V).$$

Take now $y \in W \cap F(x)$. The expression above yields $y \in \liminf_{V \in \mathcal{U}_x} F(x_V)$. Because W is a neighborhood of y, we conclude that $F(x_V) \cap W \neq \emptyset$ eventually in $\mathcal{U}_x$. This fact clearly contradicts (2.15) and hence the conclusion must be true. Conversely, assume that the statement on neighborhoods holds, and take a net $\{x_i\}_{i \in I}$ converging to x. We must prove that $F(x) \subset \liminf_{i \in I} F(x_i)$. Take $y \in F(x)$ and denote by $\mathcal{U}_y$ the family of all neighborhoods of y. Fix an open neighborhood $W \in \mathcal{U}_y$. We claim that $W \cap F(x_i) \neq \emptyset$ eventually in I. Because $y \in F(x)$ and $W \in \mathcal{U}_y$ we have that $W \cap F(x) \neq \emptyset$. By our assumption, the set $F^{-1}(W)$ is a neighborhood of x. Using now the fact that the net $\{x_i\}_{i \in I}$ converges to x we conclude that $x_i \in F^{-1}(W)$ eventually in I. In other words, $W \cap F(x_i) \neq \emptyset$ eventually in I. Our claim holds and the proof of the first statement is complete. The second statement in (b) follows directly from the first one.   □

Definition 2.5.10. *We say that $F : X \rightrightarrows Y$ is* closed-valued *if $F(x)$ is a closed subset of Y for all $x \in X$.*

Let us recall the definition of locally compact topology (see, e.g., [63, Definition 2.1.1]).

Definition 2.5.11. *A topological space Z is* locally compact *if each $x \in Z$ has a compact neighborhood. In other words, when for each point we can find a compact set V and an open set U such that $x \in U \subset V$.*

We call to mind that when the space is Hausdorff and locally compact, then the family of compact neighborhoods of a point is a base of neighborhoods of that point. As a consequence, every neighborhood of a point contains a compact neighborhood of that point. (see, e.g., [115, Ch. 5, Theorems 17 and 18]). These properties are used in the result below, where we state the semicontinuity properties in terms of closed sets.

Proposition 2.5.12.

(a) *F is ISC if and only if $F^{+1}(R)$ is closed for all closed $R \subset Y$.*

(b) F is USC if and only if $F^{-1}(R)$ is closed for all closed $R \subset Y$.

(c) If F is OSC, then $F^{-1}(K)$ is closed for every compact set $K \subset Y$. If F is closed-valued and X is locally compact, then the converse of the last statement is also true.

Proof. Items (a) and (b) follow from Proposition 2.5.9 and the fact that $F^{+1}(A) = (F^{-1}(A^c))^c$ for all $A \subset Y$. For proving (c), assume first that F is OSC and take a compact set $K \subset Y$. In order to prove that $F^{-1}(K)$ is closed, take a net $\{x_i\}_{i \in I} \subset F^{-1}(K)$, converging to x. By Theorem 2.1.6 it is enough to prove that $x \in F^{-1}(K)$. Because $F(x_i) \cap K \neq \emptyset$ for all $i \in I$, there exists a net $\{y_i\}_{i \in I}$ such that $y_i \in F(x_i) \cap K$ for all $i \in I$. Because $\{y_i\} \subset K$ and K is compact, by Theorem 2.1.9(a), there exists a cluster point $\bar{y}$ of $\{y_i\}$ belonging to K. Therefore, for every neighborhood W of $\bar{y}$ we have that $y_i \in W$ frequently. This implies that $F(x_i) \cap W \neq \emptyset$ frequently. In other words, $\bar{y} \in \lim \mathrm{ext}_i F(x_i)$. By outer-semicontinuity of F, we get $\bar{y} \in F(x)$, or, equivalently, $x \in F^{-1}(K)$. For proving the converse statement, take a net $\{(x_i)\}_{i \in I}$ such that $x_i \to x$. We must prove that

$$\lim_{i \in I} \mathrm{ext}\, F(x_i) \subset F(x).$$

Take $y \in \lim \mathrm{ext}_{i \in I} F(x_i)$ and assume that $y \notin F(x)$. Because $F(x)$ is closed and X is locally compact, there exists a compact neighborhood V of y such that

$$V \cap F(x) = \emptyset. \tag{2.16}$$

The fact that $y \in \lim \mathrm{ext}_{i \in I} F(x_i)$ implies that $V \cap F(x_i) \neq \emptyset$ frequently. Therefore, for a cofinal subset J_0 of I we have that $V \cap F(x_i) \neq \emptyset$ for all $i \in J_0$. In other words, the subnet $\{x_i\}_{i \in J_0}$ is contained in $F^{-1}(V)$ and converges to x. By assumption, the set $F^{-1}(V)$ is closed, so we must have $x \in F^{-1}(V)$, or, equivalently, $F(x) \cap V \neq \emptyset$, which contradicts (2.16). The claim holds and $y \in F(x)$. $\square$

Remark 2.5.13. If F is not closed-valued, the last statement in Proposition 2.5.12(c) cannot hold. By Remark 2.5.2, outer-semicontinuity forces F to be closed-valued. The statement on compact sets in Proposition 2.5.12(c) may hold for nonclosed-valued mappings. Consider $F : \mathbb{R} \rightrightarrows \mathbb{R}$ defined by $F(x) := [0, 1)$ for all $x \in \mathbb{R}$. It is easy to see that for all compact $K \subset \mathbb{R}$, either $F^{-1}(K) = \emptyset$ or $F^{-1}(K) = \mathbb{R}$, always closed sets. Because its graph is not closed, F is not OSC. See Figure 2.4.

The three concepts of semicontinuity are independent, in the sense that a point-to-set mapping can satisfy any one of them without satisfying the remaining two. This is shown in the following example.

Example 2.5.14. Consider the point-to-set mappings $F_1, F_2, F_3 : \mathbb{R} \rightrightarrows \mathbb{R}$ defined as:

$$F_1(x) := \begin{cases} [0, 1] & \text{if } x \neq 0, \\ [0, 1/2] & \text{if } x = 0. \end{cases} \qquad F_2(x) := \begin{cases} [0, 1/2] & \text{if } x \neq 0, \\ [0, 1) & \text{if } x = 0. \end{cases}$$

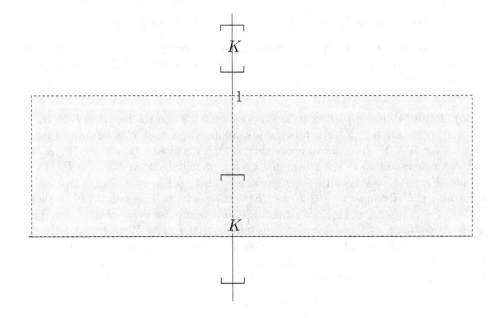

Figure 2.4. *Counterexample for outer-semicontinuity condition*

$$F_3(x) := \begin{cases} [1/x, +\infty) & \text{if } x > 0, \\ (-\infty, 0] & \text{if } x = 0, \\ [-1/x, 0] & \text{if } x < 0. \end{cases}$$

It is easy to check that F_1 is only ISC, F_2 is only USC, and F_3 is only OSC (see Figure 2.5).

Example 2.5.8 justifies the introduction of outer-semicontinuity in addition to upper-semicontinuity. There are some cases, however, in which one of these kinds of semicontinuity implies the other. In order to present these cases, we need some definitions.

Definition 2.5.15. *Let X be a topological space and Y a metric space. A point-to-set mapping $F : X \rightrightarrows Y$ is said to be* locally bounded at $x \in D(F)$ *if there exists a neighborhood U of x such that $F(U) := \cup_{z \in U} F(z)$ is bounded, and it is said to be* locally bounded *when it is locally bounded at every $x \in D(F)$.*

2.5.1   Weak and weak* topologies

We collect next some well-known facts concerning weak and weak star topologies in Banach spaces (see e.g., [22], [197] for a more detailed study).

Let X be a Banach space. The space of all bounded (i.e., continuous) linear functionals defined on X is called the *dual space* of X and denoted X^*. Let $\langle \cdot, \cdot \rangle :$

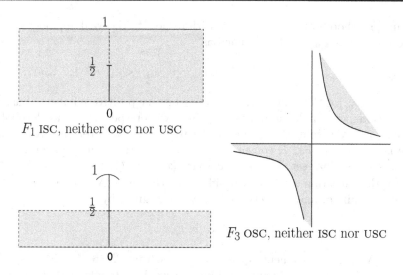

F_1 ISC, neither OSC nor USC

F_2 USC, neither ISC nor OSC

F_3 OSC, neither ISC nor USC

Figure 2.5. *Inner-, upper-, and outer-semicontinuous mappings*

$X^* \times X \to \mathbb{R}$ be the duality pairing in $X \times X^*$; that is, the linear functional $v \in X^*$ takes the value $\langle v, x \rangle \in \mathbb{R}$ at the point $x \in X$. The *bidual* of X, denoted X^{**}, is the set of all continuous linear functionals defined on X^*. An element $x \in X$ can be identified with the linear functional

$$\langle x, \cdot \rangle : \quad X^* \to \mathbb{R}, \\ v \mapsto \langle v, x \rangle.$$

With this identification, we have that $X \subset X^{**}$. When the opposite inclusion holds (i.e., when $X = X^{**}$) we say that X is reflexive. The space X^* is also a normed Banach space, with the norm given by

$$\|v\| := \sup_{\|x\| \leq 1} |\langle v, x \rangle|, \tag{2.17}$$

for all $v \in X^*$. It is clear that all the elements of X^* are continuous functions from X to $\mathbb{R}$ with respect to the norm topology in X. In other words, for every open set $U \subset \mathbb{R}$, the sets $v^{-1}(U) := \{x \in X : \langle v, x \rangle \in U\}$ are open with respect to the norm topology in X. However, it is also important to consider the weakest topology that makes every element of X^* continuous. By construction, this topology is contained in the original (norm) topology of X. For this reason, it is called the *weak* topology of X. Because it is the weakest tolopology that makes all elements of X^* continuous, the weak topology in X consists of arbitrary unions of finite intersections of sets of the form $v^{-1}(U)$, with $v \in X^*$ and $U \subset \mathbb{R}$ an open set. A basis of neighborhoods of a point $x_0 \in X$ with respect to the weak topology consists of the sets $V(x_0)_{[v_1,\ldots,v_m,\varepsilon_1,\ldots,\varepsilon_m]}$ defined as

$$V(x_0)_{[v_1,\ldots,v_m,\varepsilon_1,\ldots,\varepsilon_m]} := \{x \in X \; : \; |\langle v_i, x \rangle - \langle v_i, x_0 \rangle| < \varepsilon_i, \; i = 1,\ldots,m\}, \quad (2.18)$$

where $\{v_1,\ldots,v_m\} \subset X^*$ and $\varepsilon_1,\ldots,\varepsilon_m$ are arbitrary positive numbers. Based on these definitions, we can define convergence in X with respect to the weak topology. A net $\{x_i\}_{i \in I} \subset X$ is said to *converge weakly* to $x \in X$ if for every $v \in X^*$ we have that the net $\{\langle x_i, v \rangle\}_{i \in I} \subset \mathbb{R}$ converges to $\langle v, x \rangle$. In other words, for every $\varepsilon > 0$ there exists a terminal set $J \subset I$ such that $|\langle x_i, v \rangle - \langle x, v \rangle| < \varepsilon$ for all $i \in J$. We denote this situation $x_i \xrightarrow{w} x$. Consider now the space X^*. For every $x \in X$, consider the family of linear functions $\Lambda_x : X^* \to \mathbb{R}$ given by

$$\Lambda_x(v) = \langle v, x \rangle,$$

for all $v \in X^*$. Every element Λ_x is continuous from X^* to $\mathbb{R}$ with respect to the norm topology in X. Again consider the weakest topology that makes Λ_x continuous for every $x \in X$. This topology is called the *weak* topology* in X^*. A basis of neighborhoods of a point $v_0 \in X^*$ with respect to the weak* topology is defined in a way similar to (2.18):

$$V(v_0)_{[x_1,\ldots,x_m,\varepsilon_1,\ldots,\varepsilon_m]} := \{v \in X^* \; : \; |\langle x_i, v \rangle - \langle x_i, v_0 \rangle| < \varepsilon_i, \; i = 1,\ldots,m\}, \quad (2.19)$$

where $\{x_1,\ldots,x_m\} \subset X$ and $\varepsilon_1,\ldots,\varepsilon_m$ are positive numbers. Hence we can define the weak* convergence in the following way. A net $\{v_i\}_{i \in I} \subset X^*$ is said to *converge weakly** to $v \in X^*$ when for every $x \in X$ we have that the net $\{\langle v_i, x \rangle\}_{i \in I} \subset \mathbb{R}$ converges to $\langle v, x \rangle$. In other words, for every $\varepsilon > 0$ there exists a terminal set $J \subset I$ such that $|\langle v_i, x \rangle - \langle x, v \rangle| < \varepsilon$ for all $i \in J$. We denote this situation $v_i \xrightarrow{w^*} v$.

An infinite-dimensional Banach space X with the weak topology is not N_1, and the same holds for its dual X^* with the weak* topology.

We recall next, for future reference, one of the most basic theorems in functional analysis: the so-called Bourbaki–Alaoglu Theorem.

Theorem 2.5.16. *Let X be a Banach space. Then,*

 (i) *A subset $A \subset X^*$ is weak* compact if and only if it is weak* closed and bounded.*

 (ii) *If $A \subset X^*$ is convex, bounded, and closed in the strong topology, it is weak* compact.*

As a consequence of (i), every weak* convergent sequence is bounded.

Proof. See [74, 1, p, 424] for (i), and p. 248 in Volume I of [125] for (ii). The last statement follows from the fact that the set consisting of the union of a weak* convergent sequence and its weak* limit is a weak* compact set. Hence by (i) it must be bounded. $\square$

As we see from the last statement of Theorem 2.5.16, weak* convergent sequences are bounded. This property is no longer true for nets, which might be weak* convergent and unbounded. This is shown in the following example, taken from [24, Exercise 5.2.19].

Example 2.5.17. Let H be a separable Hilbert space with orthonormal basis $\{e_n\}$ and consider the sequence of sets $\{A_n\}$ defined as

$$A_n := \{e_n + n\, e_{n+p} : p \in \mathbb{N}\}.$$

Using the sequence of sets $\{A_n\}$, we construct a net $\{\xi_i\}_{i\in I} \subset H$ that is weakly convergent to 0 and *eventually unbounded*; that is, for every $\theta > 0$ there exists a terminal subset $J \subset I$ such that $\|\xi_i\| > \theta$ for all $i \in J$. Our construction is performed in two steps. Denote $\mathcal{U}_0$ the family of all weak neighborhoods of 0.

Step 1. In this step we prove that $0 \in w - \liminf_n A_n$, where $w-$ means that the $\liminf$ is taken with respect to the weak topology in H. In order to prove this fact, we consider the family of sets

$$W(x_0, \varepsilon) := \{u \in H : |\langle u, x_0 \rangle| < \varepsilon\},$$

for every given $x_0 \in H$, $\varepsilon > 0$. We claim that for every $x_0 \in H$ and $\varepsilon > 0$, there exists $n_0 = n_0(x_0, \varepsilon)$ such that

$$W(x_0, \varepsilon) \cap A_n \neq \emptyset \quad \text{for all } n \geq n_0. \tag{2.20}$$

Fix $x_0 \in H$ and $\varepsilon > 0$. Because $x_0 \in H$, there exists $n_1 = n_1(x_0, \varepsilon)$ such that

$$|\langle e_n, x_0 \rangle| < \frac{\varepsilon}{2} \quad \text{for all } n \geq n_1. \tag{2.21}$$

Indeed, this follows from the fact that $\|x_0\|^2 = \sum_n |\langle e_n, x_0 \rangle|^2 < +\infty$. For every fixed $n \geq n_1$ there exists $j_n \in \mathbb{N}$ such that

$$|\langle e_{n+p}, x_0 \rangle| < \frac{\varepsilon}{2\,n} \quad \text{for all } p \geq j_n.$$

Altogether, for every $n \geq n_1 = n_1(x_0, \varepsilon)$ we have

$$x_{n,j_n} := e_n + n\, e_{n+j_n} \in W(x_0, \varepsilon) \cap A_n,$$

which proves (2.20) for $n_0 := n_1(x_0, \varepsilon)$. From (2.18), we see that finite intersections of sets of the form $W(x_0, \varepsilon)$ make a base of $\mathcal{U}_0$ (i.e., the family of sets $\{W(x_0, \varepsilon)\}_{x_0 \in H, \varepsilon > 0}$ forms a subbase of $\mathcal{U}_0$). We claim that the argument used above for proving that (2.20) holds for $W(x_0, \varepsilon)$, can be extended to a finite intersection of these sets. Indeed, let $W \in \mathcal{U}_0$, so $W \supset U$, where U is given by

$$U := \{u \in H : |\langle u, x_i \rangle| < \varepsilon, \ i = 1, \ldots, q\} = \cap_{i=1}^q W(x_i, \varepsilon).$$

Choose $n_0 := \max\{n_1(x_1, \varepsilon), \ldots, n_q(x_q, \varepsilon)\}$, where each $n_i(x_i, \varepsilon)$ is obtained as in (2.21). For each fixed $n \geq n_0$, there exists $j_n \in \mathbb{N}$ such that

$$|\langle e_{n+p}, x_i \rangle| < \frac{\varepsilon}{2\,n} \quad \text{for all } p \geq j_n, \ i = 1, \ldots, q.$$

Altogether, for every $n \geq n_0$ we have

$$x_{n,j_n} := e_n + n\, e_{n+j_n} \in U \cap A_n \subset W \cap A_n,$$

which proves that $0 \in w - \liminf_n A_n$. A consequence of this fact is that, for every $W \in \mathcal{U}_0$, there exists a natural number $n = n(W)$ such that

$$W \cap A_n \neq \emptyset \quad \text{for all } n \geq n(W). \tag{2.22}$$

Without loss of generality, we can always choose $n(W)$ such that

$$\begin{aligned} n(W) &\geq n(W') \quad \text{whenever} \quad W \subset W' \\ n(W) &> n(W') \quad \text{whenever} \quad W \subsetneq W'. \end{aligned} \tag{2.23}$$

This choice implies that the set $Q := \{n(W) \in \mathbb{N} : W \in \mathcal{U}_0\}$ must be infinite.

Step 2. There exists a net $\{\xi_W\}_{W \in \mathcal{U}_0} \subset H$ weakly convergent to 0 and eventually unbounded. Take $W \in \mathcal{U}_0$. Using (2.22) we can choose $\xi_W \in W \cap A_{n(W)}$. Because $\xi_W \in W$ for all $W \in \mathcal{U}_0$ it is clear that ξ_W converges weakly to 0. We prove now that $\{\xi_W\}_{W \in \mathcal{U}_0}$ is eventually unbounded. Take $\theta > 0$. The set Q is infinite, thus there exists $W_\theta \in \mathcal{U}_0$ such that $n(W_\theta) > \theta$. Consider the (terminal) set $J_\theta := \{W \in \mathcal{U}_0 : W \subset W_\theta\}$. By (2.23) we have that $n(W) \geq n(W_\theta) > \theta$ for all $W \in J_\theta$. Because $\xi_W \in A_{n(W)}$, there exists $p \in \mathbb{N}$ such that $\xi_W = e_{n(W)} + n(W)\, e_{n(W)+p}$, and hence we get

$$\begin{aligned} \|\xi_W\|^2 &= \|e_{n(W)} + n(W)\, e_{n(W)+p}\|^2 = \|e_{n(W)}\|^2 + n(W)^2 \|e_{n(W)+p}\|^2 \\ &= 1 + n(W)^2 > 1 + \theta^2 > \theta^2, \end{aligned}$$

which gives $\|\xi_W\| > \theta$ for all $W \in J_\theta$, establishing that $\{\xi_W\}_{W \in \mathcal{U}_0}$ is eventually unbounded.

A topological space is called *sequentially compact* whenever every sequence has a convergent subsequence. When a Banach space is *separable* (i.e., when there is a countable dense set in X) then the conclusion of Bourbaki–Alaoglu theorem can be strengthened to sequential compactness. The most common examples of separable Banach spaces are ℓ_p and $L^p[a,b]$, for $1 \leq p < \infty$, whereas ℓ_∞ and $L^\infty[a,b]$ are not separable (see, e.g., Theorems 3.2 and Theorem 4.7(d) in [135]).

The result stating this stronger conclusion follows (see Theorem 3.17 in [197]).

Theorem 2.5.18. *If X is a separable topological vector space and if $\{v_n\} \subset X^*$ is a bounded sequence, then there is a subsequence $\{v_{n_k}\} \subset \{v_n\}$ and there is a $v \in X^*$ such that $v_{n_k} \xrightarrow{w^*} v$.*

When X is a reflexive Banach space (i.e., when the weak and weak* topologies in X^* coincide), we can quote a stronger fact (see [231, Sec. 4 in the appendix to Chapter V]). The spaces ℓ_p and $L^p[a,b]$, for $1 < p < \infty$ are reflexive, whereas ℓ_1, ℓ_∞ and $L^1[a,b], L^\infty[a,b]$ are not.

Theorem 2.5.19. *A Banach space X is reflexive if and only if every bounded sequence contains a subsequence converging weakly to an element of X.*

Let X be a Banach space. A subset of X is weakly closed when it contains the limits of all its weakly convergent nets. Because every strongly convergent net also converges weakly, we have that the strong closure of a set is contained in the weak closure. Denote $\overline{A}$ the strong closure and by $\overline{A}^w$ the weak closure. Our remark is expressed as $\overline{A} \subset \overline{A}^w$. When the set A is convex, then the converse inclusion holds.

Theorem 2.5.20. *If X is a Banach space and $A \subset X$ is convex, then the strong and the weak closure of A coincide. In particular, if A is weakly closed and convex, then it is strongly closed.*

Proof. See, for example, page 194 in [70]. $\square$

Let X be a Banach space and $F : X \rightrightarrows X^*$ a point-to-set mapping. We can consider in X the strong topology (induced by the norm) or the weak topology, as in (2.18). In X^* we can also consider the strong topology (induced by the norm defined in (2.17)), or the weak* topology, as in (2.19). Therefore, when studying semicontinuity of a point-to-set mapping $F : X \rightrightarrows X^*$, we must specify which topology is used both in X and in X^*. Denote (sw*) the product topology in $X \times X^*$ where X is endowed with the strong topology and X^* is endowed with the weak* topology.

Proposition 2.5.21. *Let X and Y be topological spaces and $F : X \rightrightarrows Y$ a point-to-set mapping.*

(i) *If Y is regular (i.e., (T3)), F is USC and $F(x)$ is closed for all $x \in X$, then F is OSC.*

(ii) *If X is a Banach space, $Y = X^*$, and $F : X \rightrightarrows X^*$ is locally bounded and (sw*) OSC at x, then F is (sw*) USC at x.*

(iii) *If F is USC and single-valued in a point $x \in D(F)^\circ$, then F is ISC at x.*

(iv) *Assume F is USC and compact-valued in a point x_0 (i.e., $F(x_0)$ is compact in Y). Let $\{x_i\}_{i \in I} \subset X$ be a net converging to x_0 and consider a net $\{y_i\}_{i \in I} \subset Y$ such that $y_i \in F(x_i)$ for all $i \in I$. Then, there exists $y_0 \in F(x_0)$ such that y_0 is a cluster point of $\{y_i\}_{i \in I}$.*

Proof.
(i) Take a net $\{x_i\}_{i \in I}$ converging to x. We must prove that

$$\limext_{i \in I} F(x_i) \subset F(x).$$

If the above inclusion does not hold, there exists $y \in \limext_{i \in I} F(x_i)$ with $y \notin F(x)$. Because $F(x)$ is closed and Y is regular, there exist open sets $V, W \subset Y$ such that $y \in V$, $F(x) \subset W$, and $V \cap W = \emptyset$. Because F is USC, there exists a neighborhood $U \subset X$ of x such that $F(U) \subset W$. Using the fact that $\{x_i\}_{i \in I}$ converges to x, we have that $x_i \in U$ eventually. So for all j in a terminal subset $J \subset I$ we have

$$F(x_j) \subset F(U) \subset W. \tag{2.24}$$

Because $y \in \lim\text{ext}_{i \in I} F(x_i)$ and V is a neighborhood of y we have that $V \cap F(x_i) \neq \emptyset$ for all i in a cofinal subset $K \subset I$. Therefore we can find

$$y_i \in F(x_i) \cap V, \tag{2.25}$$

for all i in the cofinal subset $K \cap J \subset I$. Using the fact that $V \cap W = \emptyset$, we get that (2.24) contradicts (2.25).

(ii) If the result does not hold, then there exist $x \in X$ and a set $W \subset Y$, open in the weak* topology and containing $F(x)$, such that for each $r > 0$ we have $F(B(x,r)) \not\subset W$. This means that we can find a sequence $\{x_n\} \subset X$ and $\{y_n\} \subset Y$ such that $\lim x_n = x$, $y_n \in F(x_n)$ for all $n \in \mathbb{N}$, and $\{y_n\} \subset W^c$. Let $U \subset X$ be a neighborhood of x such that $F(U)$ is bounded, whose existence is guaranteed by the local boundedness of F. Without loss of generality, we may assume that $\{x_n\} \subset U$. Because $F(U)$ is bounded, it is contained in some strongly closed ball B of X^*. By Theorem 2.5.16(ii), B is weak* compact. Note that $y_n \in F(x_n) \subset F(U) \subset B$, so $\{y_n\} \subset X^*$ is a net contained in the weak* compact set B. By Theorem 2.1.9(a) we have that there exists a weak* cluster point y of $\{y_n\}$. Moreover, there is a subnet (not necessarily a subsequence) of $\{y_n\}$ that converges in the weak* topology to y. Because $\{y_n\} \subset W^c$, and W^c is weak* closed, we conclude that $y \in W^c$. Thus,

$$y \notin F(x), \tag{2.26}$$

because $F(x) \subset W$. Using also the subnet of $\{y_n\} \subset W^c$ which converges to y, we obtain a subnet of $\{(x_n, y_n)\}$ contained in the graph of F that is (sw*)-convergent to (x, y). Because F is (sw*)-OSC at x, we must have $y \in F(x)$, contradicting (2.26).

(iii) Assume that $F(x)$ is a singleton, and let $\{x_i\}_{i \in I}$ be a net converging to x. We must prove that $F(x) \in \lim\text{int}_i F(x_i)$. Suppose that this is not true. Then there exists an open set $W \subset Y$ such that $F(x) \in W$ and

$$W \cap F(x_i) = \emptyset, \tag{2.27}$$

for all $i \in J_0$, with J_0 a cofinal subset of I. Because $F(x) \in W$ and F is upper-semicontinuous at x, there exists an open neighborhood U' of x such that $F(U') \subset W$. Because $x \in D(F)^\circ$, we can assume that $U' \subset D(F)$. Using the fact that $\{x_i\}_{i \in I}$ converges to x, there exists a terminal set $I_1 \subset I$ such that $x_i \in U'$ for all $i \in I_1$. For $i \in J_0 \cap I_1$ we have $\emptyset \neq F(x_i) \subset F(U') \subset W$, which contradicts (2.27).

(iv) Suppose that the conclusion is not true. Hence, no $z \in F(x_0)$ is a cluster point of $\{y_i\}_{i \in I}$. Therefore, for every $z \in F(x_0)$ we can find a neighborhood $W \in \mathcal{U}_z$ for which the set $J_z := \{i \in I : y_i \in W_z\}$ is not cofinal. By Theorem 2.1.2(vii), we have that

$$(J_z)^c = \{i \in I : y_i \notin W_z\} \quad \text{is terminal.} \tag{2.28}$$

Because $F(x_0) \subset \cup_{z \in F(x_0)} W_z$, by compacity there exist $z_1, \ldots, z_p \in F(x_0)$ such that $F(x_0) \subset \cup_{l=1}^p U_l$, with $U_l := W_{z_l}$ for all $l = 1, \ldots, p$. Call $\bar{U} := \cup_{l=1}^p U_l$

and $I_l := \{i \in I : y_i \notin U_l\}$. By (2.28), we know that I_l is terminal. We claim that the set $\bar{I} := \{i \in I : y_i \in \bar{U}\}$ is not terminal. By Theorem 2.1.2(vi), it is enough to check that $(\bar{I})^c$ is cofinal. Indeed,

$$(\bar{I})^c = \{i \in I : y_i \notin \bar{U}\} = \cap_{l=1}^{p}\{i \in I : y_i \notin U_l\} = \cap_{l=1}^{p} I_l,$$

so $(\bar{I})^c$ is terminal, because it is a finite intersection of terminal sets. In particular, it is cofinal as claimed. So our claim is true and $\bar{I}$ is not terminal. On the other hand, $F(x_0) \subset \bar{U}$ and by upper-semicontinuity of F, there exists $V \in \mathcal{U}_{x_0}$ such that $F(V) \subset \bar{U}$. Because the net $\{x_i\}_{i \in I}$ converges to x_0, the set

$$I_0 := \{i \in I : x_i \in V\},$$

is terminal. Altogether, for $i \in I_0$

$$y_i \in F(x_i) \subset F(V) \subset \bar{U},$$

which yields $\bar{I} = \{i \in I : y_i \in \bar{U}\} \supset I_0$, and so $\bar{I}$ must be terminal. This contradiction yields the conclusion. □

The result below is similar to the one in Proposition 2.5.21(ii).

Proposition 2.5.22. *Let X be a topological space, Y be a metric space and $F : X \rightrightarrows Y$ be a point-to-set mapping. If Y is compact (with respect to the metric topology) and F is OSC, then F is USC.*

Proof. Because Y is compact, it is locally compact. The compactness of Y also implies that $F(U) \subset Y$ is bounded for all $U \subset X$. Hence F is locally bounded. From this point the proof follows steps similar to those in Proposition 2.5.21(ii) and it is left as an exercise. □

The following proposition connects the concepts of semicontinuity with set convergence. The tools for establishing this connection are Proposition 2.2.14 and its corollary.

Proposition 2.5.23. *Let $F : X \rightrightarrows Y$ be a point-to-set mapping.*

(i) If F is OSC at x then for any net $\{x_i\}_{i \in I}$ such that

$$x_i \to x$$
$$F(x_i) \to D,$$

it holds that $D \subset F(x)$. When X is N_1 and Y is N_2, the converse of the previous statement holds, with sequences substituting for nets. Namely, suppose that for any sequence $\{x_n\}_{n \in \mathbb{N}}$ such that

$$x_n \to x$$
$$F(x_n) \to D,$$

it holds that $D \subset F(x)$; then F is OSC.

(ii) F is ISC at x if and only if for any net $\{x_i\}$ such that

$$x_i \to x$$
$$F(x_i) \to D,$$

it holds that $D \supset F(x)$.

Proof.

(i) Suppose that F is OSC at x and consider a net $\{x_i\}$ satisfying the assumption. Because F is OSC at x we have $\lim \mathrm{ext}_i\, F(x_i) \subset F(x)$. Using the fact that $F(x_i) \to D$, we conclude that $D = \lim \mathrm{int}_i\, F(x_i) \subset \lim \mathrm{ext}_i\, F(x_i) \subset F(x)$. Let us prove now the second part of (i). Assume that the statement on sequences holds and take a sequence $\{(x_n, y_n)\}$ that satisfies

$$x_n \to x,$$
$$y_n \in F(x_n),$$
$$y_n \to y.$$

We must prove that $y \in F(x)$. Because $y \in \lim \mathrm{int}_n\, F(x_n)$, we get that $\lim \mathrm{ext}_n\, F(x_n) \neq \emptyset$. By Proposition 2.2.14 there exist $D \neq \emptyset$ and an index set $J \in \mathcal{N}_\sharp$ such that $F(x_j) \to_J D$. Because $y_j \in F(x_j)$ for all $j \in J$ and $y_j \to_J y$, we get $y \in \lim \mathrm{int}_J\, F(x_j) = D$. By the assumption, $D \subset F(x)$, and hence $y \in F(x)$.

(ii) Suppose that F is ISC at x and consider a net $\{x_i\}$ satisfying the assumption. Take $y \in F(x)$. By inner-semicontinuity of F, we have that $y \in \lim \mathrm{int}_i\, F(x_i) = D$, as required. Conversely, assume that the statement on nets holds for $x \in D(F)$ and suppose F is not ISC at x. In other words, there exists a net $x_i \to x$ such that $y \in F(x)$ with $y \notin \lim \mathrm{int}_i\, F(x_i)$. Therefore there exist a neighborhood W of y and a cofinal subset $K \subset I$ such that

$$W \cap F(x_i) = \emptyset,$$

for all $i \in K$. Consider now the net $\{F(x_i)\}_{i \in K}$. By Theorem 2.2.12 there exists a subnet $\{z_j\}_{j \in J}$ of $\{x_i\}_{i \in K}$ such that $\{F(z_j)\}_{j \in J}$ is a convergent subnet of $\{F(x_i)\}_{i \in K}$. By definition of subnet, there exists a function $\varphi : J \to K$ such that $x_{\varphi(j)} = z_j$ for all $j \in J$, where φ and J verify condition (SN2) of the definition of subnet. Let D be the limit of the subnet $\{F(z_j)\}_{j \in J}$. The assumption on F implies $F(x) \subset D$. Hence $y \in D$. In particular, we have that $y \in \lim \mathrm{int}_j\, F(z_j)$ and therefore there exists a terminal subset $\tilde{J} \subset J$ such that

$$W \cap F(z_j) \neq \emptyset,$$

for all $j \in \tilde{J}$. On the other hand, we have $W \cap F(z_j) = W \cap F(x_{\varphi(j)}) = \emptyset$ by the definition of K. This contradiction implies that F must be ISC at x.   □

Now we state and prove some metric properties associated with outer-semicontinuit in the context of Banach spaces.

Proposition 2.5.24. *Let X, Y be metric spaces and $F : X \rightrightarrows Y$ a point-to-set mapping that is locally bounded at $\bar{x} \in D(F)$.*

(i) *Let Y be finite-dimensional. Assume that F is* OSC *with respect to the metric topology in X at $\bar{x}$. Then for all $\varepsilon > 0$ there exists $\delta > 0$ such that whenever $d(x, \bar{x}) < \delta$ it holds that $F(x) \subset F(\bar{x}) + B(0, \varepsilon)$.*

(ii) *Let Y be finite-dimensional. If $F(\bar{x})$ is closed, then the ε–δ statement in (i) implies outer-semicontinuity of F at $\bar{x}$ with respect to the metric topologies in X and Y.*

(iii) *Let X be a Banach space and $Y = X^*$ with the weak* topology. Assume that $F(\bar{x})$ is weak* closed, then the ε–δ statement in (i) implies (sw*) outer-semicontinuity of F at $\bar{x}$.*

Proof.

(i) Suppose that F is outer-semicontinuous at $\bar{x}$ and assume that the assertion on ε and δ is not true for F. Then there exists $\varepsilon_0 > 0$ and a sequence $\{x_n\}$ such that $\lim_n x_n = \bar{x}$ and $F(x_n) \not\subset F(\bar{x}) + B(0, \varepsilon_0)$. This implies the existence of a sequence $\{y_n\}$ such that $y_n \in F(x_n)$ for all n and $y_n \notin F(\bar{x}) + B(0, \varepsilon_0)$. In other words,

$$d(y_n, F(\bar{x})) \geq \varepsilon_0 \quad \text{for all } n. \tag{2.29}$$

Let U be a neighborhood of $\bar{x}$ such that $F(U)$ is bounded and take $n_0 \in \mathbb{N}$ such that $x_n \in U$ for all $n \geq n_0$. By the assumption on U, there exists a subsequence $\{y_{n_k}\}$ of $\{y_n\}$ converging to some $\bar{y}$. Using (2.29) with $n = n_k$ and taking limits we get

$$d(\bar{y}, F(\bar{x})) \geq \varepsilon_0. \tag{2.30}$$

By outer-semicontinuity of F at $\bar{x}$, we have $\bar{y} \in F(\bar{x})$, which contradicts (2.30).

(ii) Take a sequence $\{(x_n, y_n)\} \subset \mathrm{Gph}(F)$ such that $\lim_n x_n = \bar{x}$ and $\lim_n y_n = \bar{y}$. Assume that $\bar{y} \notin F(\bar{x})$. Hence there exists $\varepsilon_0 > 0$ such that $d(\bar{y}, F(\bar{x})) \geq 2\varepsilon_0$. Let $\delta_0 > 0$ be such that $F(x) \subset F(\bar{x}) + B(0, \varepsilon_0/2)$ whenever $d(x, \bar{x}) < \delta_0$. Fix n_0 such that $d(x_n, \bar{x}) < \delta_0$ and $d(y_n, \bar{y}) < \varepsilon_0/2$ for all $n \geq n_0$. Our assumption implies that $F(x_n) \subset F(\bar{x}) + B(0, \varepsilon_0/2)$ for all $n \geq n_0$. But $y_n \in F(x_n)$, so for $n \geq n_0$ we have $y_n \in F(\bar{x}) + B(0, \varepsilon_0/2)$. Altogether,

$$
\begin{aligned}
2\varepsilon_0 \leq d(\bar{y}, F(\bar{x})) &\leq d(\bar{y}, F(x_n)) + d(F(x_n), F(\bar{x})) \\
&\leq d(\bar{y}, y_n) + d(F(x_n), F(\bar{x})) \\
&\leq \varepsilon_0/2 + \varepsilon_0/2 = \varepsilon_0,
\end{aligned}
$$

a contradiction. Therefore F is OSC at $\bar{x}$.

(iii) By Theorem 2.5.4, it is enough to prove (sw*)-closedness of $\mathrm{Gph}(F)$. Take a net $\{(x_i, y_i)\} \subset \mathrm{Gph}(F)$ that verifies $y_i \overset{w^*}{\to} \bar{y}$ and $\lim_i x_i = \bar{x}$. Assume again that $\bar{y} \notin F(\bar{x})$. Because $F(\bar{x})$ is weak* closed, it must be strongly closed (because every net that converges strongly converges also weakly*), and hence there exists $\varepsilon_0 > 0$ such that

$$d(\bar{y}, F(\bar{x})) > \varepsilon_0. \tag{2.31}$$

For this ε_0, choose $\delta_0 > 0$ such that whenever $\|x - \bar{x}\| < \delta_0$ we have $F(x) \subset F(\bar{x}) + B(0, \varepsilon_0/2)$. Fix now a terminal subset $J \subset I$ such that $\|x_i - x\| < \delta_0$ for all $i \in J$. Then our assumption yields $F(x_i) \subset F(\bar{x}) + B(0, \varepsilon_0/2)$. Therefore $\{y_i\} \subset F(\bar{x}) + B(0, \varepsilon_0/2)$. The closed ball $B(0, \varepsilon)$ is weak* compact by Theorem 2.5.16(ii) and $F(\bar{x})$ is weak* closed. By Theorem 2.5.27(iv), $F(\bar{x}) + B(0, \varepsilon)$ is weak* closed. Using the fact that $w^* - \lim_n y_i = \bar{y}$, we get $\bar{y} \in F(\bar{x}) + B(0, \varepsilon_0/2)$. This contradicts (2.31).   □

Remark 2.5.25. When Y is infinite-dimensional, item (i) of Proposition 2.5.24 may fail to hold. Indeed, let H be a Hilbert space and $\{e_n\}$ an orthonormal basis of H. Take $F : H \to H$ defined as

$$F(x) = \begin{cases} e_n & \text{if } x \in \{t e_n \ : \ 0 \neq t \in \mathbb{R}\}, \\ 0 & \text{if } x = 0, \\ \emptyset & \text{otherwise.} \end{cases}$$

This F is locally bounded at $\bar{x} := 0$, and it is also OSC at $\bar{x} := 0$, from the strong topology to the weak one. However, if $\varepsilon < 1$ and $\{x_n\}$ is a nontrivial sequence converging strongly to 0, then we get $y_n := e_n \in F(x_n)$. In this case, $d(y_n, F(\bar{x})) = d(y_n, 0) = 1 > \varepsilon$ and hence the ε–δ assertion is not valid. Local boundedness at $\bar{x}$ in this item is also necessary, as we show next. Consider $F : \mathbb{R} \rightrightarrows \mathbb{R}$ given by

$$F(x) = \begin{cases} \{1/x\} & \text{if } x \neq 0 \\ 0 & \text{if } x = 0. \end{cases}$$

Clearly, F is OSC at $\bar{x} := 0$ and the ε–δ assertion does not hold for $\bar{x} := 0$. If $F(\bar{x})$ is not closed, item (ii) may fail to hold. The construction of a suitable counterexample is left to the reader (cf. Exercise 16).

The geometric interpretation of outer-semicontinuity stated above is not suitable for point-to-set mappings with unbounded images. The (continuous) mapping of Example 2.5.8, for instance, does not satisfy the requirements of Proposition 2.5.24. For addressing such cases, it becomes necessary to "control" the size of the sets $F(\cdot)$ by considering its intersections with compact sets. We need now some well-known facts from the theory of topological vector spaces, which have been taken from [26, Chapter IX, § 2].

Definition 2.5.26. *A vector space Z together with a topology is said to be a topological vector space when the following conditions are satisfied.*

(1) The function

$$\sigma : \begin{array}{ccc} Z \times Z & \to & Z \\ (x, y) & \mapsto & x + y, \end{array}$$

is (jointly) continuous; in other words, for each neighborhood W of $z_1 + z_2$, there exist neighborhoods V_1 of z_1 and V_2 of z_2 such that $V_1 + V_2 \subset W$.

(2) The function

$$\tau : \quad \mathbb{R} \times Z \quad \to \quad Z$$
$$(\lambda, y) \quad \mapsto \quad \lambda y,$$

is (jointly) continuous; in other words, for each neighborhood W of $\lambda_0\, z_0$, there exists a neighborhood V_0 of z_0 and $r > 0$ such that

$$B(\lambda_0, r)\, V_0 := \{\lambda z \mid \lambda \in B(\lambda_0, r)\, , z \in V_0\} \subset W.$$

Theorem 2.5.27. *[26, Chapter IX, § 2]. Let Z be a topological vector space.*

(i) *There exists a basis of neighborhoods $\mathcal{U}_S$ of zero such that every element $V \in \mathcal{U}_S$ is symmetric; that is,*

$$x \in V \quad \Longrightarrow \quad -x \in V.$$

(ii) *For each neighborhood U of 0, there exists a neighborhood V of 0 such that $V + V \subset U$.*

(iii) *If $\{z_i\}_{i \in I}$ is a net converging to z, then for every $\alpha \in \mathbb{R}$ and every $\bar{z} \in Z$ the net $\{\alpha z_i + \bar{z}\}$ converges to $\alpha z + \bar{z}$.*

(iv) *If $F \subset Z$ is closed and $K \subset Z$ is compact, then $F + K$ is closed.*

Proposition 2.5.28. *Let X be a topological space, Y a topological vector space, and $F : X \rightrightarrows Y$ a point-to-set mapping.*

(i) *If F is OSC in X at $\bar{x}$, then for every pair $V, W \subset Y$ of neighborhoods of 0, with V compact, there exists a neighborhood $U \subset X$ of $\bar{x}$ such that whenever $x \in U$, it holds that $F(x) \cap V \subset F(\bar{x}) + W$.*

(ii) *Assume that X and Y are topological spaces and that Y is locally compact. If $F(\bar{x})$ is closed, then the statement on neighborhoods in (i) implies outer-semicontinuity of F at $\bar{x}$.*

Proof.
    (i) Assume that there exists a pair $V_0, W_0 \subset Y$ of neighborhoods of 0, with V_0 compact, such that for every neighborhood U of $\bar{x}$, the conclusion of the statement does not hold. Then there exists a standard net $\{x_U\}_{U \in \mathcal{U}_{\bar{x}}}$ (hence converging to $\bar{x}$), such that

$$F(x_U) \cap V_0 \not\subset F(\bar{x}) + W_0.$$

This allows us to construct a net $\{y_U\}$ such that $y_U \in F(x_U) \cap V_0$ for all U and

$$y_U \notin F(\bar{x}) + W_0. \tag{2.32}$$

By compactness of V_0, there exists a cluster point $\bar{y}$ of the net $\{y_U\}_{U \in \mathcal{U}_{\bar{x}}}$ with $\bar{y} \in V_0$. Hence for every neighborhood W of $\bar{y}$ we have $y_U \in W$ frequently in

$\mathcal{U}_{\bar{x}}$, and because $y_U \in F(x_U)$ for all $U \in \mathcal{U}_{\bar{x}}$, we conclude that $W \cap F(x_U) \neq \emptyset$ frequently in $\mathcal{U}_{\bar{x}}$. This yields $\bar{y} \in \lim \text{ext}_i F(x_i)$. By outer-semicontinuity of F, we have that $\bar{y} \in F(\bar{x})$. Consider now the neighborhood of $\bar{y}$ given by $W_1 := \bar{y} + W_0$. By definition of cluster point we must have

$$y_U \in W_1, \tag{2.33}$$

for all U in some cofinal set $\mathcal{U}_0$ of $\mathcal{U}_{\bar{x}}$. Because $\bar{y} \in F(\bar{x})$ we have

$$W_1 = \bar{y} + W_0 \subset F(\bar{x}) + W_0.$$

Combining the expression above with (2.33) we get $y_U \in F(\bar{x}) + W_0$ for all $U \in \mathcal{U}_0$, contradicting (2.32). Therefore the conclusion in the statement of (i) holds.

(ii) Assume that the statement on neighborhoods holds. Take a net $\{(x_i, y_i)\} \subset \text{Gph}(F)$ converging to $(\bar{x}, \bar{y})$. Suppose that $\bar{y} \notin F(\bar{x})$. Because $F(\bar{x})$ is closed, there exists a neighborhood V of $\bar{y}$, which we can assume to be compact, such that $V \cap F(\bar{x}) = \emptyset$. Take a neighborhood V_0 of 0 such that $V = V_0 + \bar{y}$. By Theorem 2.5.27(i) we can assume that V_0 is symmetric. Because V is compact, V_0 is also compact. We claim that $\bar{y} \notin F(\bar{x}) + V_0$. Indeed, if there exist $z \in F(\bar{x})$ and $v_0 \in V_0$ such that $\bar{y} = z + v_0$ then $z = \bar{y} - v_0 \in \bar{y} + V_0 = V$, which is a contradiction, because $z \in F(\bar{x})$ and $V \cap F(\bar{x}) = \emptyset$. Consider now a compact neighborhood $\widehat{W}$ of 0 such that $\widehat{W} \supset \bar{y} + V_0$. The assumption implies that there exists a neighborhood U of $\bar{x}$ such that whenever $x \in U$, it holds that $F(x) \cap \widehat{W} \subset F(\bar{x}) + V_0$. Because $\{(x_i, y_i)\}$ converges to $(\bar{x}, \bar{y})$, we have that $y_i \in \bar{y} + V_0 \subset \widehat{W}$ and $x_i \in U$ for all $i \in J$, where J is a terminal subset of I. Therefore we have $y_i \in F(x_i) \cap \widehat{W} \subset F(\bar{x}) + V_0$ for all $i \in J$, and hence $y_i \in F(\bar{x}) + V_0$ for all $i \in J$. Note that V_0 is compact and $F(\bar{x})$ is closed, so that the set $F(\bar{x}) + V_0$ is closed by Theorem 2.5.27(iv). Using now the fact that $y_i \to \bar{y}$ we conclude that $\bar{y} \in F(\bar{x}) + V_0$, contradicting the assumption on V_0.   □

The result for inner-semicontinuity, analogous to Proposition 2.5.28, is stated next.

Proposition 2.5.29. *Let X be N_1, Y locally compact and N_1 and $F : X \rightrightarrows Y$ a point-to-set mapping.*

(i) *If $F(\bar{x})$ is closed and F is ISC in X at $\bar{x}$, then for every pair $V, W \subset Y$ of neighborhoods of 0, with V compact, there exists a neighborhood $U \subset X$ of $\bar{x}$ such that $F(\bar{x}) \cap V \subset F(x) + W$ for all $x \in U$.*

(ii) *The statement on neighborhoods in (i) implies inner-semicontinuity of F at $\bar{x}$.*

Proof.

(i) If the statement on neighborhoods were not true, we could find neighborhoods $V_0, W_0 \subset Y$, with V_0 compact, and a sequence $\{x_n\}$ converging to $\bar{x}$ such that $F(\bar{x}) \cap V_0 \not\subset F(x_n) + W_0$. Define a sequence $\{z_n\}$ such that $z_n \in F(\bar{x}) \cap V_0$ and $z_n \notin F(x_n) + W_0$. By compactness of V_0 and closedness of $F(\bar{x})$, there exists a subsequence $\{z_{n_k}\}$ of $\{z_n\}$ converging to some $\bar{z} \in F(\bar{x}) \cap V_0$. By inner-semicontinuity and Theorem 2.5.6, we can find a sequence $\{y_n \in F(x_n)\}$ also converging to $\bar{z}$. Then $\lim_k z_{n_k} - y_{n_k} = 0$, so that for large enough k we have that $z_{n_k} - y_{n_k} \in W_0$. Altogether, we conclude that $z_{n_k} = y_{n_k} + [z_{n_k} - y_{n_k}] \in F(x_{n_k}) + W_0$ for large enough k, but this contradicts the definition of $\{z_n\}$.

(ii) We use again Theorem 2.5.6. Take a sequence $\{x_n\}$ converging to $\bar{x}$ and $\bar{y} \in F(\bar{x})$. We claim that for every neighborhood W of 0 in Y, it holds that $\bar{y} \in F(x_n) + W$ for large enough n. Note that if this claim is true, then we can take a nested sequence of neighborhoods $\{W_n\}$ of 0 such that $\bar{y} \in F(x_n) + W_n$. Thus, there exist sequences $\{y_n\}$, $\{z_n\}$ such that $y_n \in F(x_n)$, $z_n \in W_n$, and $\bar{y} = y_n + z_n$ for all n. The sequence $\{W_n\}$ is nested, therefore $\{z_n\}$ converges to 0, and hence we get that $\{y_n\}$ converges to $\bar{y}$. So it is enough to prove the stated claim. If the claim is not true we can find a neighborhood W_0 and $J \in \mathcal{N}_\sharp$ such that $\bar{y} \notin F(x_n) + W_0$ for $n \in J$. Take a compact neighborhood V_0 of $\bar{y}$. By the statement on neighborhoods with $V = V_0$ and $W = W_0$, there exists a neighborhood U of $\bar{x}$ such that

$$F(\bar{x}) \cap V_0 \subset F(x) + W_0$$

for all $x \in U$. Thus, $\bar{y} \in F(\bar{x}) \cap V_0 \subset F(x_n) + W_0$ for large enough n, which contradicts the assumption on W_0, establishing the claim. $\square$

We can derive from the above proposition the following geometric characterization of continuity.

Corollary 2.5.30. *Let X be N_1, Y N_1 and locally compact and $F : X \rightrightarrows Y$ a point-to-set mapping. Assume that F is closed-valued at $\bar{x} \in D(F)$. Then F is continuous at $\bar{x}$ if and only if for every pair V, $W \subsetneq Y$ of neighborhoods of 0, with V compact, there exists a neighborhood $U \subset X$ of $\bar{x}$ such that*

$$x \in U \implies \begin{cases} F(\bar{x}) \cap V \subset F(x) + W, \text{ and} \\ F(x) \cap V \subset F(\bar{x}) + W. \end{cases}$$

Lipschitz continuity can also be expressed in a similar way, via compact neighborhoods of 0. However, we only need the following stronger notion.

Definition 2.5.31. *Let X and Y be Banach spaces and $F : X \rightrightarrows Y$ a point-to-set mapping. Let U be a subset of $D(F)$ such that F is closed-valued on U. The mapping F is said to be* Lipschitz continuous *on U if there exists a Lipschitz constant $\kappa > 0$ such that for all x, $x' \in U$ it holds that*

$$F(x) \subset F(x') + \kappa \|x - x'\| B(0,1).$$

In other words, for every $x \in U$, $v \in F(x)$, and $x' \in U$, there exists $v' \in F(x')$ such that

$$\|v - v'\| \leq \kappa \|x - x'\|.$$

The next results present cases in which Lipschitz continuity implies continuity.

Proposition 2.5.32. *Let X and Y be Banach spaces and $F : X \rightrightarrows Y$ a point-to-set mapping. Let U be a subset of $D(F)$ such that F is Lipschitz continuous and closed-valued on U. Then,*

(i) If U is open, then F is (strongly) ISC in U.

(ii) If $Y = X^*$, then F is (sw*)-ISC in U.

(iii) If $Y = X^*$ and $F(x)$ is weak* closed then F is (sw*)-OSC in U.

In particular, when $Y = X^*$ and $F(x)$ is weak* closed, Lipschitz continuity of F on U yields continuity of F on U.

Proof.
    (i) Take a net $\{x_i\}_{i \in I}$ strongly convergent to $\bar{x} \in U$. We must show that $F(\bar{x}) \subset \liminf_i F(x_i)$. Because U is open we can assume that $\{x_i\}_{i \in I} \subset U$. By Lipschitz continuity of F, there exists $\kappa > 0$ such that

$$F(\bar{x}) \subset F(x_i) + B(0, \kappa \|\bar{x} - x_i\|). \tag{2.34}$$

Fix $v \in F(\bar{x})$ and take an arbitrary $r > 0$. There exists $J \subset I$ a terminal set such that $\|\bar{x} - x_i\| < r/(2\kappa)$ for all $i \in J$. Using (2.34) for $i \in J$ we can define a net $\{v_i\}_{i \in J}$ such that $v_i \in F(x_i)$ and $v = v_i + u$ with $\|v - v_i\| = \|u\| \leq \kappa \|\bar{x} - x_i\| < r/2$ for all $i \in J$. Therefore $v_i \in B(v, r/2)$ for all $i \in J$, which gives $F(x_i) \cap B(v, r/2) \neq \emptyset$ for all $i \in J$. Because J is terminal and $r > 0$ is arbitrary, we get the desired inner-semicontinuity.
    Item (ii) follows from (i) and the fact that the strong topology is finer that the weak* one. In other words, for every weak* neighborhood W of $v \in X^*$ there exists $r > 0$ such that $v \in B(v, r) \subset W$, so we get $(strong) - \liminf_i F(x_i) \subset (w^*) - \liminf_i F(x_i)$.
    Item (iii) follows from Proposition 2.5.24(iii) and the fact that Lipschitz continuity at $\bar{x} \in U$ implies the ε–δ statement in this proposition. $\square$

Exercises

2.1. Prove that for a point-to-point mapping $F : X \to Y$, the concepts of USC, ISC are equivalent to continuity. However, OSC can hold for discontinuous functions. Hint: take $F : \mathbb{R} \to \mathbb{R}$ given by

$$F(x) := \begin{cases} 1/x & \text{if } x \neq 0, \\ 0 & \text{if } x = 0. \end{cases}$$

2.2. Prove that the point-to-set mapping F defined in Example 2.5.8 is continuous, but not USC at any $\alpha \in [0, 2\pi]$.

2.3. Prove that F is OSC if and only if F^{-1} is OSC. Give an example of an ISC point-to-set mapping for which F^{-1} is not ISC. Hint: take $F : \mathbb{R} \rightrightarrows \mathbb{R}$ defined by $F(x) := \{x + 1\}$ if $x < -1$, $F(x) := \{x - 1\}$ if $x > 1$, and $F(x) = \{0\}$ if $x \in [-1, 1]$.

2.4. Prove that the sum of ISC point-to-set mappings is ISC. The sum of OSC point-to-set mappings may fail to be OSC. Hint for the last statement: take $F_1, F_2 : \mathbb{R} \rightrightarrows \mathbb{R}$, where $F_1(x) := \{1/x\}$ if $x \neq 0$, $F_1(0) := \{0\}$, and $F_2(x) := \{x - (1/x)\}$ if $x \neq 0$, $F_2(0) := [1, +\infty)$, and prove that $F_1 + F_2$ is not OSC at 0.

2.5. Prove that if F is OSC, then λF is OSC for all $\lambda \in \mathbb{R}$. Idem for inner-semicontinuity.

2.6. Let $F : X \rightrightarrows Y$ and $G : Y \rightrightarrows Z$ be point-to-set mappings. Prove that

(i) If F is ISC (respectively, USC) at x and G is ISC (respectively, USC) in the set $F(x)$, then $G \circ F$ is ISC (respectively, USC) at x. In other words, the composition of ISC (respectively, USC) mappings is ISC (respectively, USC).

(ii) The composition of OSC mappings in general is not OSC. Hint: take $F(x) = \{1/x\}$ for $x \neq 0$, $F(0) = \{0\}$, and $G(x) = [1, x]$ for all x.

(iii) If Y is a metric space, F is OSC and locally bounded at x, and G is OSC in the set $F(x)$, then $G \circ F$ is OSC at x.

2.7. Prove items (a) and (b) of Proposition 2.5.9. Hint: for the "if" part of item (b), assume that F is not ISC at x, and use Proposition 2.2.9(a) for concluding the existence of an open set W for which the statement in the latter proposition does not hold. This will lead to a contradiction.

2.8. Prove items (a) and (b) of Proposition 2.5.12.

2.9. Prove that F is OSC if and only if for all $y \notin F(x)$, there exists $W \in \mathcal{U}_y$ and $V \in \mathcal{U}_x$ such that $V \cap F^{-1}(W) = \emptyset$.

2.10. Prove that F is ISC if and only if for all $y \in F(x)$ and for all $W \in \mathcal{U}_y$, there exists $V \in \mathcal{U}_x$ such that $V \subset F^{-1}(W)$.

2.11. Prove the statements in Remark 2.5.13.

2.12. Prove the statements in Example 2.5.14.

2.13. Let X, Y, and Z be metric spaces, $F : X \rightrightarrows Y$, and $\varphi : Y \to Z$ an homeomorphism (i.e., φ is a bijection such that φ and φ^{-1} are continuous). If F is USC then $\varphi \circ F : X \rightrightarrows Z$ is USC (see the proof of Theorem 2.7.7(III) for the corresponding results for ISC and OSC).

2.14. Let X and Y be metric spaces. Given $F : X \rightrightarrows Y$, define $\bar{F} : X \rightrightarrows Y$ as $\bar{F}(x) = \overline{F(x)}$.

(i) Prove that if $\bar{F}$ is ISC then F is ISC (see the proof of Theorem 2.7.7(III) for the converse statement).

(ii) Prove that if F is OSC, then $F(x)$ is closed. Use this fact for proving that when F is OSC, then $\bar{F}$ is also OSC. Give an example in which $\bar{F}$ is OSC and F is not.

(iii) Prove that $\mathrm{Gph}(F) \subset \mathrm{Gph}(\bar{F}) \subset \overline{\mathrm{Gph}(F)}$.

(iv) Give examples for which the inclusions in item (iii) are strict.

2.15. Give an example of a mapping $F : X \rightrightarrows Y$, satisfying the ε-δ statement of Proposition 2.5.24(i), but not the conclusion of Proposition 2.5.24(ii), in a case in which $F(\bar{x})$ is not closed.

2.16. Prove all the assertions of Remark 2.5.25.

2.6   Semilimits of point-to-set mappings

In this section we connect the continuity concepts for point-to-set mappings introduced in the previous section with the notions of set convergence given in Chapter 2. We consider point-to-set mappings $F : X \rightrightarrows Y$ where X and Y are metric spaces. Because metric spaces are N_1, the topological properties can be studied using sequences.

In point-to-point analysis, upper-semicontinuity and lower-semicontinuity are connected with upper and lower limits. Similarly, outer- and inner-semicontinuity can be expressed in terms of semilimits of the kind given in Definition 2.2.3(i)–(ii). For a definition that is independent of the particular sequence, denote by

$$x' \xrightarrow{F} x, \tag{2.35}$$

the fact that $x' \in D(F)$ and converges to x. We point out that when $x \in D(F)$, then $x' \equiv x$ satisfies (2.35).

Definition 2.6.1. *Given $x \in D(F)$, the* outer limit *of F at x is the point-to-set mapping given by*

$$\ell_+ F(x) := \{y \in Y \ : \liminf_{x' \xrightarrow{F} x} d(y, F(x')) = 0\} = \operatorname*{lim\,ext}_{x' \xrightarrow{F} x} F(x'), \tag{2.36}$$

and the inner limit *of F at x is the point-to-set mapping given by*

$$\ell_- F(x) := \{y \in Y \ : \limsup_{x' \xrightarrow{F} x} d(y, F(x')) = 0\}$$

$$= \{y \in Y \ : \lim_{x' \xrightarrow{F} x} d(y, F(x')) = 0\} = \operatorname*{lim\,int}_{x' \xrightarrow{F} x} F(x'). \tag{2.37}$$

In the same way as outer and inner limits of sequences of sets, the point-to-set mappings $\ell_+ F(\cdot)$ and $\ell_- F(\cdot)$ are closed-valued. It also holds that

$$\ell_- F(x) \subset \overline{F(x)} \subset \ell_+ F(x), \tag{2.38}$$

for any $x \in D(F)$.

The following result gives a useful characterization of $\ell_+ F(\cdot)$ and $\ell_- F(\cdot)$. The proof is a direct consequence of Definition 2.6.1 and is omitted.

Proposition 2.6.2.

(a) $y \in \ell_+ F(x)$ if and only if $(x, y) \in \overline{\mathrm{Gph}(F)}$, or, in other words,

$$\ell_+ F(x) = \{y \in Y : \exists\{(x_n, y_n)\} \subset \mathrm{Gph}(F) \text{ with } \lim_{n \to \infty}(x_n, y_n) = (x, y)\}.$$

(b) $y \in \ell_- F(x)$ if and only if

$$\forall\{x_n\} \text{ such that } x_n \xrightarrow{F} x \ \exists \ y_n \in F(x_n) \text{ such that } y_n \to y.$$

Proposition 2.6.2 allows us to express outer- and inner-semicontinuity in terms of outer and inner limits of F.

Corollary 2.6.3.

(a) F is OSC at x if and only if $\ell_+ F(x) \subset F(x)$.

(b) F is ISC at x if and only if $F(x) \subset \ell_- F(x)$.

Remark 2.6.4. As a consequence (see (2.38)), when F is continuous, it is closed-valued and $\ell_- F(x) = F(x) = \ell_+ F(x)$, resembling the analogous fact for point-to-point mappings. Also, F is continuous if and only if $\mathrm{Gph}(F)$ is closed and F is ISC.

Example 2.6.5. With the notation of Example 2.5.14, it is easy to check that

$$\ell_- F_1(0) = F_1(0) = [0, 1/2] \subset \ell_+ F_1(0) = [0, 1],$$

$$\ell_- F_2(0) = [0, 1/2] \subset F_2(0) = [0, 1) \subset \ell_+ F_2(0) = [0, 1],$$

$$\ell_- F_3(0) = \emptyset \subset F_3(0) = (-\infty, 0] = \ell_+ F_3(0).$$

Remark 2.6.6. (*Outer- and inner-semicontinuity in the context of point-to-point mappings*). Looking at a point-to-point mapping $F : X \to Y$ as a point-to-set one with values $\{F(x)\}$, it is clear that a point x belongs to $D(F)$ if and only if $F(x)$ exists. It holds that

(i) F is continuous at x (in the classical sense) if and only if $\ell_- F(x) = \ell_+ F(x) = \{F(x)\}$.

(ii) F is discontinuous at x (in the classical sense) if and only if $\ell_- F(x) = \emptyset$.

So we see that continuity in the sense of Definition 2.5.6 coincides with the classical one for point-to-point mappings.

Exercises

2.1. Prove Proposition 2.6.2.

2.2. Prove that

  (i) $\lim \operatorname{ext}_{y \to x} F(y)$ and $\lim \operatorname{int}_{y \to x} F(y)$ are closed,

  (ii) $\lim \operatorname{int}_{y \to x} F(y) \subset \overline{F(x)} \subset \lim \operatorname{ext}_{y \to x} F(y)$, which explains the notation of "interior" and "exterior" limits.

2.3. Prove that

  (i) $\lim \operatorname{ext}_{y \to x} F(y) = \bigcap_{\eta > 0} \overline{\bigcup_{y \in B(x,\eta) \cap D(F)} F(y)} = \bigcap_{\varepsilon > 0} \left[\bigcap_{\eta > 0} \bigcup_{y \in B(x,\eta) \cap D(F)} B(F(y), \varepsilon) \right]$.

  (ii) $\lim \operatorname{int}_{y \to x} F(y) = \bigcap_{\varepsilon > 0} \left[\bigcup_{\eta > 0} \bigcap_{y \in B(x,\eta) \cap D(F)} B(F(y), \varepsilon) \right]$.

2.4. Take $\ell_+ F$, $\ell_- F$ as in Definition 2.6.1. Let $\bar{F}$ be defined as $\bar{F}(x) := \overline{F(x)}$. Prove that

  (i) $\operatorname{Gph}(\ell_- F) \subset \operatorname{Gph}(\bar{F})$.

  (ii) $\operatorname{Gph}(\ell_+ F) = \overline{\operatorname{Gph}(F)}$.

2.5. Let X and Y be metric spaces and $F : X \rightrightarrows Y$. Define the set of *cluster values of F at x*, denoted $\operatorname{cv}(F, x)$, as the point-to-set mapping

$$\operatorname{cv}(F, x) := \left\{ y \in Y : \begin{array}{c} \liminf \\ x' \to x \\ x' \neq x \end{array} d(y, F(x')) = 0 \right\}$$

$$= \left\{ y \in Y : \forall \varepsilon > 0 \sup_{r > 0} \inf_{\substack{x' \in B(x,r) \\ x' \neq x}} d(y, F(x')) < \varepsilon \right\}.$$

Prove that

  (i) $\ell_+ F(x) = \operatorname{cv}(F, x) \cup \{F(x)\}$.

  (ii) F is continuous at x if and only $\operatorname{cv}(F, x) = \{F(x)\}$.

2.6. Prove the statements in Example 2.6.5.

2.7. Prove items (i) and (ii) in Remark 2.6.6.

2.7   Generic continuity

Recall that a point-to-set mapping $F : X \rightrightarrows Y$ is said to be continuous when it is OSC and ISC at any $x \in D(F)$. When X is a complete metric space and Y is a complete and separable metric space, then under mild conditions semicontinuity implies continuity on a dense subset of X. Some classical definitions are necessary. Let X be a topological space and let $A \subset X$. Given $U \subset A$, recall that the family of open sets $\{G \cap A \mid G \text{ open in } X\}$ is a topology in A, called the *relative topology* in A. Given $U \subset A$, the *interior of U* relative to A, denoted $(U)_A^o$, is the set of points $x \in U$ for which there exists an open set $V \subset X$ such that $x \in V \cap A \subset U$. Analogously, the *closure of U* relative to A, denoted $(\overline{U})_A$, is the set of points $y \in U$ such that for every open set $V \subset X$ such that $x \in V \cap A$, it holds that $(V \cap A) \cap U \neq \emptyset$.

Definition 2.7.1. *A set $V \subset X$ is* nowhere dense *when $(\overline{V})^o = \emptyset$. For $A \subset X$, we say that a set $V \subset A$, which is closed relative to A, is* nowhere dense *in A when $(V)_A^o = \emptyset$.*

Note that a closed set V is nowhere dense if and only if V^c is a dense open set. A typical example of a nowhere dense set is $\partial L = L \setminus L^o$, where L is a closed set. We use this fact in the proof of Theorem 2.7.7.

Definition 2.7.2. *$S \subset X$ is* meager *if and only if $S = \cup_{n=1}^{\infty} V_n$, where $\{V_n\} \subset X$ is closed and nowhere dense for all $n \in \mathbb{N}$. For $A \subset X$, we say that a set $T \subset A$ is* meager *in A when $T = \cup_{n=1}^{\infty} S_n$, where $\{S_n\} \subset A$ is closed relative to A and nowhere dense in A for all $n \in \mathbb{N}$.*

Definition 2.7.3. *A set $Q \subset X$ is* residual *in X, or second category if and only if $Q \supset S^c$ for some meager subset S of X. It is* first category *when its complement is residual. For $A \subset X$, we say that a set $P \subset A$ is* residual *in A when $P \supset T^c$, where T is meager in A.*

A crucial result involving residual sets in metric spaces is Baire's theorem (1899): if X is a complete metric space and R is residual in X, then R is dense in X (see, e.g., Theorem 6.1 in [22]).

Two direct and very useful consequences of Baire's theorem are stated below. The proofs can be found, for example, in [63, Chapter 3].

Corollary 2.7.4. *Let X be a complete metric space and $\{D_n\}$ a countable family of dense open subsets of X. Then $\cap_n D_n$ is dense in X.*

Taking complements in the above statement we get at once the following.

Corollary 2.7.5. *Let X be a complete metric space and $\{F_n\}$ a family of closed subsets of X such that $X = \cup_n F_n$. Then there exists n_0 such that $F_n^o \neq \emptyset$.*

Definition 2.7.6. *A topological space is said to be a* Baire space *if the intersection of any countable family of open dense sets is dense.*

It is known that an open subset of a Baire is also a Baire space, with respect to the relative topology. As a consequence, if X is complete and $D(F) \subset X$ is open, then $D(F)$ is a Baire space and the conclusion of Corollary 2.7.4 holds with respect to the topology induced in $D(F)$.

When a property holds in a residual set, it is said to be *generic*. The next result asserts that a semicontinuous point-to-set mapping is generically continuous; that is, continuous in a residual set.

Theorem 2.7.7. *Let X and Y be complete metric spaces. Assume that Y satisfies N_2 and consider $F : X \rightrightarrows Y$.*

(I) If F is USC and $F(x)$ is closed for all $x \in X$, then there exists $R \subset X$, which is a residual set in $D(F)$, such that F is continuous on R.

(II) If F is ISC and $F(x)$ is compact for all $x \in X$, then there exists a residual $R \subset X$ such that F is K-continuous on R.

(III) If F is ISC and $F(x)$ is closed for all $x \in X$, then there exists a residual $R \subset X$ such that F is continuous on R.

Proof. Let $V = \{V_n\}_{n \in \mathbb{N}}$ be the countable basis of open sets of Y. Without loss of generality we may assume that for any $y \in Y$ and for any neighborhood U of y, there exists $n \in \mathbb{N}$ such that $y \in \overline{V}_n \subset U$, and also that V is closed under finite unions; which means that for all $n_1, \ldots, n_p \in \mathbb{N}$ there exists $m \in \mathbb{N}$ such that $\cup_{i=1}^p V_{n_i} = V_m$.

(I) Define the sets

$$L_n := F^{-1}(\overline{V_n}) = \{x \in X \ : \ F(x) \cap \overline{V_n} \neq \emptyset\}.$$

Because F is USC, from Proposition 2.5.12(b) we have that L_n is closed relative to $D(F)$. It is clear that $D(F) = \cup_{n=1}^\infty L_n$. We use the sequence $\{L_n\}$ to construct the residual set in $D(F)$ where F is continuous. Because L_n is closed in $D(F)$, $(\partial L_n)^c \cap D(F)$ is open and dense in $D(F)$. Then the set

$$R := \bigcap_{n=1}^\infty [(\partial L_n)^c \cap D(F)]$$

is a residual set in $D(F)$. Note that, by Proposition 2.5.21(i), F is OSC. Hence it is enough to show that F is ISC on R. By Proposition 2.5.9(b), the latter fact will hold at $x \in R$ if we prove that, whenever $U \subset X$ is an open set such that $U \cap F(x) \neq 0$, then the set $F^{-1}(U)$ is a neighborhood of x in $D(F)$. Indeed, take $x \in R$ and U such that $U \cap F(x) \neq 0$. Take $V_n \in V$ such that $\overline{V}_n \subset U$ and $V_n \cap F(x) \neq 0$. This means that $x \in L_n$. Because $x \in R$, for every n such that $x \in L_n$, we get $x \notin \partial L_n$. Hence we must have $x \in (L_n)^o_{D(F)}$, where the latter set denotes the interior of L_n relative to $D(F)$. By the definition of the interior relative to $D(F)$, this means that there exists a ball $B(x, \eta) \subset X$ such that $B(x, \eta) \cap D(F) \subset L_n = F^{-1}(\overline{V_n}) \subset F^{-1}(U)$. Therefore, $F^{-1}(U)$ is a neighborhood of x in $D(F)$, as we wanted to prove. Hence F is ISC on R.

(II) Again, we construct the residual set R by means of the countable basis $V = \{V_n\}$. Define

$$K_n := F^{+1}(\overline{V_n}) = \{x \in X \ : \ F(x) \subset \overline{V_n}\}.$$

By Proposition 2.5.12(a), these sets are closed. We claim that $D(F) \subset \cup_{n=1}^{\infty} K_n$. Indeed, take $x \in D(F)$ and $y \in F(x)$. We know that there exists $n_y \in \mathbb{N}$ such that $y \in V_{n_y} \subset \overline{V_{n_y}}$, so that $F(x) \subset \cup_{y \in F(x)} V_{n_y}$. By compactness of $F(x)$ there exists a finite subcovering $V_{n_1}, \ldots, V_{n_p}$ such that $F(x) \subset \cup_{i=1}^{p} V_{n_i}$. Because $\mathcal{V}$ is closed under finite unions, there exists $m \in \mathbb{N}$ such that $\cup_{i=1}^{p} V_{n_i} = V_m \subset \overline{V_m}$. Thus, $x \in F^{+1}(\overline{V_m}) = K_m$. As before, $(\partial K_n)^c$ is a dense open set, implying that the set $R := \cap_{n=1}^{\infty} (\partial K_n)^c$ is residual. We claim that F is USC on R. Indeed, take $x \in R$ and let $U \subset Y$ be an open set such that $F(x) \subset U$. For every $y \in F(x)$, there exists $n_y \in \mathbb{N}$ such that $y \in V_{n_y} \subset \overline{V_{n_y}} \subset U$. In this way we obtain a covering $F(x) \subset \cup_{y \in F(x)} V_{n_y}$, which by compactness has a finite subcovering $F(x) \subset \cup_{i=1}^{p} V_{n_i} \subset U$. Because $\{V_n\}$ is closed under finite unions, there exists $V_m \in \mathcal{V}$ such that $F(x) \subset \cup_{i=1}^{p} V_{n_i} = V_m \subset \overline{V_m} \subset U$. Then $x \in K_m$. Because $x \in R$, in the same way as in item (I), we must have $x \in (K_m)^o$. Thus, there exists $\eta > 0$ such that $B(x, \eta) \subset K_m$. In other words, $F(x') \subset \overline{V_m} \subset U$ for any $x' \in B(x, \eta)$, which implies that F is USC at x. Therefore, F is USC on R and inasmuch as it is also ISC, it is K-continuous in R.

(III) We use the fact that if a metric space Y satisfies N_2 then it is homeomorphic to a subset Z_0 of a compact metric space Z (see, e.g., Proposition 5 in [134]). Call $\varphi : Y \to Z_0 \subset Z$ the above mentioned homeomorphism and consider the point-to-set mappings $\varphi \circ F : X \to Z_0 \subset Z$ and $G := \overline{\varphi \circ F} : X \to Z$. We claim that $\varphi \circ F$ and G are ISC. Because $G = \overline{\varphi \circ F}$, it is enough to prove inner-semicontinuity of $\varphi \circ F$. Take $y \in \varphi \circ F(x)$ and a sequence $x_n \to x$. Then $\varphi^{-1}(y) \in F(x)$ and by inner-semicontinuity of F there exists a sequence $\{z_n\}$ such that $z_n \to \varphi^{-1}(y)$ with $z_n \in F(x_n)$. Because φ is continuous, we have that $\varphi(z_n) \to y$ with $\varphi(z_n) \in \varphi \circ F(x_n)$. We have proved that $\varphi \circ F$ is ISC. Because $G(x)$ is a closed subset of Z, it is compact. So G satisfies all the assumptions of item (II), which implies the existence of a residual set $R \subset X$ such that G is USC in R. Using now Proposition 2.5.21(i) and the fact that G is closed-valued, we conclude that G is in fact OSC in R. Note that the topology of $Z_0 \subset Z$ is the one induced by Z, and recall that the induced closure of a set $A \subset Z_0$ is given by $\overline{A} \cap Z_0$, where $\overline{A}$ is the closure in Z. This implies that a subset $A \subset Z_0$ is closed for the induced topology if and only if $A = \overline{A} \cap Z_0$. We apply this fact to the set $A := (\varphi \circ F)(x)$. Because $F(x)$ is closed and φ is an homeomorphism, we have that $(\varphi \circ F)(x)$ is closed in Z_0. Hence, $(\varphi \circ F)(x) = \overline{(\varphi \circ F)(x)} \cap Z_0$. Take now $x \in R$. We claim that $\varphi \circ F$ is OSC at x. In order to prove this claim, consider a sequence $\{(x_n, y_n)\} \subset \text{Gph}(\varphi \circ F) \subset X \times Z_0$ such that $x_n \to^X x$ and $y_n \to^{Z_0} y$, where we are emphasizing the topologies with respect to which each sequence converges. Because the topology in Z_0 is the one induced by Z, we have that $(x_n, y_n) \to^{X \times Z} (x, y)$. Observing now that $\{(x_n, y_n)\} \in \text{Gph}(\varphi \circ F) \subset \text{Gph } G$ and using the fact that G is OSC at $x \in R$, we conclude that $y \in G(x) \cap Z_0 = \overline{(\varphi \circ F)(x)} \cap Z_0 = (\varphi \circ F)(x)$, as mentioned above. This yields $y \in (\varphi \circ F)(x)$ and hence outer-semicontinuity of $\varphi \circ F$ in R is established. Finally, we establish outer semicontinuity of F in R. Consider a sequence $\{(x_n, y_n)\} \subset \text{Gph } F$ such that $(x_n, y_n) \to (x, y)$. Because φ is an homeomorphism, $(x_n, \varphi(y_n)) \to (x, \varphi(y))$, with $\{(x_n, \varphi(y_n))\} \subset \text{Gph } \varphi \circ F$. By outer-semicontinuity of $\varphi \circ F$ in R, we have that $\varphi(y) \in (\varphi \circ F)(x)$. This implies that $y \in F(x)$ and therefore F is OSC in R. $\square$

Remark 2.7.8. The compactness of $F(x)$ is necessary in Theorem 2.7.7(II). The point-to-set mapping of Example 2.5.8 is ISC, but there exists no $\alpha \in [0, 2\pi]$ such that F is USC at α. Also, the closedness of $F(x)$ is necessary in Theorem 2.7.7(III). The point-to-set mapping of Remark 2.5.13 is ISC, but there exists no $x \in \mathbb{R}$ such that F is OSC at x.

2.8   The closed graph theorem for point-to-set mappings

The classical closed graph theorem for linear operators in Banach spaces states that such an operator is continuous if its graph is closed. Informally, lack of continuity of a (possibly nonlinear) mapping $F : X \to Y$ at $\bar{x} \in X$ can occur in two ways: existence of a sequence $\{x^n\} \subset X$ converging to $\bar{x}$ such that $\{\|F(x^n)\|\}$ is unbounded, or such that $\{F(x^n)\}$ converges to some $\bar{y} \neq F(\bar{x})$. Closedness of $\mathrm{Gph}(F)$ excludes the second alternative. The theorem under consideration states that the first one is impossible when F is linear. In this section we extend this theorem to point-to-set mappings. The natural extension of linear operators to the point-to-set setting is the notion of a *convex process*, which we define next.

Definition 2.8.1. *Let X and Y be Banach spaces. A mapping $F : X \rightrightarrows Y$ is*

(i) *Convex if $\mathrm{Gph}(F)$ is convex*

(ii) *Closed if $\mathrm{Gph}(F)$ is closed*

(iii) *a Process if $\mathrm{Gph}(F)$ is a cone*

Hence a closed convex process is a set-valued map whose graph is a closed convex cone. Most of the properties of continuous linear operators are shared by closed convex processes. The one under consideration here is the closed graph theorem mentioned above. An important example of closed convex processes are the derivatives of point-to-set mappings introduced in Section 3.7. We start with an introductory result.

Proposition 2.8.2. *$F : X \rightrightarrows Y$ is convex if and only if $\alpha F(x_1) + (1 - \alpha)F(x_2) \subset F(\alpha x_1 + (1 - \alpha)x_2)$ for all $x_1, x_2 \in D(F)$ and all $\alpha \in [0, 1]$.*
It is a process if and only if $0 \in F(0)$ and $\lambda F(x) = F(\lambda x)$ for all $x \in X$ and all $\lambda > 0$.
It is a convex process if and only if it is a process satisfying $F(x_1) + F(x_2) \subset F(x_1 + x_2)$ for all $x_1, x_2 \in X$.

Proof. Elementary. $\square$

We remark that if F is a closed convex process then both $D(F)$ and $R(F)$ are convex cones, not necessarily closed.

We present now an important technical result, which has among its consequences both the closed graph theorem and the open map one. It was proved independently by Robinson and by Ursescu (see [179] and [222]).

Theorem 2.8.3. *If X and Y are Banach spaces such that X is reflexive, and $F : X \rightrightarrows Y$ is a closed convex mapping, then for all $y_0 \in (R(F))^o$ there exists $\gamma > 0$ such that F^{-1} is Lipschitz continuous in $B(y_0, \gamma)$ (cf. Definition 2.5.31).*

Proof. Fix $y_0 \in (R(F))^o$ and $x_0 \in F^{-1}(y_0)$. Define $\rho : Y \to \mathbb{R} \cup \{\infty\}$ as $\rho(y) := d\left(x_0, F^{-1}(y)\right)$. We claim that ρ is convex and lsc. The convexity is an easy consequence of the fact that F^{-1} has a convex graph, which follows from the fact that $\mathrm{Gph}(F)$ is convex. We proceed to prove the lower-semicontinuity of ρ. We must prove that the sets $S_\rho(\lambda) := \{y \in Y \mid \rho(y) \le \lambda\}$ are closed. Take a sequence $\{y_n\} \subset S_\rho(\lambda)$ converging to some $\bar{y}$. Fix any $\varepsilon > 0$. Note that $\rho(y_n) \le \lambda < \lambda + \varepsilon$ for all n, in which case there exists $x_n \in F^{-1}(y_n)$ such that $\|x_n - x_0\| < \lambda + \varepsilon$; that is, $\{x_n\} \subset B(x_0, \lambda + \varepsilon)$. By reflexivity of X, $B(x_0, \lambda + \varepsilon)$ is weakly compact (cf. Theorem 2.5.19), so that there exists a subsequence $\{x_{n_k}\}$ of $\{x_n\}$ weakly convergent to some $\bar{x} \in B(x_0, \lambda + \varepsilon)$. Thus, $\{(x_{n_k}, y_{n_k})\} \subset \mathrm{Gph}(F)$ is weakly convergent to $(\bar{x}, \bar{y})$. Because $\mathrm{Gph}(F)$ is closed and convex, $\mathrm{Gph}(F)$ is weakly closed by Corollary 3.4.16, and hence $(\bar{x}, \bar{y})$ belongs to $\mathrm{Gph}(F)$. Thus, noting that $\bar{x} \in B(x_0, \lambda + \varepsilon)$, we have

$$\rho(\bar{y}) = \inf_{z \in F^{-1}(\bar{y})} \|z - x_0\| \le \|\bar{x} - x_0\| \le \lambda + \varepsilon.$$

We conclude that $\rho(\bar{y}) \le \lambda + \varepsilon$ for all $\varepsilon > 0$, and hence $\bar{y} \in S_\rho(\lambda)$, establishing the claim. Note that

$$\mathrm{Dom}(\rho) = R(F) = \cup_{n=1}^{\infty} S_\rho(n).$$

Because $(R(F))^o \ne \emptyset$ and the sublevel sets $S_\rho(n)$ are all closed, there exists n_0 such that $(S_\rho(n_0))^o \ne \emptyset$; that is, there exists $r > 0$ and $y' \in S_\rho(n_0)$ such that $B(y', r) \subset S_\rho(n_0)$. We have seen that ρ is bounded above in a neighborhood of some point $y' \in \mathrm{Dom}(\rho)$. We invoke Theorem 3.4.22 for concluding that ρ is locally Lipschitz in the interior of its domain $R(F)$. Hence there exist $\theta > 0$ and $L > 0$ such that

$$|\rho(y) - \rho(y_0)| = |\rho(y)| = \rho(y) \le L \|y - y_0\|$$

for all $y \in B(y_0, \theta)$. It follows that $\rho(y) < (L+1) \|y - y_0\|$, and now the definition of ρ yields the existence of $x \in F^{-1}(y)$ such that $\|x - x_0\| < (L + 1) \|y - y_0\|$. Hence F^{-1} is Lipschitz continuous in $B(y_0, \theta)$. $\square$

We mention that the result of Theorem 2.8.3 also holds when X is not reflexive (see [179, 222]). Next we obtain several corollaries of this theorem, including the open map theorem and the closed graph one.

Corollary 2.8.4. *If X and Y are Banach spaces such that X is reflexive, and $F : X \rightrightarrows Y$ is a closed convex mapping, then F^{-1} is inner-semicontinuous in*

$(R(F))^o$.

Proof. The result follows from the fact that local Lipschitz continuity implies inner-semicontinuity.   ☐

Note that F is a closed convex process if and only if F^{-1} is a closed and convex process.

Corollary 2.8.5 (Open map theorem). *If X and Y are Banach spaces, X is reflexive, $F : X \rightrightarrows Y$ is a closed convex process, and $R(F) = Y$, then F^{-1} is Lipschitz continuous; that is, there exists $L > 0$ such that for all $x_1 \in F^{-1}(y_1)$ and all $y_2 \in Y$, there exists $x_2 \in F^{-1}(y_2)$ such that $\|x_1 - x_2\| \le L \|y_1 - y_2\|$.*

Proof. Note that $0 \in (R(F))^o$ because $R(F) = Y$. Also $0 \in F(0)$ because $\mathrm{Gph}(F)$ is a closed cone. Thus, we can use Theorem 2.8.3 with $(x_0, y_0) = (0, 0)$, and conclude that there exists $\theta > 0$ and $\sigma > 0$ such that

$$|\rho(y) - \rho(0)| = \rho(y) \le \sigma\|y\| < (\sigma + 1)\|y\|$$

for all $y \in B(0, \theta)$. Hence, there exists $x \in F^{-1}(y)$ such that

$$\|x - x_0\| = \|x\| < (\sigma + 1) \|y\| . \qquad (2.39)$$

Take $y_1, y_2 \in Y$ and $x_1 \in F^{-1}(y_1)$. Because $R(F) = Y$, there exists $z \in F^{-1}(y_2 - y_1)$. If $y_1 = y_2$, then take $x_1 = x_2$ and the conclusion holds trivially. Otherwise, let $\tilde{y} := \lambda(y_2 - y_1)$, with $\lambda := \theta/(2\|y_2 - y_1\|)$ Using (2.39) with $y := \tilde{y}$ and $x := \tilde{x} \in F^{-1}(\tilde{y})$, we get that $\|\tilde{x}\| \le (\sigma + 1) \|\tilde{y}\|$. Take $x_2 := x_1 + \lambda^{-1}\tilde{x}$. We claim that $x_2 \in F^{-1}(y_2)$. Indeed,

$$
\begin{aligned}
x_2 &= x_1 + \lambda^{-1}\tilde{x} \in F^{-1}(y_1) + \lambda^{-1}F^{-1}(\tilde{y}) \\
&= F^{-1}(y_1) + F^{-1}(\lambda^{-1}\tilde{y}) = F^{-1}(y_1) + F^{-1}(y_2 - y_1) \subset F^{-1}(y_2)
\end{aligned}
$$

because F^{-1} is a convex process. It follows that

$$\|x_1 - x_2\| = \lambda^{-1} \|\tilde{x}\| \le \lambda^{-1}(\sigma + 1) \|\tilde{y}\| = L \|y_1 - y_2\| ,$$

with $L = \sigma + 1$.   ☐

Now we apply our results to the restriction of continuous linear maps to closed and convex sets.

Corollary 2.8.6. *Let X and Y be Banach spaces such that X is reflexive, $A : X \rightrightarrows Y$ a continuous linear mapping, and $K \subset X$ a closed and convex set. Take $x_0 \in K$ such that Ax_0 belongs to $(A(K))^o$. Then there exist positive constants $\theta > 0$ and $L > 0$ such that for all $y \in B(Ax_0, \theta)$ there exists a solution $x \in K$ to the equation $Ax = y$ satisfying $\|x - x_0\| \le L \|y - Ax_0\|$.*

Proof. The result follows from Theorem 2.8.3, defining $F : X \rightrightarrows Y$ as the restriction of A to K, which is indeed closed and convex: indeed, $\mathrm{Gph}(F) =$

Gph(A) $\cap$ $K \times Y$ which is closed and convex, because Gph(A) is closed by continuity of A, and convex by linearity of A. □

When K is a cone, the restriction of a continuous linear operator A to K is a convex process, and we get a slightly stronger result.

Corollary 2.8.7. *Let X and Y be Banach spaces such that X is reflexive, $A : X \rightrightarrows Y$ a continuous linear mapping, and $K \subset X$ a closed and convex cone such that $A(K) = Y$. Then the set-valued map $G : Y \rightrightarrows X$ defined as $G(y) := A^{-1}(y) \cap K$ is Lipschitz continuous.*

Proof. We apply Corollary 2.8.5 to $A_{/K}$, which is onto by assumption. □

We close this section with the extension of the closed graph theorem to point-to-set closed convex processes, and the specialization of this result to point-to-point linear maps, for future reference.

Corollary 2.8.8 (Closed graph theorem). *Take Banach spaces X and Y such that Y is reflexive. If $F : X \rightrightarrows Y$ is a closed convex process and $D(F) = X$, then F is Lipschitz continuous; that is, there exists $L > 0$ such that $F(x_1) \subset F(x_2) + L \|x_1 - x_2\| B(0,1)$ for all $x_1, x_2 \in X$.*

Proof. The result follows from Corollary 2.8.5 applied to the closed convex process F^{-1}. □

Corollary 2.8.9. *Take Banach spaces X and Y such that Y is reflexive. If $A : X \rightrightarrows Y$ is a linear map such that $D(A) = X$ and Gph(A) is closed, then A is Lipschitz continuous.*

Proof. The graph of a linear map is always a convex cone. Under the hypothesis of this corollary, A is a closed convex process, and hence Corollary 2.8.8 applies. □

Exercises

2.1. Prove that F is a convex closed process if and only if F^{-1} is a closed and convex process.

2.2. Prove Proposition 2.8.2.

2.9   Historical notes

The notion of semilimits of sequences of sets was introduced by P. Painlevé in 1902 during his lectures on analysis at the University of Paris, according to his student L. Zoretti [239]. The first published reference seems to be Painlevé's comment in [165]. Hausdorff [90] and Kuratowski [128] included this notion of convergence in their books, where they developed the basis of calculus of limits of set-valued mappings, and as a consequence this concept is known as Kuratowski–Painlevé's convergence. We mention parenthetically that in these earlier references the semilimits were called "upper" and "lower". We prefer instead "exterior" and "interior", respectively, following the trend started in [94] and [194].

Upper- and lower-semicontinuity of set-valued maps appeared for the first time in 1925, as part of the thesis of F. Vasilesco [224], a student of Lebesgue who considered only the case of point-to-set maps $S : \mathbb{R} \rightrightarrows \mathbb{R}$. The notion was extended to more general settings in [35, 127], and [128].

Set-valued analysis received a strong impulse with the publication of Berge's topology book [26] in 1959, which made the concept known outside the pure mathematics community. Nevertheless, until the late 1980s, the emphasis in most publications on the subject stayed within the realm of topology. More recently, the book by Beer [24] and several survey articles (e.g., [212] and [137]) shifted the main focus to more applied concerns (e.g., optimization, random sets, economics). The book by Aubin and Frankowska [18] has been a basic milestone in the development of the theory of set-valued analysis. The more recent book by Rockafellar and Wets [194] constitutes a fundamental addition to the literature on the subject.

The compactness result given in Theorem 2.2.12 for nets of sets in Hausdorff spaces was obtained by Mrowka in 1970 [163], and the one in Proposition 2.2.14 for sequences of sets in N_2 spaces was published by Zarankiewicz in 1927 [233].

The generic continuity result in Theorem 2.7.7 is due to Kuratowski [127] and [128]. It was later extended by Choquet [62], and more recently by Zhong [236].

The notion of convex process presented in Section 2.8 was introduced independently, in a finitely dimensional framework, by Rockafellar ([185], with further developments in [187]) and Makarov–Rubinov (see [138] and references therein). The latter reference is, to the best of our knowledge, the first one dealing with the infinite dimensional case.

The extension of the classical closed graph theorem to the set-valued framework was achieved independently by Robinson [179] and Ursescu [222].

Chapter 3

Convex Analysis and Fixed Point Theorems

3.1 Lower-semicontinuous functions

In this chapter we establish several results on fixed points of point-to-set mappings. We start by defining the notion of a fixed point in this setting.

Definition 3.1.1. *Let X be a topological space and $G : X \rightrightarrows X$ a point-to-set mapping.*

(a) $\bar{x} \in X$ is a fixed point of G if and only if $\bar{x} \in G(\bar{x})$.

(b) Given $K \subset X$, $\bar{x}$ is a fixed point in K if and only if $\bar{x}$ belongs to K and $\bar{x} \in G(\bar{x})$.

We also use the related concept of equilibrium point.

Definition 3.1.2. *Given topological vector spaces X and Y, a point-to-set mapping $G : X \rightrightarrows Y$, and a subset $K \subset X$, $\bar{x}$ is an equilibrium point in K if and only if $\bar{x}$ belongs to K and $0 \in G(\bar{x})$.*

We first establish a result on existence of "unconstrained" fixed points, as in Definition 3.1.1(a), namely Caristi's theorem, and then one on "constrained" fixed points, as in Definition 3.1.1(b), namely Kakutani's one. The first requires the notion of lower-semicontinuous functions, to which this section is devoted. Kakutani's theorem requires convex functions, which we study in some detail in Section 3.4.

In the remainder of this section, X is a Hausdorff topological space. Consider a real function $f : X \to \mathbb{R} \cup \{\infty, -\infty\}$. The extended real line $\mathbb{R} \cup \{\infty, -\infty\}$ allows us to transform constrained minimization problems into unconstrained ones: given $K \subset X$, the constrained problem consisting of minimizing a real-valued $\varphi : X \to \mathbb{R}$ subject to $x \in K$ is equivalent to the unconstrained one of minimizing $\varphi_K : X \to \mathbb{R} \cup \{\infty\}$, with φ_K given by

$$\varphi_K(x) = \begin{cases} \varphi(x) & \text{if } x \in K, \\ \infty & \text{otherwise.} \end{cases} \qquad (3.1)$$

We also need the notions of an indicator function of a set and of a domain in the setting of the extended real line.

Definition 3.1.3. *Given $K \subset X$, the* indicator function *of K is $\delta_K : X \to \mathbb{R} \cup \{\infty\}$ given by*

$$\delta_K(x) = \begin{cases} 0 & \text{if } x \in K, \\ \infty & \text{otherwise.} \end{cases}$$

It follows from Definition 3.1.3 that $\varphi_K = \varphi + \delta_K$.

Definition 3.1.4. *Given $f : X \to \mathbb{R} \cup \{\infty, -\infty\}$, its domain $\mathrm{Dom}(f)$ is defined as $\mathrm{Dom}(f) = \{x \in X : f(x) < \infty\}$. The function f is said to be* strict *if $\mathrm{Dom}(f)$ is nonempty, and* proper *if it is strict and $f(x) > -\infty$ for all $x \in X$.*

We introduce next the notion of epigraph, closely linked to the study of lower-semicontinuous functions.

Definition 3.1.5. *Given $f : X \to \mathbb{R} \cup \{\infty, -\infty\}$ its* epigraph $\mathrm{Epi}(f) \subset X \times \mathbb{R}$ *is defined as $\mathrm{Epi}(f) := \{(x, \lambda) \in X \times \mathbb{R} : f(x) \leq \lambda\}$.*

Observe that f is strict if and only if $\mathrm{Epi}(f) \neq \emptyset$, and f is proper if and only if $\mathrm{Epi}(f) \neq \emptyset$ and it does not contain "vertical" lines (i.e., a set of the form $\{(x, \lambda) : \lambda \in \mathbb{R}\}$ for some $x \in X$).

The projections of the epigraph onto X and $\mathbb{R}$ give rise to two point-to-set mappings that are also of interest and deserve a definition. These point-to-set mappings, as well as the epigraph of f, are depicted in Figure 3.1.

Definition 3.1.6. *Given $f : X \to \mathbb{R} \cup \{\infty, -\infty\}$,*

(a) Its level mapping $S_f : \mathbb{R} \rightrightarrows X$, *is defined as $S_f(\lambda) = \{x \in X : f(x) \leq \lambda\}$.*

(b) Its epigraphic profile $E_f : X \rightrightarrows \mathbb{R}$ *is defined as $E_f(x) = \{\lambda \in \mathbb{R} : f(x) \leq \lambda\}$.*

Proposition 3.1.7. *Take $f : X \to \mathbb{R}$. Then*

(i) E_f is OSC if and only if $\mathrm{Epi}(f)$ is closed.

(ii) S_f is OSC if and only if $\mathrm{Epi}(f)$ is closed.

Proof.

(i) Follows from the fact that $\mathrm{Gph}(E_f) = \mathrm{Epi}(f)$.

Item (ii) follows from (i), after observing that $(\lambda, x) \in \mathrm{Gph}(S_f)$ if and only if $(x, \lambda) \in \mathrm{Gph}(E_f)$. $\square$

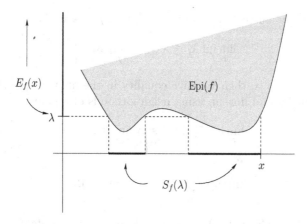

Figure 3.1. *Epigraph, level mapping, and epigraphic profile*

Next we introduce lower-semicontinuous functions.

Definition 3.1.8. *Given $f : X \to \mathbb{R} \cup \{\infty, -\infty\}$ and $\bar{x} \in \mathbb{R}$, we say that*

(a) *f is lower-semicontinuous at $\bar{x}$ (lsc at $\bar{x}$) if and only if for all $\lambda \in \mathbb{R}$ such that $f(\bar{x}) > \lambda$, there exists a neighborhood U of $\bar{x}$ such that $f(x) > \lambda$ for all $x \in U$.*

(b) *f is upper-semicontinuous at $\bar{x}$ (Usc at $\bar{x}$) if and only if the function $-f$ is lower-semicontinuous at $\bar{x}$.*

(c) *f is lower-semicontinuous (respectively, upper-semicontinuous) if and only if f is lower-semicontinuous at x (respectively, upper-semicontinuous at x) for all $x \in X$.*

It is well-known that f is continuous at x if and only if it is both lsc and Usc at x.

The next elementary proposition, whose proof is left to the reader, provides an equivalent definition of lower-semicontinuity. For $x \in X$, $\mathcal{U}_x \subset \mathcal{P}(X)$ denotes the family of neighborhoods of x.

For a net $\{x_i\} \subset X$ and a function $f : X \to \mathbb{R}$, we can define

$$\liminf_{i \in I} f(x_i) := \sup_{\substack{J \subset I, \\ J \text{ terminal}}} \inf_{j \in J} f(x_j),$$

and

$$\limsup_{i \in I} f(x_i) := \inf_{\substack{J \subset I, \\ J \text{ terminal}}} \sup_{j \in J} f(x_j).$$

Recall that for every cluster point λ of a net $\{\lambda_i\}_{i\in I} \subset \mathbb{R}$ we have

$$\liminf_{i\in I} \lambda_i \leq \lambda \leq \limsup_{i\in I} \lambda_i.$$

If the net converges to λ, then we have equality in the above expression. We also need to define lim inf and lim sup using neighborhoods of a point:

$$\liminf_{x\to\bar{x}} f(x) := \sup_{U\in\mathcal{U}_{\bar{x}}} \inf_{x\in U} f(x),$$

and

$$\limsup_{x\to\bar{x}} f(x) := \inf_{U\in\mathcal{U}_{\bar{x}}} \sup_{x\in U} f(x).$$

Proposition 3.1.9. *Given $f : X \to \mathbb{R} \cup \{\infty\}$, the following statements are equivalent.*

(i) *f is lsc at $\bar{x}$.*

(ii) *$f(\bar{x}) \leq \liminf_{x\to\bar{x}} f(x)$.*

(iii) *$f(\bar{x}) \leq \liminf_i f(x_i)$ for all net $\{x_i\} \subset X$ such that $\lim_i x_i = \bar{x}$.*

Proof. The proofs of (i)→(ii) and (iii)→(i) are a direct consequence of the definitions, and (ii)→(iii) uses the fact that $\liminf_{x\to\bar{x}} f(x) \leq \liminf_i f(x_i)$ for every net $\{x_i\} \subset X$ such that $\lim_i x_i = \bar{x}$. The details are left to the reader. □

The following proposition presents an essential property of lower semicontinuous functions.

Proposition 3.1.10. *Given $f : X \to \mathbb{R} \cup \{\infty\}$, the following statements are equivalent.*

(i) *f is lsc.*

(ii) *$\mathrm{Epi}(f)$ is closed.*

(iii) *E_f is osc.*

(iv) *S_f is osc.*

(v) *$S_f(\lambda)$ is closed for all $\lambda \in \mathbb{R}$.*

Proof.

(i) $\Rightarrow$ (ii) Take any convergent net $\{(x_i, \lambda_i)\} \subset \mathrm{Epi}(f)$ with limit $(x, \lambda) \in X \times \mathbb{R}$. It suffices to prove that $(x, \lambda) \in \mathrm{Epi}(f)$. By Definition 3.1.5, for all $i \in I$,

$$f(x_i) \leq \lambda_i. \tag{3.2}$$

Taking "limites inferiores" in both sides of (3.2) we get, using Proposition 3.1.9(iii),

$$f(x) \leq \liminf_i f(x_i) \leq \liminf_i \lambda_i = \lambda,$$

so that $f(x) \leq \lambda$; that is, (x, λ) belongs to $\mathrm{Epi}(f)$.

(ii) $\Rightarrow$ (iii) The result follows from Proposition 3.1.7(i).

(iii) $\Rightarrow$ (iv) Follows from Proposition 3.1.7.

(iv) $\Rightarrow$ (v) Follows from the fact that S_f is OSC (cf. Exercise 2.14(ii) of Section 2.5).

(v) $\Rightarrow$ (i) Note that (v) is equivalent to the set $S_f(\lambda)^c$ being open, which is precisely the definition of lower-semicontinuity.

$\square$

Corollary 3.1.11. *A subset K of X is closed if and only if its indicator function δ_K is lsc.*

Proof. By Definition 3.1.3, $\mathrm{Epi}(\delta_K) = K \times [0, \infty]$, which is closed if and only if K is closed. On the other hand, by Proposition 3.1.10, δ_K is lsc if and only if $\mathrm{Epi}(\delta_K)$ is closed. $\square$

Classical continuity of f is equivalent to the continuity of the point-to-set mapping E_f.

Proposition 3.1.12. *Given $f : X \to \mathbb{R} \cup \{\infty, -\infty\}$ and $x \in X$,*

(a) E_f is ISC at x if and only if f is USC at x.

(b) E_f is continuous at x if and only if f is continuous at x.

Proof.

(a) Assume that E_f is ISC at x. If the conclusion does not hold, then there exist a net $\{x_i\}$ converging to x and some $\gamma \in \mathbb{R}$ such that $f(x) < \gamma < \limsup_i f(x_i)$. It follows that $\gamma \in E_f(x)$ and because $\{x_i\}$ converges to x, by inner-semicontinuity of E_f we have that $E_f(x) \subset \liminf_i E_f(x_i)$. This means that for every $\varepsilon > 0$ there exists a terminal set $J_\varepsilon \subset I$ such that $B(\gamma, \varepsilon) \cap E_f(x_i) \neq \emptyset$ for all $i \in J_\varepsilon$. In other words, for every $i \in J_\varepsilon$ there exists $\gamma_i \in E_f(x_i)$ with $\gamma - \varepsilon < \gamma_i < \gamma + \varepsilon$. Hence we have

$$\limsup_i f(x_i) \leq \sup_{i \in J_\varepsilon} f(x_i) \leq \gamma_i < \gamma + \varepsilon.$$

Because $\varepsilon > 0$ is arbitrary, the above expression yields $\limsup_i f(x_i) \leq \gamma$, contradicting the assumption on γ.

(b) The result follows from (a) and the equivalence between items (i) and (iii) of
   Proposition 3.1.10.

   □

Remark 3.1.13. Inner-semicontinuity of S_f neither implies nor is implied by
semicontinuity of f. The function $f : \mathbb{R} \to \mathbb{R}$ depicted in Figure 3.2 is continuous,
and S_f is not ISC at λ. Indeed, taking λ and x as in Figure 3.2 and a sequence
$\lambda_n \to \lambda$ such that $\lambda_n < \lambda$ for all n, there exists no sequence $\{x_n\}$ converging to x
such that $x_n \in S_f(\lambda_n)$. The function $f : \mathbb{R} \to \mathbb{R}$ defined by

$$f(x) = \begin{cases} 0 & \text{if } x \in \mathbb{Q} \\ 1 & \text{otherwise} \end{cases}$$

is discontinuous everywhere, and S_f is ISC.

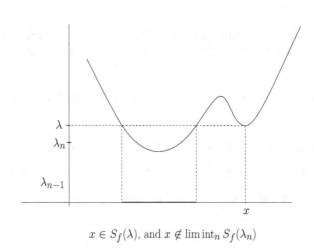

$x \in S_f(\lambda)$, and $x \notin \liminf_n S_f(\lambda_n)$

Figure 3.2. *Continuous function with non inner-semicontinuous level mapping*

The following proposition deals with operations that preserve lower-semicontinuity

Proposition 3.1.14.

(i) *If* $f, g : X \to \mathbb{R} \cup \{\infty\}$ *are* lsc *then* $f + g$ *is* lsc.

(ii) *If* $f_i : X \to \mathbb{R} \cup \{\infty, -\infty\}$ *is* lsc *for* $1 \le i \le m$ *then* $\inf_{1 \le i \le m} f_i$ *is* lsc.

(iii) *For an arbitrary set* I, *if* $f_i : X \to \mathbb{R} \cup \{\infty, -\infty\}$ *is* lsc *for all* $i \in I$ *then*
   $\sup_{i \in I} f_i$ *is* lsc.

(iv) If $f : X \to \mathbb{R} \cup \{\infty, -\infty\}$ is lsc then αf is lsc for all real $\alpha > 0$.

(v) If X and Y are topological spaces, $g : Y \to X$ is continuous and $f : X \to \mathbb{R} \cup \{\infty, -\infty\}$ is lsc, then $f \circ g : Y \to \mathbb{R} \cup \{\infty, -\infty\}$ is lsc.

Proof.

(i) Use Proposition 3.1.9(iii), noting that

$$\liminf_i f(x_i) + \liminf_i g(x_i) \leq \liminf_i [f(x_i) + g(x_i)],$$

if the sum on the left is not $\infty - \infty$ (see, e.g., [194, Proposition 1.38]).

(ii) and (iii) Use Proposition 3.1.10(ii), noting that $\mathrm{Epi}(\inf_{1 \leq i \leq m} f_i) = \cup_{i=1}^m \mathrm{Epi}(f_i)$, $\mathrm{Epi}(\sup_{i \in I} f_i) = \cap_{i \in I} \mathrm{Epi}(f_i)$.

(iv) Use Proposition 3.1.10(v), noting that $S_{\alpha f}(\lambda) = S_f(\lambda/\alpha)$.

(v) Note that $S_{f \circ g}(\lambda) = g^{-1}(S_f(\lambda))$. $S_f(\lambda)$ is closed by Proposition 3.1.10(v) because f is lsc, so that $g^{-1}(S_f(\lambda))$ is closed by continuity of g, and the result follows.

$\square$

The next proposition contains a very important property of lsc functions.

Proposition 3.1.15. *If $f : X \to \mathbb{R} \cup \{\infty\}$ is lsc, K is a compact subset of X and $\mathrm{Dom}(f) \cap K \neq \emptyset$, then f is bounded below on K and it attains its minimum on K, which is finite.*

Proof. For every $n \in \mathbb{N}$, define the sets

$$L_n := \{x \in X : f(x) > -n\}.$$

Note that $L_n \subset L_{n+1}$. All these sets are open by lower-semicontinuity of f and $K \subset \cup_n L_n$. Because K is compact and the sets are nested, there exists $n_0 \in \mathbb{N}$ such that $K \subset L_{n_0}$. Therefore for every $x \in K$ we have $f(x) > -n_0$ and hence $\alpha := \inf_{x \in K} f(x) \geq -n_0$ which yields f bounded below in K with $\alpha \in \mathbb{R}$ (note that $\alpha < \infty$ because $\mathrm{Dom}(f) \cap K \neq \emptyset$). To finish the proof, we must show that there exists $\bar{x} \in K$ such that $f(\bar{x}) = \alpha$. If this is not the case, then K can be covered by the family of sets defined by

$$A_n := \{x \in X : f(x) > \alpha + \frac{1}{n}\}.$$

We have $A_n \subset A_{n+1}$ and all these sets are open by lower-semicontinuity of f. Using compacity again we conclude that there exists $n_1 \in \mathbb{N}$ such that $K \subset A_{n_1}$. In other words,

$$\alpha = \inf_{x \in K} f(x) \geq \alpha + \frac{1}{n_1} > \alpha,$$

a contradiction. Therefore f must attain its minimum value α on K.   □

Exercises

3.1. Let X, Y be metric spaces. Given $F : X \rightrightarrows Y$, for each $y \in Y$ define
$\varphi_y : X \to \mathbb{R} \cup \{\infty\}$ as $\varphi_y(x) = d(y, F(x))$. Prove that

   (i) F is ISC at x if and only if φ_y is USC at x for all $y \in Y$.

   (ii) Assume that $F(x)$ is closed. If φ_y is USC at x for all $y \in Y$, then F is
   OSC at x.

   (iii) Assume that Y is finite-dimensional. If F is OSC at x then φ_y is USC at
   x for all $y \in Y$. Hint: use Definition 2.5.6 and Proposition 2.2.11.

3.2. Given $\{f_i\}$ $(i \in I)$, let $\bar{f} := \sup_{i \in I} f_i$ and $\underline{f} := \inf_{i \in I} f_i$.

   (i) Prove that $\text{Epi}(\bar{f}) = \cap_{i \in I} \text{Epi}(f_i)$.

   (ii) Prove that if I is finite then $\text{Epi}(\underline{f}) = \cup_{i \in I} \text{Epi}(f_i)$.

   (iii) Give an example for which $\cup_{i \in I} \text{Epi}(f_i) \subsetneq \text{Epi}(\underline{f})$. Hint: consider $\{f_n\}_{n \in \mathbb{N}}$
   defined as $f_n(x) = x^2 + 1/n$.

3.3. Take $\bar{f}$, $\underline{f}$ as in Exercise 2 and S_f as in Definition 3.1.6(a).

   (i) Prove that $S_{\bar{f}}(\lambda) = \cap_{i \in I} S_{f_i}(\lambda)$.

   (ii) Prove that $S_{\underline{f}}(\lambda) \supset \cap_{i \in I} S_{f_i}(\lambda)$, and find an example for which the
   inclusion is strict.

3.4. Prove Proposition 3.1.9.

3.5. Take E_f as in Definition 3.1.6(b). Prove that $\text{Gph}(E_f) = \text{Epi}(f)$ and that
$D(E_f) = \text{Dom}(f)$.

3.6. Prove the statements in Remark 3.1.13.

3.2   Ekeland's variational principle

Ekeland's variational principle is the cornerstone of our proof of the first fixed point
theorem in this chapter. In its proof, we use the following well-known result on
complete metric spaces. We recall that the diameter $\text{diam}(A)$ of a subset A of a
metric space X is defined as $\text{diam}(A) = \sup\{d(x, y) : x, y \in A\}$.

Proposition 3.2.1. *A metric space X is complete if and only if for every se-*
quence $\{F_n\}$ of nonempty closed sets, such that $F_{n+1} \subset F_n$ for all $n \in \mathbb{N}$ and
$\lim_n \text{diam}(F_n) = 0$, *there exists $\bar{x} \in X$ such that $\cap_{n=1}^{\infty} F_n = \{\bar{x}\}$.*

Proof. See Theorem 5.1 in [22].   □

The next theorem is known as Ekeland's variational principle, and was published for the first time in [80].

Theorem 3.2.2. *Let X be a complete metric space and $f : X \to \mathbb{R} \cup \{\infty\}$ a strict, nonnegative, and lsc function. For all $\hat{x} \in \mathrm{Dom}(f)$ and all $\varepsilon > 0$ there exists $\bar{x} \in X$ such that*

(a) $f(\bar{x}) + \varepsilon d(\hat{x}, \bar{x}) \leq f(\hat{x})$.

(b) $f(\bar{x}) < f(x) + \varepsilon d(x, \bar{x})$ for all $x \neq \bar{x}$.

Proof. Clearly, it suffices to prove the result for $\varepsilon = 1$, because if it holds for this value of ε, then it holds for any value, applying the result to the function $\tilde{f} = f/\varepsilon$, which is lsc by Proposition 3.1.14(iv).

Define the point-to-set mapping $F : X \rightrightarrows X$ as $F(x) = \{y \in X : f(y) + d(x, y) \leq f(x)\}$. Because $d(x, \cdot)$ is continuous and f is lsc, we get from Proposition 3.1.14(i) that $f(\cdot) + d(x, \cdot)$ is lsc, and then from Proposition 3.1.10(v) that $F(x)$ is closed for all $x \in X$. We claim that F satisfies:

$$\text{if } x \in \mathrm{Dom}(f) \quad \text{then} \quad x \in F(x) \subset \mathrm{Dom}(f), \tag{3.3}$$

$$\text{if } y \in F(x) \text{ then } F(y) \subset F(x). \tag{3.4}$$

Statement (3.3) is straightforward from the definition of $F(x)$; in order to prove (3.4), take any $z \in F(y)$. Using the triangular property of d, and the facts that $z \in F(y)$, $y \in F(x)$, we get

$$f(z) + d(x, z) \leq f(z) + d(x, y) + d(y, z) \leq f(y) + d(x, y) \leq f(x),$$

which establishes the claim.

Define now $\psi : X \to \mathbb{R} \cup \{\infty\}$ as $\psi(y) = \inf_{z \in F(y)} f(z)$. For all $y \in \mathrm{Dom}(f)$, we have, by (3.3) and nonnegativity of f, that

$$0 \leq \psi(y) < \infty. \tag{3.5}$$

It follows from the definition of ψ that

$$d(x, y) \leq f(x) - f(y) \leq f(x) - \psi(x) \tag{3.6}$$

for all $x \in \mathrm{Dom}(f)$, $y \in F(x)$, which allows us to estimate the diameter of each $F(x)$, according to

$$\mathrm{diam}(F(x)) = \sup\{d(u, v) : u, v \in F(x)\} \leq 2[f(x) - \psi(x)], \tag{3.7}$$

because it follows from (3.6) and the triangular property of d that $d(u, v) \leq d(u, x) + d(x, v) \leq 2[f(x) - \psi(x)]$. Given $\hat{x} \in \mathrm{Dom}(f)$, we define a sequence $\{x_n\} \subset X$ from which we in turn construct a sequence of closed sets

$F_n \subset X$ satisfying the hypotheses of Proposition 3.2.1. Let $x_0 = \hat{x}$, and, given $x_n \in \mathrm{Dom}(f)$, take $x_{n+1} \in F(x_n)$ so that

$$f(x_{n+1}) \le \psi(x_n) + 2^{-n}. \tag{3.8}$$

The definition of ψ and (3.5) ensure existence of x_{n+1}. Define $F_n := F(x_n)$. Note that $\emptyset \subsetneq F_n \subset \mathrm{Dom}(f)$ by (3.3), and that $F_{n+1} \subset F_n$ for all $n \in \mathbb{N}$ by (3.4). These facts imply that

$$\psi(x_n) \le \psi(x_{n+1}) \tag{3.9}$$

for all $n \in \mathbb{N}$. If we prove that $\lim_n \mathrm{diam}(F_n) = 0$, then the sequence $\{F_n\}$ will satisfy the hypotheses of Proposition 3.2.1. We proceed to establish this fact. By (3.8), the definition of ψ, and (3.9), we have

$$\psi(x_{n+1}) \le f(x_{n+1}) \le \psi(x_n) + 2^{-n} \le \psi(x_{n+1}) + 2^{-n}. \tag{3.10}$$

By the definition of F_n, (3.7), and (3.10), $\mathrm{diam}(F_{n+1}) \le 2[f(x_{n+1}) - \psi(x_{n+1})] \le 2^{-n+1}$, so that $\lim_n \mathrm{diam}(F_n) = 0$, and we proceed to apply Proposition 3.2.1 to $\{F_n\}$, concluding that there exists $\bar{x} \in X$ such that $\{\bar{x}\} = \cap_{n=0}^{\infty} F_n$. Because $\bar{x} \in F(x_0) = F(\hat{x})$, we get that $f(\bar{x}) + d(\bar{x}, \hat{x}) \le f(\hat{x})$, establishing (a). The latter inequality also yields $\bar{x} \in \mathrm{Dom}(f)$. Because $\bar{x} \in F_n = F(x_n)$ for all $n \in \mathbb{N}$, we conclude from (3.4) that $F(\bar{x}) \subset F(x_n) = F_n$ for all $n \in \mathbb{N}$, so that $F(\bar{x}) \subset \cap_{n=0}^{\infty} F_n = \{\bar{x}\}$, and henceforth, in view of (3.3), $F(\bar{x}) = \{\bar{x}\}$. Thus, $x \notin F(\bar{x})$ for all $x \ne \bar{x}$; that is, $f(x) + d(x, \bar{x}) > f(\bar{x})$ for all $x \ne \bar{x}$, establishing (b). $\square$

The following is an easy consequence of the previous theorem, and is also found in the literature as Ekeland's variational principle. Its proof is direct from the previous result.

Corollary 3.2.3. *Let X be a Banach space and $f : X \to \mathbb{R} \cup \{\infty\}$ a strict, lsc, and bounded below function. Suppose that $x_0 \in X$ and $\varepsilon > 0$ are such that*

$$f(x_0) < \inf_{x \in X} f(x) + \varepsilon.$$

Then for all $\lambda \in (0, 1)$ there exists $\bar{x} \in \mathrm{Dom}(f)$ such that

(a) $\lambda \|\bar{x} - x_0\| \le \varepsilon$.

(b) $\lambda \|\bar{x} - x_0\| \le f(x_0) - f(\bar{x})$.

(c) $\lambda \|\bar{x} - x\| > f(\bar{x}) - f(x)$, *for all $x \ne \bar{x}$.*

3.3   Caristi's fixed point theorem

Next we present our main result on existence of unconstrained fixed points of point-to-set mappings, an extension of Caristi's theorem for the point-to-point case (see [58]). We follow here, for the point-to-set case, the presentation in [16], taken from [201].

Theorem 3.3.1. *Let X be a complete metric space and $G : X \rightrightarrows X$ a point-to-set mapping. If there exists a proper, nonnegative, and lsc function $f : X \to \mathbb{R} \cup \{\infty\}$ such that for all $x \in X$ there exists $y \in G(x)$ satisfying*

$$f(y) + d(x, y) \le f(x), \qquad (3.11)$$

then G has a fixed point; that is, there exists $\bar{x} \in X$ such that $\bar{x} \in G(\bar{x})$. If the stronger condition $G(x) \subset F(x)$ holds for all $x \in X$, with $F(x) = \{y \in X : f(y) + d(x, y) \le f(x)\}$, then there exists $\bar{x} \in X$ such that $G(\bar{x}) = \{\bar{x}\}$.

Proof. Fix $\varepsilon \in (0, 1)$. By Theorem 3.2.2, there exists $\bar{x} \in X$ such that

$$f(\bar{x}) < f(x) + \varepsilon d(x, \bar{x}) \qquad (3.12)$$

for all $x \neq \bar{x}$. For $F : X \rightrightarrows X$ as defined in the statement of the theorem, we observe that (3.11) is equivalent to $G(x) \cap F(x) \neq \emptyset$ for all $x \in X$. Thus, $G(\bar{x}) \cap F(\bar{x}) \neq \emptyset$, or equivalently, there exists $\bar{y} \in G(\bar{x})$ such that

$$f(\bar{x}) \ge f(\bar{y}) + d(\bar{y}, \bar{x}). \qquad (3.13)$$

We claim that $\bar{y} = \bar{x}$. Otherwise, by (3.12) with $x = \bar{y}$, we have that $f(\bar{x}) < f(\bar{y}) + \varepsilon d(\bar{y}, \bar{x})$, which, together with (3.13), implies that $0 < d(\bar{x}, \bar{y}) < \varepsilon d(\bar{x}, \bar{y})$, which is a contradiction. Thus, the claim holds and $\bar{y} = \bar{x}$. Because $\bar{y} \in G(\bar{x})$, it follows that $\bar{x}$ is a fixed point of G, proving the first statement of the theorem. Under the stronger hypothesis, (3.13) holds for all $y \in G(\bar{x})$ and the same argument shows that $\bar{x} = y$ for any such y. Thus, $G(\bar{x}) = \{\bar{x}\}$. $\square$

Next we present a related result, where the assumption of lower-semicontinuity of f is replaced by outer-semicontinuity of G.

Theorem 3.3.2. *Let X be a complete metric space, and $G : X \rightrightarrows X$ an OSC point-to-set mapping. If there exists a nonnegative $f : X \to \mathbb{R} \cup \{\infty\}$ such that $G(x) \cap F(x) \neq \emptyset$ for all $x \in X$, with $F(x) = \{y \in X : f(y) + d(x, y) \le f(x)\}$, then G has a fixed point.*

Proof. Take an arbitrary $x_0 \in \text{Dom}(f)$. We define inductively $\{x_n\} \subset X$ as follows. Given $x_n \in \text{Dom}(f)$ take $x_{n+1} \in G(x_n)$ such that

$$d(x_{n+1}, x_n) \le f(x_n) - f(x_{n+1}). \qquad (3.14)$$

This sequence is well defined because $G(x) \cap F(x) \neq \emptyset$ for all $x \in X$. It follows from (3.14) that $\{f(x_n)\}$ is nonincreasing, and bounded below by nonnegativity of f, so that $\{f(x_n)\}$ converges. Summing (3.14) with n between p and $q - 1$ we get

$$d(x_p, x_q) \le \sum_{n=p}^{q-1} d(x_{n+1}, x_n) \le f(x_p) - f(x_q). \qquad (3.15)$$

Inasmuch as $\{f(x_n)\}$ converges, it follows from (3.15) that $\{x_n\}$ is a Cauchy sequence, which thus converges, by completeness of X, say to $\bar{x} \in X$. By definition of

$\{x_n\}$, we have that $(x_n, x_{n+1}) \in \mathrm{Gph}(G)$ for all $n \in \mathbb{N}$. Because $\lim_n (x_n, x_{n+1}) = (\bar{x}, \bar{x})$ and G is OSC, we conclude that $(\bar{x}, \bar{x}) \in \mathrm{Gph}(G)$, in other words, $\bar{x} \in G(\bar{x})$.
□

Remark 3.3.3. In the case of Theorem 3.3.2, the stronger condition $G(x) \subset F(x)$ for all $x \in X$ is not enough for guaranteeing that the fixed point $\bar{x}$ satisfies $\{\bar{x}\} = G(\bar{x})$, because f may fail to be lsc, and thus Ekeland's variational principle cannot be invoked.

3.4   Convex functions and conjugacy

We start with some basic definitions. Let X be a real vector space.

Definition 3.4.1. *A subset K of X is* convex *if and only if $\lambda x + (1 - \lambda)y$ belongs to K for all $x, y \in K$ and all $\lambda \in [0, 1]$.*

Definition 3.4.2.

(a) *A function $f : X \to \overline{\mathbb{R}} := \mathbb{R} \cup \{\infty\}$ is* convex *if and only if $f(\lambda x + (1 - \lambda)y) \leq \lambda f(x) + (1 - \lambda)f(y)$ for all $x, y \in X$ and all $\lambda \in [0, 1]$.*

(b) *A function $f : X \to \overline{\mathbb{R}}$ is* concave *if and only if the function $-f$ is convex.*

(c) *A function $f : X \to \overline{\mathbb{R}}$ is* strictly convex *if and only if $f(\lambda x + (1 - \lambda)y) < \lambda f(x) + (1 - \lambda)f(y)$ for all $x, y \in X, x \neq y$, and all $\lambda \in (0, 1)$.*

The following elementary proposition, whose proof is left to the reader, presents a few basic properties of convex functions.

Proposition 3.4.3.

(i) *If $f, g : X \to \overline{\mathbb{R}}$ are convex, then $f + g : X \to \overline{\mathbb{R}}$ is convex.*

(ii) *If $f : X \to \overline{\mathbb{R}}$ is convex, then $\alpha f : X \to \overline{\mathbb{R}}$ is convex for all real $\alpha > 0$.*

(iii) *If $f_i : X \to \overline{\mathbb{R}}$ are convex for all $i \in I$, where I is an arbitrary index set, then $\sup_{i \in I} f_i : X \to \overline{\mathbb{R}}$ is convex.*

(iv) *$f : X \to \overline{\mathbb{R}}$ is convex if and only if $\mathrm{Epi}(f) \subset X \times \mathbb{R}$ is convex.*

Although convex functions can be defined in purely algebraic terms, as in Definition 3.4.2, they become interesting only through their interplay with continuity and duality notions. Thus, in the remainder of this chapter we assume that X is a Banach space and X^* its dual; that is, the set of linear and continuous real functionals defined on X, endowed with the weak topology, as defined in Section

2.5.1. We recall that $\langle \cdot, \cdot \rangle : X^* \times X \to \mathbb{R}$ denotes the duality pairing (i.e., $\langle p, x \rangle$ means just the value $p(x)$ of $p \in X^*$ at $x \in X$).

We start with a basic topological property of convex sets, whose proof is left to the reader.

Proposition 3.4.4. *If $K \subset X$ is convex then its closure $\overline{K}$ and its interior K^o are also convex.*

We introduce next the very important concept of conjugate functions, originally developed by Fenchel in [84].

Definition 3.4.5. *Given $f : X \to \overline{\mathbb{R}}$, we define its* conjugate function $f^* : X^* \to \overline{\mathbb{R}}$ *as $f^*(p) = \sup_{x \in X} \{ \langle p, x \rangle - f(x) \}$.*

Proposition 3.4.6. *For any $f : X \to \overline{\mathbb{R}}$, its conjugate f^* is convex and lsc.*

Proof. For fixed $x \in X$ the function $\varphi_x : X^* \to \mathbb{R}$ defined as $\varphi_x(p) = \langle p, x \rangle$ is continuous, by definition of the dual topology, and henceforth lsc. It is also trivially convex, by linearity. Constant functions are also convex and lsc, therefore the result follows from Definition 3.4.5 and Propositions 3.1.14(i), 3.1.14(iii), 3.4.3(i), and 3.4.3(iii). □

Example 3.4.7 (Normalized duality mapping). *Let $g : X \to \mathbb{R}$ be defined by $g(x) = \frac{1}{2}\|x\|^2$. Then $g^* : X^* \to \overline{\mathbb{R}}$ is given by $g^*(v) = \frac{1}{2}\|v\|^2$ for all $v \in X^*$. Indeed,*

$$g^*(v) = \sup_{x \in X} \left\{ \langle v, x \rangle - \frac{1}{2}\|x\|^2 \right\} = \sup_{r \geq 0} \sup_{\|x\|=r} \left\{ \langle v, x \rangle - \frac{1}{2}\|x\|^2 \right\}$$

$$= \sup_{r \geq 0} \left\{ r\|v\| - \frac{r^2}{2} \right\} = \frac{1}{2}\|v\|^2.$$

Remark 3.4.8. Note that if f is strict (i.e., if $\mathrm{Dom}(f) \neq \emptyset$), then $f^*(p) > -\infty$ for all $p \in X^*$.

Definition 3.4.9. *Assume that $f : X \to \overline{\mathbb{R}}$ is strict. We define the* biconjugate $f^{**} : X^{**} \to \overline{\mathbb{R}}$ *as $f^{**} = (f^*)^*$; that is, $f^{**}(\xi) = \sup_{p \in X^*} \{ \langle \xi, p \rangle - f^*(p) \}$.*

It is important to determine conditions under which $f^{**} = f$. In view of Definition 3.4.9, strictly speaking, reflexivity of X is a necessary condition, because f^{**} is defined on $X^{**} = (X^*)^*$. But, because the application $\mathcal{I} : X \to X^{**}$ defined as $[\mathcal{I}(x)](p) = \langle p, x \rangle$ is always one-to-one, we can see X as a subspace of X^{**}, identifying it with the image of $\mathcal{I}$, in which case the issue becomes the equality of f and $f^{**}_{/X}$, meaning the restriction of f^{**} to $X \subset X^{**}$. According to Proposition 3.4.6, convexity and lower-semicontinuity of f are necessary conditions; it turns out that they are also sufficient. For the proof of this result, we need one of the basic tools of convex analysis, namely the convex separation theorem, which is itself a consequence of the well-known Hahn–Banach theorem on extensions of dominated linear functionals defined on subspaces of Banach spaces. We thus devote some space to these analytical preliminaries.

Definition 3.4.10. *Let X be a Banach space. A function $h : X \to \mathbb{R}$ is said to be a* sublinear functional *if and only if*

(i) $h(x + y) \leq h(x) + h(y)$ *for all $x, y \in X$ (i.e., h is* subadditive*), and*

(ii) $h(\lambda x) = \lambda h(x)$ *for all $x \in X$ and all real $\lambda > 0$ (i.e., h is* positively homogeneous*).*

Theorem 3.4.11 (Hahn–Banach theorem). *Let X be a Banach space, $V \subset X$ a linear subspace, $h : X \to \mathbb{R}$ a sublinear functional, and $g : V \to \mathbb{R}$ a linear functional such that $g(x) \leq h(x)$ for all $x \in V$. Then there exists a linear $\bar{g} : X \to \mathbb{R}$ such that $\bar{g}(x) = g(x)$ for all $x \in V$ and $\bar{g}(x) \leq h(x)$ for all $x \in X$.*

Proof. See Theorem 11.1 in [22]. ☐

A variant of the Hahn–Banach theorem, also involving the behavior of g and h in a convex set $C \subset X$ is due to Mazur and Orlicz [146]. We state it next, and we use it later on for establishing Brøndsted–Rockafellar's lemma; that is, Theorem 5.3.15.

Theorem 3.4.12. *If X is a Banach space and $h : X \to \mathbb{R}$ a sublinear functional, then there exists a linear functional $g : X \to \mathbb{R}$ such that $g(x) \leq h(x)$ for all $x \in X$ and $\inf_{x \in C} g(x) = \inf_{x \in C} h(x)$, where C is a convex subset of X.*

Proof. See Theorem 1.1 in [206]. ☐

Before introducing the separation version of Hahn–Banach theorem, we need some "separation" notation.

Definition 3.4.13. *Given subsets A and B of a Banach space X, $p \in X^*$, and $\alpha \in \mathbb{R}$, we say that*

(a) *The hyperplane $H_{p,\alpha} := \{x \in X : \langle p, x \rangle = \alpha\}$ separates A and B if and only if $\langle p, x \rangle \leq \alpha$ for all $x \in A$ and $\langle p, x \rangle \geq \alpha$ for all $x \in B$.*

(b) *The hyperplane $H_{p,\alpha}$ strictly separates A and B if and only if there exists $\epsilon > 0$ such that $\langle p, x \rangle \leq \alpha - \epsilon$ for all $x \in A$ and $\langle p, x \rangle \geq \alpha + \epsilon$ for all $x \in B$.*

Theorem 3.4.14. *Let X be a Banach space and A, B two nonempty, convex, and disjoint subsets of X.*

(i) *If $A^o \neq \emptyset$ then there exist $p \in X^*$ and $\alpha \in \mathbb{R}$ such that the hyperplane $H_{p,\alpha}$ separates A and B.*

(ii) *If x is a boundary point of A and $A^o \neq \emptyset$, then there exists a hyperplane H such that $x \in H$ and A lies on one side of H.*

(iii) *If A is closed and B is compact then there exist $p \in X^*$ and $\alpha \in \mathbb{R}$ such that the hyperplane $H_{p,\alpha}$ strictly separates A and B.*

Proof. This theorem follows from Theorem 3.4.11, and is in fact almost equivalent to it; see, for example, Theorem 1.7 in [38]. $\square$

We state for future reference an elementary consequence of Theorem 3.4.14.

Proposition 3.4.15. *Let X be a Banach space and A a nonempty, convex, and open subset of X. Then $\overline{A}^o = A$.*

Proof. Clearly $A = A^o \subset \overline{A}^o$. Suppose that the converse inclusion fails; that is, there exists $x \in \overline{A}^o \setminus A$. By Theorem 3.4.14(i) with $B = \{x\}$, there exists a hyperplane H separating x from A, so that A lies on one of the two closed halfspaces defined by H, say M. It follows that $\overline{A} \subset M$, so that $\overline{A}^o \subset M^o$, implying that $x \in M^o$, contradicting the fact that x belongs to the remaining halfspace defined by H. $\square$

Given $C \subset X$, denote $\overline{C}^w$ the weak closure of C. It is clear from the definitions of weak and strong topology that $\overline{A}^w \supset \overline{A}$ for all $A \subset X$. A classical application of Theorem 3.4.14(iii) is the fact that the opposite inclusion holds for convex sets.

Corollary 3.4.16. *Let X be a Banach space. If A is convex, then $\overline{A}^w = \overline{A}$.*

Proof. Suppose that there exists $x \in \overline{A}^w$ such that $x \notin \overline{A}$. By Theorem 3.4.14(iii) there exist $p \in X^*$ and $\alpha \in \mathbb{R}$ such that $\langle p, x \rangle < \alpha < \langle p, z \rangle$ for all $z \in \overline{A}$. Because $x \in \overline{A}^w$, there exists a net $\{a_i\}_{i \in I} \subset A$ weakly converging to x. Therefore, $\lim_i \langle p, a_i \rangle = \langle p, x \rangle$, which contradicts the separation property. $\square$

Our next use of the convex separation theorem establishes that conjugates of proper, convex, and lsc functions are proper.

Proposition 3.4.17. *Let X be a Banach space. If $f : X \to \mathbb{R} \cup \{\infty\}$ is proper, convex, and lsc, then f^* is proper.*

Proof. Take $\bar{x} \in \text{Dom}(f)$ and $\bar{\lambda} < f(\bar{x})$. We apply Theorem 3.4.14(iii) in $X \times \mathbb{R}$ with $A = \text{Epi}(f)$ and $B = \{(\bar{x}, \bar{\lambda})\}$. Because f is lsc and convex with nonempty domain, A is closed, nonempty, and convex by Propositions 3.1.10(ii) and 3.4.3(iv). The set B is trivially convex and compact, and thus there exists a hyperplane $H_{\pi,\alpha}$ strictly separating A and B. Observe that π belongs to $(X \times \mathbb{R})^* = X^* \times \mathbb{R}$; that is, it is of the form $\pi = (p, \mu)$ with $p \in X^*$, so that $\langle \pi, (x, \lambda) \rangle = \langle p, x \rangle + \mu\lambda$. By Definition 3.4.13(b),

$$\langle p, x \rangle + \mu\lambda > \alpha \qquad \forall (x, \lambda) \in \text{Epi}(f), \tag{3.16}$$

$$\langle p, \bar{x} \rangle + \mu\bar{\lambda} < \alpha. \tag{3.17}$$

Taking $x \in \text{Dom}(f)$ and $\lambda = f(x)$ in (3.16), we get

$$\langle p, x \rangle + \mu f(x) > \alpha \qquad \forall x \in \text{Dom}(f), \tag{3.18}$$

and thus, using (3.17) and (3.18),

$$\langle p, \bar{x} \rangle + \mu f(\bar{x}) > \langle p, \bar{x} \rangle + \mu\bar{\lambda}. \tag{3.19}$$

We conclude from (3.19) that μ is positive, so that, multiplying (3.18) by $-\mu^{-1}$, we obtain $\langle -\mu^{-1}p, x \rangle - f(x) < -\mu^{-1}\alpha$ for all $x \in \text{Dom}(f)$. It follows from Definition 3.4.5 that $f^*(-\mu^{-1}p) \leq -\mu^{-1}\alpha < \infty$. In view of Remark 3.4.8, we conclude that f^* is proper. □

With the help of Theorem 3.4.14(iii), we establish now the relation between f and $f_{/X}^{**}$.

Theorem 3.4.18. *Assume that $f : X \to \mathbb{R} \cup \{\infty\}$ is strict. Then,*

(i) $f_{/X}^{**}(x) \leq f(x)$ *for all $x \in X$.*

(ii) The following statements are equivalent.

    (a) f is convex and lsc.

    (b) $f = f_{/X}^{**}$.

Proof.
    (i) It follows from Definition 3.4.9, and the form of the immersion $\mathcal{I}$ of X in X^{**}, that

$$f_{/X}^{**}(x) = \sup_{p \in X^*} \{\langle p, x \rangle - f^*(p)\}. \tag{3.20}$$

By the definition of f^*, $f^*(p) \geq \langle p, x \rangle - f(x)$, or equivalently $f(x) \geq \langle p, x \rangle - f^*(p)$ for all $x \in X$, $p \in X^*$, so that the result follows from (3.20).
    (ii) (a)$\Rightarrow$ (b) We consider first the case of nonnegative f: $f(x) \geq 0$ for all $x \in X$. In view of item (i), it suffices to prove that $f(x) \leq f_{/X}^{**}(x)$ for all $x \in X$. Assume, by contradiction, that there exists $\bar{x} \in X$ such that $f_{/X}^{**}(\bar{x}) < f(\bar{x})$. We apply Theorem 3.4.14(iii) in $X \times \mathbb{R}$ with $A = \text{Epi}(f)$ and $B = \{(\bar{x}, f_{/X}^{**}(\bar{x}))\}$. As in the proof of Proposition 3.4.17, we get a separating hyperplane $H_{\pi, \alpha}$ with $\pi = (p, \mu)$ and $\langle \pi, (x, \lambda) \rangle = \langle p, x \rangle + \mu\lambda$. Thus, we get from Definition 3.4.13(b) that

$$\langle p, x \rangle + \mu\lambda > \alpha \qquad \forall (x, \lambda) \in \text{Epi}(f), \tag{3.21}$$

$$\langle p, \bar{x} \rangle + \mu f^{**}_{/X}(\bar{x}) < \alpha. \tag{3.22}$$

Taking $x \in \mathrm{Dom}(f)$ and making $\lambda \to \infty$ in (3.21), we conclude that $\mu \geq 0$. Non-negativity of f and the definition of epigraph imply that (3.21) can be rewritten as $\langle p, x \rangle + (\mu + \varepsilon)f(x) > \alpha$ for all $x \in \mathrm{Dom}(f)$ and all real $\varepsilon > 0$, and therefore, in view of Definition 3.4.5, we have

$$f^*[-(\mu + \varepsilon)^{-1}p] \leq -(\mu + \varepsilon)^{-1}\alpha. \tag{3.23}$$

Using now (3.23) and (3.20) we get

$$f^{**}_{/X}(\bar{x}) \geq \langle -(\mu + \varepsilon)^{-1}p, \bar{x} \rangle - f^*[-(\mu + \varepsilon)^{-1}p] \geq$$
$$\langle -(\mu + \varepsilon)^{-1}p, \bar{x} \rangle + (\mu + \varepsilon)^{-1}\alpha. \tag{3.24}$$

It follows from (3.24) that $\langle p, \bar{x} \rangle + (\mu + \varepsilon)f^{**}_{/X}(\bar{x}) \geq \alpha$ for all $\varepsilon > 0$, contradicting (3.22). The case of nonnegative f is settled; we address next the general case. Take $\bar{p} \in \mathrm{Dom}(f^*)$, which is nonempty by Proposition 3.4.17. Define $\hat{f} : X \to \mathbb{R} \cup \{\infty\}$ as

$$\hat{f}(x) = f(x) - \langle \bar{p}, x \rangle + f^*(\bar{p}). \tag{3.25}$$

Because $-\langle \bar{p}, \cdot \rangle + f^*(\bar{p})$ is trivially convex and lsc, we get from Propositions 3.1.14(i) and 3.4.3(i) that $\hat{f}$ is convex and lsc. The definition of f^* and (3.25) also imply that $\hat{f}(x) \geq 0$ for all $x \in X$. Thus we apply the already established result for nonnegative functions to $\hat{f}$, concluding that $\hat{f} = \hat{f}^{**}_{/X}$. It is easy to check that $\hat{f}^*(p) = f^*(p + \bar{p}) - f^*(\bar{p})$, $\hat{f}^{**}_{/X}(x) = f^{**}_{/X}(x) - \langle \bar{p}, x \rangle + f^*(\bar{p})$. The latter equality, combined with (3.25) yields $f = f^{**}_{/X}$.

(b)$\Rightarrow$ (a) The result follows from the definition of $f^{**}_{/X}$ and Proposition 3.4.6.
$\square$

We define next the support function of a subset of X.

Definition 3.4.19. *Let X be a Banach space and K a nonempty subset of X. The support function $\sigma_K : X^* \to \mathbb{R} \cup \{\infty\}$ is defined as $\sigma_K(p) = \sup_{x \in K} \langle p, x \rangle$.*

Example 3.4.20. Let X be a reflexive Banach space, so that $\mathcal{I}$ can be identified with the identity, and $f^{**}_{/X}$ with f^{**}. Given $K \subset X$, consider the indicator function δ_K of K. We proceed to compute $\delta_K^*, \delta_K^{**}$. Observe that

$$\delta_K^*(p) = \sup_{x \in X}\{\langle p, x \rangle - \delta_K(x)\} = \sup_{x \in K} \langle p, x \rangle = \sigma_K(p),$$

$$\delta_K^{**}(x) = \sup_{p \in X^*}\{\langle p, x \rangle - \sigma_K(p)\}.$$

Note that $\delta_K^{**}(x) \geq 0$ for all $x \in X$. By Theorem 3.4.18(ii), $\delta_K^{**} = \delta_K$ if and only if δ_K is convex and lsc, and it is easy to check that this happens if and only if K is closed and convex. It is not difficult to prove, using Theorem 3.4.14(iii), that for a closed and convex K it holds that

$$K = \{x \in X : \langle p, x \rangle \leq \sigma_K(p) \ \forall p \in X^*\}. \tag{3.26}$$

The equality above allows us to establish that $\delta_K^{**}(x) = 0$ whenever $x \in K$ and that $\delta_K^{**}(x) = \infty$ otherwise. Hence we conclude that K is closed and convex if and only if $\delta_K^{**} = \delta_K$.

The following lemma is used for establishing an important minimax result, namely Theorem 5.3.10. The proof we present here is taken from Lemma 2.1 in [206]. Given $\alpha, \beta \in \mathbb{R}$, we define $\alpha \vee \beta = \max\{\alpha, \beta\}$, $\alpha \wedge \beta = \min\{\alpha, \beta\}$.

Lemma 3.4.21. *Let X be a Banach space. Consider a nonempty and convex subset $Z \subset X$.*

(a) *If $f_1, \ldots, f_p : X \to \overline{\mathbb{R}}$ are convex functions on Z, then there exist $\alpha_1, \ldots, \alpha_p \geq 0$ such that $\sum_{i=1}^{p} \alpha_i = 1$ and*

$$\inf_{x \in Z} \{f_1(x) \vee f_2(x) \vee \cdots \vee f_p(x)\}$$

$$= \inf_{x \in Z} \{\alpha_1 f_1(x) + \alpha_2 f_2(x) + \cdots + \alpha_p f_p(x)\}.$$

(b) *If $g_1, \ldots, g_p : X \to \mathbb{R} \cup \{-\infty\}$ are concave functions on Z, then there exist $\alpha_1, \ldots, \alpha_p \geq 0$ such that $\sum_{i=1}^{p} \alpha_i = 1$ and*

$$\sup_{x \in Z} \{g_1(x) \wedge g_2(x) \wedge \cdots \wedge g_p(x)\} =$$

$$\sup_{x \in Z} \{\alpha_1 g_1(x) + \alpha_2 g_2(x) + \cdots + \alpha_p g_p(x)\}.$$

Proof.
(a) Define $h : \mathbb{R}^p \to \mathbb{R}$ as $h(a_1, a_2, \ldots, a_p) = a_1 \vee \cdots \vee a_p$. Note that h is sublinear in $\mathbb{R}^p$. Consider also the set $C := \{(a_1, \ldots, a_p) \in \mathbb{R}^p \mid \exists x \in Z \text{ such that } f_i(x) \leq a_i (1 \leq i \leq p)\}$. It is easy to see that C is a convex subset of $\mathbb{R}^p$. By Theorem 3.4.12, we conclude that there exists a linear functional $L : \mathbb{R}^p \to \mathbb{R}$ such that $L \leq h$ in $\mathbb{R}^p$ and $\inf_{a \in C} L(a) = \inf_{a \in C} h(a)$. Because L is linear, there exist $\lambda_1, \cdots, \lambda_p \in \mathbb{R}$ such that $L(a) = \lambda_1 a_1 + \cdots + \lambda_p a_p$ for all $a = (a_1, \ldots, a_p) \in \mathbb{R}^p$. Using the fact that $L \leq h$ and the definition of h, it follows that $\lambda_i \geq 0$ for all $i = 1, \ldots, p$ and that $\sum_{i=1}^{p} \alpha_i = 1$ (Exercise 3). Define $f(x) := (f_1(x), \ldots, f_p(x))$. We claim that

$$\inf_{x \in Z} h(f(x)) = \inf_{x \in Z} \lambda_1 f_1(x) + \cdots + \lambda_p f_p(x). \tag{3.27}$$

We proceed to prove the claim. Take $\mu \in C$. Observe that

$$\inf_{x \in Z} h(f(x)) = \inf_{x \in Z} f_1(x) \vee \cdots \vee f_p(x) \leq h(\mu). \tag{3.28}$$

Taking the infimum on C in (3.28) and using the definition of L we get

$$\inf_{x \in Z} L(f(x)) \leq \inf_{x \in Z} h(f(x)) \leq \inf_{\mu \in C} h(\mu) = \inf_{\mu \in C} L(\mu). \tag{3.29}$$

Inasmuch as $f(x) \in C$ for all $x \in Z$,

$$\inf_{\mu \in C} L(\mu) \leq \inf_{x \in Z} L(f(x)). \tag{3.30}$$

Combining (3.29) and (3.30), we get (3.27), from which the result follows.

(b) The result follows from (a) by taking $f_i := -g_i$ $(1 \leq i \leq p)$. $\square$

We close this section with the following classical result on local Lipschitz continuity of convex functions, which is needed in Chapter 5. The proof below was taken from Proposition 2.2.6 in [66],

Theorem 3.4.22. *Let $f : X \to \mathbb{R}$ be a convex function and $U \subset X$. If f is bounded above in a neighborhood of some point of U, then f is locally Lipschitz at each point of U.*

Proof. Let $x_0 \in U$ be a point such that f is bounded above in a neighborhood of x_0. Replacing U by $V := U - \{x_0\}$ and $f(\cdot)$ by $g := f(\cdot + x_0)$ we can assume that the point of U at which the boundedness assumption holds for f is 0. By assumption there exists $\varepsilon > 0$ and $M > 0$ such that $B(0, \varepsilon) \subset U$ and $f(z) \leq M$ for every $z \in B(0, \varepsilon)$. Our first step is to show that f is also bounded below in the ball $B(0, \varepsilon)$. Note that $0 = (1/2)z + (1/2)(-z)$ for all $z \in B(0, \varepsilon)$, and hence $f(0) \leq (1/2)f(z) + (1/2)f(-z)$. Therefore, $f(z) \geq 2 f(0) - f(-z)$. Because $-z \in B(0, \varepsilon)$, we get $f(z) \geq 2 f(0) - M$, showing that f is also bounded below. Fix $x \in U$. We prove now that the boundedness assumption on f implies that f is bounded on some neighborhood of x. Because x is an interior point of U, there exists $\rho > 1$ such that $\rho x \in U$. If $\lambda = 1/\rho$, then $B(x, (1 - \lambda)\varepsilon)$ is a neighborhood of x. We claim that f is bounded above in $B(x, (1 - \lambda)\varepsilon)$. Indeed, note that $B(x, (1-\lambda)\varepsilon) = \{(1-\lambda)z+\lambda(\rho x)\,|\,z \in B(0, \varepsilon)\}$. By convexity, for $y \in B(x, (1-\lambda)\varepsilon)$, we have

$$f(y) \leq (1 - \lambda)f(z) + \lambda f(\rho x) \leq (1 - \lambda)M + \lambda f(\rho x) < \infty.$$

Hence f is bounded above in a neighborhood of x. It is therefore bounded below by the first part of the proof. In other words, f is locally bounded at x. Let $R > 0$ be such that $|f(z)| \leq R$ for all $z \in B(x, (1 - \lambda)\varepsilon)$. Define $r := (1 - \lambda)\varepsilon$. Let us prove that f is Lipschitz in the ball $B(x, r/2)$. Let $x_1, x_2 \in B(x, r/2)$ be distinct points and set $x_3 := x_2 + (r/2\alpha)(x_2 - x_1)$, where $\alpha := \|x_2 - x_1\|$. Note that $x_3 \in B(x, r)$. By definition of x_3 we get

$$x_2 = \frac{r}{2\alpha + r} x_1 + \frac{2\alpha}{2\alpha + r} x_3,$$

and by convexity of f we obtain

$$f(x_2) \leq \frac{r}{2\alpha + r} f(x_1) + \frac{2\alpha}{2\alpha + r} f(x_3).$$

Then

$$f(x_2) - f(x_1) \leq \frac{2\alpha}{2\alpha + r} [f(x_3) - f(x_1)]$$

$$\leq \frac{2\alpha}{r}|f(x_3) - f(x_1)| \leq \frac{2R}{r}\|x_1 - x_2\|,$$

using the definition of α, the fact that $x_1, x_3 \in B(x, r)$, and the local boundedness of f on the latter set. Interchanging the roles of x_1 and x_2 we get

$$|f(x_2) - f(x_1)| \leq \frac{2R}{r}\|x_1 - x_2\|,$$

and hence f is Lipschitz on $B(x, r/2)$. $\square$

3.5   The subdifferential of a convex function

Definition 3.5.1. *Given a Banach space B, a proper and convex $f : X \to \mathbb{R} \cup \{\infty\}$, and a point $x \in X$ we define the* subdifferential *of f at x as the subset $\partial f(x)$ of X^* given by*

$$\partial f(x) = \{v \in X^* : f(y) \geq f(x) + \langle v, y - x \rangle \quad \forall y \in X\}.$$

The elements of the subdifferential of f at x are called *subgradients* of f at x. It is clear that ∂f can be seen as a point-to-set mapping $\partial f : X \rightrightarrows X^*$. We start with some elementary properties of the subdifferential.

Proposition 3.5.2. *If $f : X \to \mathbb{R} \cup \{\infty\}$ is proper and convex, then*

(i) $\partial f(x)$ *is closed and convex for all $x \in X$.*

(ii) $0 \in \partial f(x)$ *if and only if $x \in \operatorname{argmin}_X f := \{z \in X : f(z) \leq f(y) \, \forall y \in X\}$.*

Proof. Both (i) and (ii) are immediate consequences of Definition 3.5.1. $\square$

The following proposition studies the subdifferential of conjugate functions.

Proposition 3.5.3. *If $f : X \to \mathbb{R} \cup \{\infty\}$ is proper and convex, and $x \in D(\partial f)$, then the following statements are equivalent.*

(a) $v \in \partial f(x)$.

(b) $\langle v, x \rangle = f(x) + f^*(v)$.

If f is also lsc then the following statement is also equivalent to (a) and (b).

(c) $x \in \partial f^*(v)$.

Proof.
(a)$\Rightarrow$ (b) Inasmuch as v belongs to $\partial f(x)$ we have that

$$f^*(v) \geq \langle v, x \rangle - f(x) \geq \sup_{y \in X}\{\langle v, y \rangle - f(y)\} = f^*(v), \tag{3.31}$$

so that equality holds throughout (3.31), which gives the result.

(b)⇒ (a) If (b) holds then

$$f(x) - \langle v, x \rangle = f^*(v) = \sup_{y \in X}\{\langle v, y \rangle - f(y)\} \geq \langle v, z \rangle - f(z)$$

for all $z \in X$, so that $f(x) - \langle v, x \rangle \geq \langle v, z \rangle - f(z)$ for all $z \in X$, which implies that $v \in \partial f(x)$.

(b)⇒ (c)

$$f^*(w) = \sup_{y \in X}\{\langle w, y \rangle - f(y)\} = \sup_{y \in X}\{[\langle v, y \rangle - f(y)] + \langle w - v, y \rangle\} \geq$$

$$\langle v, x \rangle - f(x) + \langle w - v, x \rangle = f^*(v) + \langle x, w - v \rangle,$$

using (b) in the last equality. It follows that $x \in \partial f^*(v)$.

(c)⇒ (b) If x belongs to $\partial f^*(v)$, then for all $w \in X^*$,

$$f^*(w) \geq f^*(v) + \langle w - v, x \rangle,$$

implying that $\langle v, x \rangle - f^*(v) \geq \langle w, x \rangle - f^*(w)$, and therefore

$$\langle v, x \rangle - f^*(v) \geq f^{**}(x) = f(x), \tag{3.32}$$

using in the last equality Theorem 3.4.18(ii), which can be applied because f is convex and lsc, and the fact that $x \in D(\partial f) \subset X$, so that $f^{**}_{/X}(x) = f^{**}(x)$. In view of (3.32), we have

$$\langle v, x \rangle - f(x) \geq f^*(v) \geq \langle v, x \rangle - f(x), \tag{3.33}$$

so that equality holds throughout (3.33) and the conclusion follows. □

Remark 3.5.4. Observe that the lower-semicontinuity of f was not used in the proof that (b) implies (c) in Proposition 3.5.3.

Corollary 3.5.5. *If X is reflexive and $f : X \to \mathbb{R} \cup \{\infty\}$ is proper, convex, and lsc then $(\partial f)^{-1} = \partial f^* : X^* \rightrightarrows X$.*

Proof. It follows immediately from the equivalence between (a) and (c) in Proposition 3.5.3. □

Next we characterize the subdifferential of indicator functions of closed and convex sets.

Proposition 3.5.6. *If $K \subset X$ is nonempty, closed, and convex, then*

(a) $D(\partial \delta_K) = K$.

(b) For every $x \in K$, we have that

$$\partial \delta_K(x) = \begin{cases} \{v \in X^* : \langle v, y - x \rangle \leq 0 \ \forall y \in K\} & \text{if } x \in K \\ \emptyset & \text{otherwise.} \end{cases} \quad (3.34)$$

Proof. By Definition 3.5.1, $v \in \partial \delta_K(x)$ if and only if

$$\langle v, y - x \rangle \leq \delta_K(y) - \delta_K(x) \quad (3.35)$$

for all $y \in X$. The image of δ_K is $\{0, \infty\}$ therefore it is immediate that $0 \in \partial \delta_k(x)$ for all $x \in K$, so that $K \subset D(\partial \delta_K)$. On the other hand, if $x \notin K$, then for $y \in K$ the right-hand side of (3.35) becomes $-\infty$, and thus no $v \in X^*$ can satisfy (3.35), so that $\partial \delta_K(x) = \emptyset$; that is, $D(\partial \delta_K) \subset K$, establishing (a). For (b), note that it suffices to consider (3.35) only for both x and y belonging to K (if $x \notin K$, then $\partial \delta_K = \emptyset$, as we have shown; if x belongs to K but y does not, then (3.35) is trivially satisfied because its right-hand side becomes ∞). Finally, note that for $x, y \in K$, (3.35) becomes (3.34), because $\delta_K(x) = \delta_K(y) = 0$. $\square$

3.5.1   Subdifferential of a sum

It follows easily from the definition of subdifferential that

$$\partial f(x) + \partial g(x) \subset \partial(f + g)(x), \quad (3.36)$$

where $f, g : X \to \mathbb{R} \cup \{\infty\}$ are proper, convex, and lsc. The fact that the converse may not hold can be illustrated by considering f and g as the indicator functions of two balls B_1 and B_2 such that $\{x\} = B_1 \cap B_2$. Using Proposition 3.5.6(b), it can be checked that $\partial \delta_{B_1 \cap B_2}(x) = \partial(\delta_{B_1} + \delta_{B_2})(x) = X \supsetneq \partial \delta_{B_1}(x) + \partial \delta_{B_1}(x)$ We must require extra conditions for the opposite inclusion in (3.36) to hold. These conditions are known as *constraint qualifications*. The most classical constraint qualification for the sum subdifferential formula is the following.
(CQ): There exist $x \in \text{Dom}(f) \cap \text{Dom}(g)$ at which one of the functions is continuous.

Theorem 3.5.7. *If $f, g : X \to \mathbb{R} \cup \{\infty\}$ are proper, convex, and lsc, and (CQ) holds for f, g, then the sum subdifferential formula holds; that is, $\partial f(x) + \partial g(x) = \partial(f + g)(x)$ for all $x \in \text{Dom}(f) \cap \text{Dom}(g)$.*

Proof. In view of (3.36), it is enough to prove that $\partial f(x) + \partial g(x) \supset \partial(f + g)(x)$. Fix $x_0 \in \text{Dom}(f) \cap \text{Dom}(g)$ and take $v_0 \in \partial(f + g)(x_0)$. Consider the functions

$$f_1(x) = f(x + x_0) - f(x_0) - \langle x, v_0 \rangle,$$

and

$$g_1(x) = f(x + x_0) - f(x_0).$$

Note that $f_1(0) = 0 = g_1(0)$. It can also be easily verified that

(i) If $v_0 \in \partial(f + g)(x_0)$, then $0 \in \partial(f_1 + g_1)(0)$.

(ii) If $0 \in \partial f_1(0) + \partial g_1(0)$ then $v_0 \in \partial f(x_0) + \partial g(x_0)$.

We claim that $0 \in \partial f_1(0) + \partial g_1(0)$. Observe that $0 \in \partial(f_1 + g_1)(0)$ by (i), and that $f_1(0) = g_1(0) = 0$. It follows that

$$(f_1 + g_1)(x) \geq (f_1 + g_1)(0) = 0. \tag{3.37}$$

By assumption, f_1 is continuous at some point of $\mathrm{Dom}(f_1) \cap \mathrm{Dom}(g_1)$, so that $\mathrm{Dom}(f_1)$ has a nonempty interior. Call $C_1 := \mathrm{Epi}(f_1)$ and $C_2 := \{(x, r) \in X \times \mathbb{R} \mid r \leq -g_1(x)\}$. By (3.37), $(C_1)^o = \{(x, \beta) \in X \times \mathbb{R} \mid f_1(x) < \beta\}$ does not intersect C_2. Using Theorem 3.4.14(i) with $A := (C_1)^o$ and $B := C_2$, we obtain a hyperplane $H := \{(x, a) \in X \times \mathbb{R} : \langle p, x \rangle + a t = \alpha\}$ separating C_1 and C_2. Because $(0, 0) \in C_1 \cap C_2$, we must have $\alpha = 0$. Using the fact that H separates C_1 and C_2 we get

$$\langle p, x \rangle + a t \geq 0 \geq \langle p, z \rangle + b t \tag{3.38}$$

for all $(x, a) \in \mathrm{Epi}(f_1)$ and all (z, b) such that $b \leq -g_1(z)$. Using (3.38) for the element $(0, 1) \in \mathrm{Epi}(f_1)$, we get $t \geq 0$. We claim that $t > 0$. Indeed, if $t = 0$ then $p \neq 0$, because $(p, t) \neq (0, 0)$. By (3.38), $\langle p, x \rangle \geq 0$ for all $x \in \mathrm{Dom}(f_1)$ and $\langle p, z \rangle \leq 0$ for all $z \in \mathrm{Dom}(g_1)$. Hence the convex sets $\mathrm{Dom}(f_1)$ and $\mathrm{Dom}(g_1)$ are separated by the nontrivial hyperplane $\{x \in X \mid \langle p, x \rangle = 0\}$. By assumption, there exists $z \in (\mathrm{Dom}(f_1))^o \cap \mathrm{Dom}(g_1) \subset \mathrm{Dom}(f_1) \cap \mathrm{Dom}(g_1)$. It follows that $\langle p, z \rangle = 0$. Because $z \in (\mathrm{Dom}(f_1))^o$, there exists $\delta > 0$ such that $z - \delta p \in \mathrm{Dom}(f_1)$. Thus, $\langle p, z - \delta p \rangle = -\delta \|p\|^2 < 0$, which contradicts the separation property. Therefore, it holds that $t > 0$ and without any loss of generality, we can assume that $t = 1$. Fix $x \in X$. The leftmost inequality in (3.38) for $(x, f_1(x))$ yields $-p \in \partial f_1(0)$, and the rightmost one for $(x, -g_1(x))$ gives $p \in \partial g_1(0)$. Therefore, $0 = -p + p \in \partial f_1(0) + \partial g_1(0)$. The claim is established. Now we invoke (ii) for concluding that $v_0 \in \partial f(x_0) + \partial g(x_0)$, completing the proof. $\square$

We leave the most important property of the subdifferential of a lsc convex function, namely its maximal monotonicity, for the following chapter, which deals precisely with maximal monotone operators.

3.6   Tangent and normal cones

Definition 3.6.1. *Let X be a Banach space and K a nonempty subset of X. The normal cone or normality operator $N_K : X \rightrightarrows X^*$ of K is defined as*

$$N_K(x) = \begin{cases} \{v \in X^* : \langle v, y - x \rangle \leq 0 \; \forall y \in K\} & \text{if } x \in K \\ \emptyset & \text{otherwise.} \end{cases} \tag{3.39}$$

The set $N_K(x)$ is the cone of directions pointing "outwards" from K at x; see Figure 3.3.

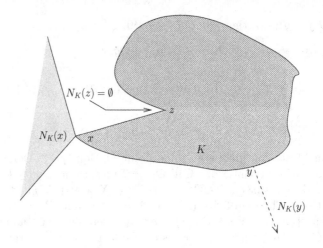

Figure 3.3. *Normal cones*

The following proposition presents some properties of the normal cone of closed and convex sets. We recall that $A \subset X$ is a *cone* if it is closed under multiplication by nonnegative scalars.

Proposition 3.6.2. *For a nonempty, closed, and convex $K \subset X$, the following properties hold.*

(i) $N_K = \partial \delta_K$.

(ii) $N_K(x)$ *is closed and convex for all $x \in X$.*

(iii) $N_K(x)$ *is a cone for all $x \in K$.*

Proof. Item (i) follows from Proposition 3.5.6(b), item (ii) follows from (i) and Proposition 3.5.2(i), and item (iii) is immediate from (3.39).   □

We introduce next three cones associated with subsets of X and X^*.

Definition 3.6.3. *Let X be a Banach space, K a nonempty subset of X, and L a nonempty subset of X^*.*

(a) *The* polar cone *$K^- \subset X^*$ of K is defined as*

$$K^- = \{p \in X^* : \langle p, x \rangle \le 0 \ \forall x \in K\}. \qquad (3.40)$$

(b) *The* antipolar cone *$L^\circ \subset X$ of L is defined as*

$$L^\circ = \{x \in X : \langle p, x \rangle \le 0 \ \forall p \in L\}. \qquad (3.41)$$

(c) *The tangent cone* $T_K : X \rightrightarrows X$ *of* K *is defined as*

$$T_K(x) = N_K(x)^\circ = \{y \in X : \langle p, y \rangle \leq 0 \ \forall p \in N_K(x)\}. \tag{3.42}$$

Remark 3.6.4. Observe that if X is reflexive then $L^\circ = L^-$ for all nonempty $L \subset X^*$. Indeed, note that in this case we have $L^- \subset X^{**} = X$ and the conclusion follows from (3.40) and (3.41).

Remark 3.6.5. As mentioned above, in geometric terms $N_K(x)$ can be thought of as containing those directions that "move away" from K at the point x, and $T_K(x)$ as containing those directions that make an obtuse angle with all directions in $N_K(x)$. We prove in Proposition 3.6.11 that $T_K(x)$ can also be seen as the closure of the cone of directions that "point inwards" to K at x.

We need in the sequel the following elementary results on polar, antipolar, tangent, and normal cones.

Proposition 3.6.6. *Let* X *be a Banach space and* $K \subset X$, $L \subset X^*$ *closed and convex cones.*

(i) $K = (K^-)^\circ$ *and* $L \subset (L^\circ)^-$.

(ii) *If* X *is reflexive, then* $K = (K^-)^-$.

Proof.

(i) The inclusions $K \subset (K^-)^\circ$ and $L \subset (L^\circ)^-$ follow from Definition 3.6.3(a) and (b). It remains to prove that $K \supset (K^-)^\circ$, which we do by contradiction, invoking Theorem 3.4.14(iii): if there exists $y \in (K^-)^\circ \setminus K$ then, because K is closed and convex, there exists $p \in X^*$ and $\alpha \in \mathbb{R}$ such that

$$\langle p, y \rangle > \alpha, \tag{3.43}$$

$$\langle p, z \rangle \leq \alpha \ \forall z \in K. \tag{3.44}$$

Because K is a cone, 0 belongs to K, so that we get $\alpha \geq 0$ from (3.44), and hence, in view of (3.43),

$$\langle p, y \rangle > 0. \tag{3.45}$$

The fact that K is a cone and (3.44) imply that, for all $\rho > 0$ and all $z \in K$, it holds that $\langle p, \rho z \rangle < \alpha$, implying that $\langle p, z \rangle < \rho^{-1}\alpha$ (i.e., $\langle p, z \rangle \leq 0$), so that $p \in K^-$, which contradicts (3.45), because y belongs to $(K^-)^\circ$.

(ii) The result follows from (i) and Remark 3.6.4. $\square$

Corollary 3.6.7. *Let* X *be a Banach space and* $K \subset X$ *a closed and convex set. Then* $N_K(x) = T_K^-(x)$ *for all* $x \in X$.

Proof. Use Proposition 3.6.6(i) and Definition 3.6.3(c) with $L := N_K(x)$ to conclude that $N_K(x) \subset (N_K(x)^\circ)^- = T_K^-(x)$. For the converse inclusion, take $p \in T_K^-(x)$, and suppose that $p \notin N_K(x)$. Then there exists $y_0 \in K$ such that

$$\langle p, y_0 - x \rangle > 0. \tag{3.46}$$

We claim that $y_0 - x \in T_K(x)$. Indeed, $\langle w, y_0 - x \rangle \leq 0$ for all $w \in N_K(x)$, and then the claim holds by Definition 3.6.3(c). Because $y_0 - x \in T_K(x)$ and $p \in T_K^-(x)$, we conclude that $\langle p, y_0 - x \rangle \leq 0$, contradicting (3.46). □

Definition 3.6.8. *Given a nonempty subset K of X, define $S_K : X \rightrightarrows X$ as*

$$S_K(x) = \begin{cases} \{(y - x)/\lambda : y \in K, \lambda > 0\} & \text{if } x \in K, \\ \emptyset & \text{otherwise.} \end{cases}$$

We mention that $S_K(x)$ is always a cone, but in general it is not closed.

Proposition 3.6.9. *If K is convex then $S_K(x)$ is convex for all $x \in K$.*

Proof. Note that $\alpha[(y - x)/\lambda] + (1 - \alpha)[(y' - x)/\lambda'] = (y'' - x)/\lambda''$, with $\lambda'' = [\alpha/\lambda + (1 - \alpha)/\lambda']^{-1}$, and $y'' = \alpha\lambda''\lambda^{-1}y + (1 - \alpha)\lambda''(\lambda')^{-1}y'$, and that $\alpha\lambda''\lambda^{-1} + (1 - \alpha)\lambda''(\lambda')^{-1} = 1$, so that y'' belongs to K whenever y and y' do. The result follows then from Definition 3.6.8. □

Example 3.6.10. Take $K = \{x \in \mathbb{R}^2 : \|x\| \leq 1, x_2 \geq 0\}$, $\bar{x} = (1, 0)$. It is easy to check that $S_K(\bar{x}) = \{(v_1, v_2) \in \mathbb{R}^2 : v_1 < 0, v_2 \geq 0\}$, $N_K(\bar{x}) = \{(v_1, v_2) \in \mathbb{R}^2 : v_1 \geq 0, v_2 \leq 0\}$. Thus, $T_K(\bar{x}) = N_K(\bar{x})^\circ = \{(v_1, v_2) \in \mathbb{R}^2 : v_1 \leq 0, v_2 \geq 0\} = \overline{S_K(\bar{x})}$. Note that $(0, 1) \in \overline{S_K(\bar{x})} \setminus S_K(\bar{x})$.

Proposition 3.6.11. *If $K \subset X$ is nonempty and convex, then $T_K(x) = \overline{S_K(x)}$ for all $x \in K$.*

Proof. In order to prove that $\overline{S_K(x)} \subset T_K(x) = N_K(x)^\circ$, it suffices to show that $S_K(x) \subset N_K(x)^\circ$, because it follows immediately from Definition 3.6.3 that antipolar cones are closed. Take $z \in S_K(x)$, so that there exist $y \in K$, $\lambda > 0$ such that $z = \lambda^{-1}(y - x)$. Observe that, for all $w \in N_K(x)$, it holds that $\langle w, z \rangle = \lambda^{-1}\langle w, y - x \rangle \leq 0$, because $y \in K$, $w \in N_K(x)$, and $\lambda > 0$. Thus, z belongs to $N_K(x)^\circ$, and the inclusion is proved.

For the converse inclusion, suppose by contradiction that there exists $w \in N_K(x)^\circ$ such that $w \notin \overline{S_K(x)}$. By Theorem 3.4.14(iii) with $A = \overline{S_K(x)}$, $B = \{w\}$, there exist $p \in X^*$ and $\alpha \in \mathbb{R}$, such that

$$\langle p, z \rangle < \alpha \qquad \forall z \in \overline{S_K(x)}, \tag{3.47}$$

$$\langle p, w \rangle > \alpha. \tag{3.48}$$

Because $0 \in \overline{S_K(x)}$, we have from (3.47) that $\alpha > 0$. Then by (3.48)

$$\langle p, w \rangle > 0. \tag{3.49}$$

Take $y \in K$ and $\lambda > 0$. By (3.47) with $z = y - x \in \overline{S_K(x)}$ it holds that

$$\langle p, y - x \rangle = \lambda \langle p, (y - x)/\lambda \rangle < \lambda \alpha. \tag{3.50}$$

Letting λ go to zero in (3.50), we get $\langle p, y - x \rangle \leq 0$. Note that $y \in K$ is arbitrary and hence $p \in N_K(x)$. Using the fact that $w \in N_K(x)^\circ$, we conclude that $\langle p, w \rangle \leq 0$, which contradicts (3.49). $\square$

We present next a property of the cones $S_K(x)$ and $T_K(x)$.

Proposition 3.6.12. *Take a convex set $K \subset X$.*

(i) *$v \in S_K(x)$ if and only if there exists $t^* > 0$ such that $x + tv \in K$ for all $t \in [0, t^*]$.*

(ii) *$v \in T_K(x)$ if and only if for all $\varepsilon > 0$ there exists u such that $\|u - v\| \leq \varepsilon$ and $t^* > 0$ such that $x + tu \in K$ for all $t \in [0, t^*]$.*

Proof.
(i) The "if" part follows readily from the definition of S_K. For the "only if" part, note that, for all $v \in S_K(x)$ there exist $t^* > 0$ and $y \in K$ such that $v = (y - x)/t^*$. Take $t \in [0, t^*]$; we can write

$$x + tv = \left(1 - \frac{t}{t^*}\right) x + \frac{t}{t^*}(x + t^* v) = \left(1 - \frac{t}{t^*}\right) x + \frac{t}{t^*} y,$$

so that $x + tv \in K$ by convexity of K.
(ii) Fix $v \in T_K(x)$ and $\varepsilon > 0$. By Proposition 3.6.11, there exists $u \in S_K(x)$ such that $\|u - v\| \leq \varepsilon$. By (i), there exists $t^* > 0$ such that $x + tu \in$ for all $t \in [0, t^*]$, proving the "only if" statement. The converse statement follows from the definition of S_K and Proposition 3.6.11. $\square$

The next result provides a useful characterization of the tangent cone T_K.

Proposition 3.6.13. *If $K \subset X$ is convex then*

$$T_K(x) = \{y \in X : x + \lambda_n y_n \in K \; \forall n \text{ for some } (y_n, \lambda_n) \to (y, 0) \text{ with } \lambda_n > 0\}. \tag{3.51}$$

Proof. Let $\hat{T}(x)$ be the right-hand side of (3.51), and let us prove that $T_K(x) \subset \hat{T}(x)$. Take $z \in T_K(x)$. By Proposition 3.6.11, z is the strong limit of a sequence $\{z_n\} \subset S_K(x)$; that is, in view of Definition 3.6.8, $z_n = \lambda_n^{-1}(y_n - x)$ with $\lambda_n >$

0, $y_n \in K$ for all $n \in \mathbb{N}$. Take $\{\gamma_n\} \subset (0,1)$ such that $\lim_n \gamma_n \lambda_n = 0$. Then $x + (\gamma_n \lambda_n) z_n = x + \gamma_n (y_n - x)$ belongs to K, because both x and y_n belong to K, which is convex. Because $\lim_n (z_n, \gamma_n \lambda_n) = (z, 0)$, we conclude that z belongs to $\hat{T}(x)$.

Next we prove that $\hat{T}(x) \subset T_K(x) = N_K(x)^\circ$. Take $z \in \hat{T}(x)$ and $v \in N_K(x)$. By definition of $\hat{T}(x)$, there exist $\{z_n\}$ and $\lambda_n > 0$ such that $z = \lim_n z_n$, and $y_n := x + \lambda_n z_n$ belongs to K for all $n \in \mathbb{N}$. Thus

$$\langle v, z \rangle = \lim_n \langle v, z_n \rangle = \lim_n \langle v, \lambda_n^{-1}(y_n - x) \rangle \leq 0, \tag{3.52}$$

using the fact that $v \in N_K(x)$ in the last inequality. It follows from (3.52) that z belongs to $N_K(x)^\circ$, establishing the required inclusion.   □

Proposition 3.6.14. *Take $K \subset X$.*

(a) *If $x \in K^\circ$ then $T_K(x) = X$, and therefore $N_K(x) = \{0\}$.*

(b) *Assume that X is finite-dimensional. If K is convex and $N_K(x) = \{0\}$, then $x \in K^\circ$.*

Proof.

(a) Take $x \in K^\circ$, so that for all $z \in X$ there exists $\rho > 0$ such that $x + \rho z$ belongs to K, and therefore $\rho z = (x + \rho z) - x$ belongs to $S_K(x)$. Because $S_K(x)$ is a cone, we conclude that $z \in S_K(x)$. We have proved that $X \subset S_K(x) \subset \overline{S_K(x)} = T_K(x)$, and thus $N_K(x) = \{0\}$ by Corollary 3.6.7.

(b) Assume that $N_K(x) = \{0\}$. Because $x \in D(N_K)$, we know that $x \in K$. Suppose, by contradiction, that x belongs to the boundary ∂K of K. Then there exists $\{x_n\} \subset K^c$ such that $\lim_n x_n = x$. Because K is convex and $x_n \in K^c$, there exists $p_n \in X^*$ such that $\langle p_n, x_n \rangle > \sigma_K(p_n) = \sup_{y \in K} \langle p_n, y \rangle$. Because σ_K is positively homogeneous, we may assume without loss of generality that $\|p_n\|_* = 1$ for all $n \in \mathbb{N}$. It follows that $\langle p_n, y - x_n \rangle < 0$ for all $y \in K$ and all $n \in \mathbb{N}$. Inasmuch as X is finite-dimensional and $\{p_n\}$ is bounded, it has a subsequence that is convergent, say to p, with $\|p\|_* = 1$. Thus, $\langle p, y - x \rangle \leq 0$, so that p is a nonnull element of $N_K(x)$, contradicting the hypothesis.   □

Remark 3.6.15. Proposition 3.6.14(b) does not hold if K is not convex: taking $K \subset \mathbb{R}^2$ defined as $K = B(0,1) \cup B((2,0),1)$, it is not difficult to check that for $x = (1,0)$, it holds that x belongs to ∂K, and $T_K(x) = \mathbb{R}^2$. Finite-dimensionality of X is also necessary for Proposition 3.6.14(b) to hold. Indeed, let X be an infinite-dimensional Hilbert space, with orthonormal basis $\{e_n\}$. Define $K := \{x \in H : -1/n \leq \langle x, e_n \rangle \leq 1/n$ for all $n \in \mathbb{N}\}$. It is a matter of routine to check that K is closed and convex. We prove next that $N_K(0) = \{0\}$: if $w \in N_K(0)$, then $\langle w, y \rangle \leq 0$ for all $y \in K$, and taking first $y = n^{-1} e_n$ and then $y = -n^{-1} e_n$, it follows easily that $w = 0$. On the other hand, $0 \notin K^\circ$, as we prove next: defining $x_n = n^{-1} e_n$,

$x'_n = 2n^{-1}e_n$, we have that $\{x_n\} \subset K$, $\{x'_n\} \subset K^c$, and both $\{x_n\}$ and $\{x'_n\}$ converge strongly to 0, implying that 0 belongs to the boundary of K.

Remark 3.6.16. $K = \{x\}$ if and only if $N_K(x) = X$, or equivalently $T_K(x) = \{0\}$.

The next two propositions, needed later on, compute normal and tangent cones of intersections of two sets.

Proposition 3.6.17. Let K, L be nonempty, closed, and convex subsets of X.

(i) $N_K(x) + N_L(x) \subset N_{K \cap L}(x)$ for all $x \in X$.

(ii) If $K^o \cap L \neq \emptyset$, then $N_{K \cap L}(x) \subset N_K(x) + N_L(x)$ for all $x \in X$.

Proof.
   (i) Note that $N_K = \partial \delta_K$, $N_L = \partial \delta_L$, and $N_{K \cap L} = \partial(\delta_K + \delta_L)$, so the inclusion follows readily from (3.36).
   (ii) The conclusion is a direct consequence of Theorem 3.5.7 and the fact δ_K is continuous at every point of the nonempty set $K^o \cap L$.   □

Proposition 3.6.18. Let K and L be nonempty subsets of a Banach space X. Then

(i) $T_{K \cap L}(x) \subset T_K(x) \cap T_L(x)$ for all $x \in X$.

(ii) If K and L are closed and convex, and $K^o \cap L \neq \emptyset$, then $T_K(x) \cap T_L(x) \subset T_{K \cap L}(x)$ for all $x \in X$.

Proof.
   (i) It follows easily from the fact that $T_A(x) \subset T_B(x)$ whenever $A \subset B$, which has the desired result as an immediate consequence.
   (ii) Take any $z \in T_K(x) \cap T_L(x)$. We must prove that z belongs to $T_{K \cap L}(x)$. By Definition 3.6.3(c) and Proposition 3.6.17(ii),

$$T_{K \cap L}(x) = N_{K \cap L}(x)^\circ = (N_K(x) + N_L(x))^\circ;$$

that is, we must establish that

$$\langle v + w, z \rangle \leq 0 \tag{3.53}$$

for all $(v, w) \in N_K(x) \times N_L(x)$. Because z belongs to $T_K(x)$ we know that $\langle v, z \rangle \leq 0$; because z belongs to $T_L(x)$, we know that $\langle w, z \rangle \leq 0$. Thus, (3.53) holds and the result follows.   □

The next proposition, also needed in the sequel, computes the tangent cone to a ball in a Hilbert space, at a point on its boundary.

Proposition 3.6.19. *Let X be a Hilbert space. Take $\bar{x} \in X$, $\rho > 0$, $K = B(\bar{x}, \rho)$, and $x \in \partial K$. Then $T_K(x) = \{v \in X : \langle v, x - \bar{x} \rangle \leq 0\}$.*

Proof. We start by proving that if $\langle v, x - \bar{x} \rangle < 0$ and $\|v\| = 1$, then $v \in T_K(x)$. The fact that $\{v \in X : \langle v, x - \bar{x} \rangle \leq 0\} \subset T_K(x)$ then follows after observing that both sets are cones and that $T_K(x)$ is closed. Let $y_n = x + n^{-1}v$. Because $\lim_n y_n = x$, Proposition 3.6.13 implies that establishing that $y_n \in K = B(\bar{x}, \rho)$ for large enough n is enough for ensuring that $v \in T_K(x)$, and henceforth that the required inclusion holds. Indeed,

$$\|y_n - \bar{x}\|^2 = \|x - \bar{x}\|^2 + n^{-2}\|v\|^2 + 2n^{-1}\langle v, x - \bar{x}\rangle =$$

$$\rho^2 + n^{-1}\left(n^{-1} + 2\langle v, x - \bar{x}\rangle\right). \tag{3.54}$$

Because $\langle v, x - \bar{x} \rangle < 0$, the rightmost expression in (3.54) is smaller than ρ^2 for large enough n, and so y_n belongs to $B(\bar{x}, \rho) = K$.

It remains to establish the converse inclusion. Take $u \in T_K(x)$. By Proposition 3.6.13 there exists a sequence $\{(u_n, \lambda_n)\} \subset X \times \mathbb{R}$ converging to $(u, 0)$ such that $y_n := x + \lambda_n u_n$ belongs to K for all $n \in \mathbb{N}$. Take now $y \in B(\bar{x}, \rho) = K$. Because $\|x - \bar{x}\| = \rho$, we have

$$\langle x - \bar{x}, y - x \rangle = -\|x - \bar{x}\|^2 + \langle x - \bar{x}, y - \bar{x} \rangle \leq$$

$$-\rho^2 + \|x - \bar{x}\|\,\|y - \bar{x}\| \leq -\rho^2 + \rho^2 = 0. \tag{3.55}$$

Because $y_n \in K$ and $u_n = (y_n - x)/\lambda_n$, putting $y = y_n$ in (3.55) and dividing by λ_n we get that $\langle u, x - \bar{x} \rangle \leq 0$, so that $u \in \{v \in X : \langle v, x - \bar{x} \rangle \leq 0\}$ as needed, and the inclusion holds. $\square$

Exercises

3.1. Prove Proposition 3.4.3.

3.2. Prove Proposition 3.4.4.

3.3. In Lemma 3.4.21 prove that the coefficients λ_i in the expression of L verify $\lambda_i \geq 0$ for all $i = 1, \ldots, p$ and $\sum_{i=1}^{p} \alpha_i = 1$.

3.4. Consider δ_K as defined in Definition 3.1.3. Prove that δ_K is convex and lsc if and only if K is closed and convex.

3.5. Let X be a Banach space and $f : X \to \mathbb{R} \cup \{\infty\}$ a convex function. Prove the following statements.

    (i) If Y is a Banach space and $Q : Y \to X$ a linear mapping, then $f \circ Q : Y \to \mathbb{R} \cup \{\infty\}$ is convex.

    (ii) If $\varphi : \mathbb{R} \to \mathbb{R}$ is convex and nondecreasing then $\varphi \circ f$ is convex.

(iii) Consider S_f as in Definition 3.1.6(a). Prove that if f is convex then $S_f(\lambda)$ is convex for all $\lambda \in \mathbb{R}$. Give an example disproving the converse statement.

3.6. Let X be a Banach space and consider $f, g : X \to \mathbb{R} \cup \{\infty\}$. Prove the following results on conjugate functions.

(i) If $f \le g$ then $g^* \le f^*$.

(ii) If $Q : X \to X$ is an automorphism, then $(f \circ Q)^* = f^* \circ (Q^*)^{-1}$, where Q^* denotes the adjoint operator of Q (cf. Definition 3.13.7).

(iii) If $h(x) = f(\lambda x)$ with $\lambda \in \mathbb{R}$, then $h^*(p) = f^*(\lambda^{-1}p)$.

(iv) If $h(x) = \lambda f(x)$ with $\lambda \in \mathbb{R}$, then $h^*(p) = \lambda f(\lambda^{-1}p)$.

3.7. Given a cone $K \subset X$, consider K^- as in Definition 3.6.3 and δ_K as in Definition 3.1.3. For a subspace V of X, let $V^{\perp} = \{p \in X^* : \langle p, x \rangle = 0 \ \forall x \in V\}$. Prove the following statements.

(i) If K is a cone, then $\delta_K^* = \delta_{K^-}$.

(ii) If V is a subspace, then $\delta_V^* = \delta_{V^{\perp}}$.

3.8. Given a subset K of X, consider σ_K as in Definition 3.4.19.

(i) Use Hahn–Banach theorem to prove that for all closed and convex K it holds that $K = \{x \in X : \langle p, x \rangle \le \sigma_K(p) \ \forall p \in X^*\}$.

(ii) Exhibit a bijection between nonempty, closed, and convex subsets of X and proper, positively homogeneous, lsc, and convex functions defined on X^*.

3.9. Prove the statements in Example 3.4.20.

3.10. Prove Proposition 3.5.2.

3.11. Prove that Proposition 3.6.17 does not hold when $K^o \cap L = \emptyset$. Hint: for $X = \mathbb{R}^2$, take $K = B((-1,0),1)$, $L = B((1,0),1)$, and $x = (0,0)$.

3.7 Differentiation of point-to-set mappings

In this section we extend to a certain class of point-to-set operators part of the theory of differential calculus for smooth maps in Banach spaces. Most of the material has been taken from [17]. Throughout this section, we deal with Banach spaces X and Y, and a point-to-set mapping $F : X \rightrightarrows Y$ that has a convex graph and is *convex-valued* (i.e., $F(x)$ is a convex subset of Y for all $x \in X$).

If we look at the case of point-to-point maps, only affine operators have convex graphs. On the other hand, we can associate with any convex function $f : X \to \mathbb{R} \cup \{\infty\}$ the epigraphic profile (cf. Definition 3.1.6(b)) $E_f : X \rightrightarrows \mathbb{R} \cup \{\infty\}$, namely

$$E_f(x) = \{t \in \mathbb{R} : t \ge f(x)\}. \tag{3.56}$$

As noted before, the graph of E_f is precisely the epigraph of f. Thus, in this section we generalize the differentiable properties of point-to-point linear mappings and convex functions. We also encompass in our analysis point-to-set extensions of convex mappings $F : X \to Y$ (where X and Y are vector spaces), with respect to a closed and convex cone $C \subset Y$, as we explain below.

Our departing point is the observation that derivatives of real-valued functions can be thought of as originating from tangents to the graph.

Definition 3.7.1.

(a) *The derivative $DF(x_0, y_0) : X \rightrightarrows Y$ of F at a point $(x_0, y_0) \in \mathrm{Gph}(F)$ is the set-valued map whose graph is the tangent cone $T_{\mathrm{Gph}(F)}(x_0, y_0)$, as introduced in Definition 3.6.3(c).*

(b) *The coderivative $DF(x_0, y_0)^* : Y^* \rightrightarrows X^*$ of F at $(x_0, y_0) \in \mathrm{Gph}(F)$ is defined as follows; $p \in DF(x_0, y_0)^*(q)$ if and only if $(p, -q) \in N_{\mathrm{Gph}(F)}(x_0, y_0)$.*

We start with some elementary properties.

Proposition 3.7.2.

(i) $D(F^{-1})(y_0, x_0) = [DF(x_0, y_0)]^{-1}$.

(ii) $D(F^{-1})(y_0, x_0)^*(p) = -[DF(x_0, y_0)^*]^{-1}(-p)$.

Proof. For (i), note that $(u, v) \in \mathrm{Gph}(D(F^{-1})(y_0, x_0)) = T_{\mathrm{Gph}(F^{-1})}(y_0, x_0)$ if and only if $(v, u) \in \mathrm{Gph}(DF(x_0, y_0)) = T_{\mathrm{Gph}(F)}(x_0, y_0)$. Item (ii) is also an elementary consequence of the definitions. $\Box$

Proposition 3.7.3. *The following statements are equivalent.*

(i) $p \in DF(x_0, y_0)^*(q)$.

(ii) $\langle q, y_0 - y \rangle \leq \langle p, x_0 - x \rangle$ *for all* $(x, y) \in \mathrm{Gph}(F)$.

(iii) $\langle p, u \rangle \leq \langle q, v \rangle$ *for all* $u \in X$ *and all* $v \in DF(x_0, y_0)(u)$.

Proof. Elementary, using Definition 3.7.1. $\Box$

Example 3.7.4. *For $K \subset X$, define $\phi_K : X \rightrightarrows Y$ as*

$$\phi_K(x) = \begin{cases} 0 & \text{if } x \in K \\ \emptyset & \text{if } x \notin K. \end{cases}$$

Then, $D\phi_K(x, 0) = \phi_{T_K}(x)$ for all $x \in K$. Indeed

$$\mathrm{Gph}(D\phi_K(x, 0)) = T_{\mathrm{Gph}(\phi_K)}(x, 0) = T_{K \times \{0\}}(x, 0)$$

$$= T_K(x) \times \{0\} = \mathrm{Gph}(\phi_{T_K(x)}).$$

Note that if $F : X \to Y$ is point-to-point and differentiable, then $DF(x_0, y_0)$ is the linear transformation $F'(x_0) : X \to Y$; that is, the Gâteaux derivative of F at x_0 (cf. Proposition 4.1.6).

Next we change our viewpoint; we move from our initial geometrical definition, in terms of tangent cones to the graph, to an analytical perspective, in which the derivative is an outer limit of incremental quotients.

Let $A \subset Y$. For a given $y \in Y$, $d(y, A)$ denotes the distance from y to A; that is, $d(y, A) = \inf_{x \in A} \|x - y\|$.

Proposition 3.7.5. *Take $F : X \rightrightarrows Y$ with convex graph and $(x_0, y_0) \in \mathrm{Gph}(F)$. Then $v_0 \in DF(x_0, y_0)(u_0)$ if and only if*

$$\liminf_{\substack{u \to u_0 \\ h > 0}} \inf d\left(v_0, \frac{F(x_0 + hu) - y_0}{h}\right) = 0, \qquad (3.57)$$

Proof. By Definition 3.7.1, if $v_0 \in DF(x_0, y_0)(u_0)$ then $(u_0, v_0) \in T_{\mathrm{Gph}(F)}(x_0, y_0)$, implying, in view of Proposition 3.6.12(ii), that for all $\varepsilon_1 > 0$, $\varepsilon_2 > 0$, there exist $u(\varepsilon_1) \in X$ and $v(\varepsilon_2) \in Y$ satisfying $\|u(\varepsilon_1)\| \le \varepsilon_1$, $\|v(\varepsilon_2)\| \le \varepsilon_2$, and $t^* > 0$ such that

$$(x_0, y_0) + t\left[(u_0, v_0) + (u(\varepsilon_1), v(\varepsilon_2))\right] \in \mathrm{Gph}(F),$$

for all $t \in [0, t^*]$. Hence,

$$v_0 \in \frac{F(x_0 + t(v_0 + u(\varepsilon_1))) - y_0}{t} - v(\varepsilon_2),$$

and thus

$$\inf_{t \in [0,t^*]} d\left(v_0, \frac{F(x_0 + t(u_0 + u(\varepsilon_1))) - y_0}{t}\right) \le \varepsilon_2. \qquad (3.58)$$

We claim that

$$\frac{F(x + su) - y}{s} \supset \frac{F(x + tu) - y}{t} \qquad (3.59)$$

for all $0 \le s \le t$ and for all $y \in F(x)$. Indeed, we get from Proposition 2.8.2

$$\frac{s}{t}F(x + tu) + \left(1 - \frac{s}{t}\right)y \subset F\left(\frac{s}{t}(x + tu) + \left(1 - \frac{s}{t}\right)x\right) = F(x + su),$$

establishing the claim. In view of (3.59), the function $\phi(t) = d(v, (F(x + tu) - y)/t)$ is increasing, in which case

$$\lim_{t \downarrow 0} d\left(v_0, \frac{F(x_0 + tu) - y_0}{t}\right) = \inf_{t > 0} d\left(v_0, \frac{F(x_0 + tu) - y_0}{t}\right). \qquad (3.60)$$

Taking now $u = u_0 + u(\varepsilon_1)$, we get from (3.58) and (3.60),

$$\varepsilon_2 \ge \inf_{t \in [0,t^*]} d\left(v_0, \frac{F(x_0 + t(u_0 + u(\varepsilon_1))) - y_0}{t}\right)$$

$$= \inf_{t>0} d\left(v_0, \frac{F(x_0 + t(u_0 + u(\varepsilon_1))) - y_0}{t}\right)$$

$$= \lim_{t\downarrow 0} d\left(v_0, \frac{F(x_0 + t(u_0 + u(\varepsilon_1))) - y_0}{t}\right). \tag{3.61}$$

It follows from (3.61) that

$$\inf_{\|u-u_0\|\le\varepsilon_1} \inf_{t>0} d\left(v_0, \frac{F(x_0 + tu) - y_0}{t}\right) \le \varepsilon_2,$$

and letting $\varepsilon_1 \to 0$ and $\varepsilon_2 \to 0$, we obtain (3.57). The proof of the converse statement is similar, and we leave it as an exercise. $\square$

This concept of derivative allows a first-order expansion for point-to-set mappings, as we show next.

Proposition 3.7.6. *Take* $F : X \rightrightarrows Y$ *with convex graph,* $(x_0, y_0) \in \mathrm{Gph}(F)$, *and* $x \in D(F)$. *Then*

$$F(x) - y_0 \subset DF(x_0, y_0)(x - x_0). \tag{3.62}$$

Proof. By convexity of $\mathrm{Gph}(F)$,

$$(1 - t)y_0 + ty \in F(x_0 + t(x - x_0))$$

for all $y \in F(x)$ and all $t \in [0, 1]$. Hence,

$$y - y_0 \in \frac{F(x_0 + t(x - x_0)) - y_0}{t}.$$

Thus

$$(x - x_0, y - y_0) \in S_{\mathrm{Gph}(F)}(x_0, y_0) \subset T_{\mathrm{Gph}(F)}(x_0, y_0);$$

that is, $y - y_0 \in DF(x_0, y_0)(x - x_0)$. Inasmuch as this inclusion holds for all $y \in F(x)$, (3.62) holds. $\square$

Next we look at the partial order "$\precsim$" induced by a closed and convex cone $C \subset Y$; that is, $y \precsim y'$ if and only if $y' - y \in C$ (cf. Section 6.8). Given $F : X \rightrightarrows Y$ and $K \subset X$, the set-valued optimization problem $\min_{\precsim} F(x)$ s.t. $x \in K$ consists of finding $x_0 \in K$ and $y_0 \in F(x_0)$ such that $y_0 \precsim y$ for all $y \in F(x)$ and all $x \in K$, which can be rewritten as

$$F(x) \subset y_0 + C \tag{3.63}$$

for all $x \in K$.

The study of optimization problems with respect to partial orders induced by cones receives the generic designation of *vector optimization*, and is currently the object of intense research (see [136]). Here we show that our concept of derivative allows us to present first-order optimality conditions for this kind of problems, when the objective F has a convex graph. As mentioned above, this includes not only the case of linear point-to-point operators $F : X \to Y$, but also the following class of nonlinear point-to-point maps. Given a cone $C \subset Y$, there is a natural extension of the definition of epigraphical profile to a C-convex map $F : X \to Y$ (see Definition 4.7.3). Indeed, we can define the point-to-set map $\hat{F} : X \rightrightarrows Y$ defined as $\hat{F}(x) := \{y \in Y : F(x) \precsim y\} = \{y \in Y : y \in F(x) + C\}$. It is easy to check that $\text{Gph}(\hat{F}) \subset X \times Y$ is convex. Thus, the first-order optimality conditions presented next include extensions to the point-to-set realm of optimization problems with C-convex point-to-point objectives (possibly nonlinear).

Definition 3.7.7. *Given a closed and convex cone $C \subset Y$, the positive polar cone is defined as $C^* \subset Y^*$ as $C^* = \{z \in Y^* : \langle z, y \rangle \geq 0 \ \forall y \in C\}$ (cf. (3.40)).*

Theorem 3.7.8. *Let $F : X \rightrightarrows Y$ be a point-to-set mapping with convex graph and fix $C \subset Y$ a convex and closed cone. Let $\precsim$ be the order induced by C in Y. Take $K \subset X$. A pair $(x_0, y_0) \in K \times F(x_0)$ is a solution of $\min_{\precsim} F(x)$ s.t. $x \in K$ if and only if one of the two following equivalent conditions holds.*

(i) $DF(x_0, y_0)(u) \subset C$ for all $u \in X$.

(ii) $0 \in DF(x_0, y_0)^*(z)$ for all $z \in C^*$.

Proof. Assume that (i) holds. By Proposition 3.7.6

$$F(x) \subset y_0 + DF(x_0, y_0)(x - x_0) \subset y_0 + C,$$

for all $x \in K$; that is, $y_0 \precsim F(x)$ for all $x \in K$, so that (x_0, y_0) solves the optimization problem. Conversely, assume that (x_0, y_0) solves (3.63). Take $u \in X$ and $v \in DF(x_0, y_0)(u)$. By Definition 3.7.1 and Proposition 3.6.12(ii), for all $\varepsilon > 0$ there exist $w \in X$, $q \in Y$, and $t^* > 0$ such that $\|w - u\| \leq \varepsilon$, $\|q - v\| \leq \varepsilon$, and

$$(x_0, y_0) + t(w, q) \in \text{Gph}(F),$$

for all $t \in [0, t^*]$. This yields $y_0 + tq \in F(x_0 + tw)$ for all $t \in [0, t^*]$. In other words,

$$q \in \frac{F(x_0 + tw) - y_0}{t}.$$

Because $v - q \in B(0, \varepsilon)$ we conclude that

$$v = q + (v - q) \in \frac{F(x_0 + tw) - y_0}{t} + B(0, \varepsilon) \subset C + B(0, \varepsilon), \qquad (3.64)$$

using (3.63) in the inclusion. Because C is closed, taking limits as $\varepsilon \to 0$ in (3.64) we obtain that v belongs to C, which establishes (i).

It remains to prove the equivalence between (i) and (ii). Note that $0 \in DF(x_0, y_0)^*(z)$ if and only if, in view of Definition 3.7.1(b), $(0, -z) \in N_{\text{Gph}(F)}(x_0, y_0) = T^-_{\text{Gph}(F)}(x_0, y_0)$, using Definition 3.6.3(c). It follows that $\langle z, v \rangle \geq 0$ whenever $(u, v) \in T_{\text{Gph}(F)}(x_0, y_0)$, or, equivalently, whenever $v \in DF(x_0, y_0)(u)$. In view of the Definition of C^*, this is equivalent to $DF(x_0, y_0) \subset C^{**} = C$ (the last equality is proved exactly as in Proposition 3.6.6(ii)).   □

Next we present the basics of calculus of derivatives and coderivatives of point-to-set maps, namely several cases for which we have a chain rule.

Proposition 3.7.9. *Consider reflexive Banach spaces X, Y, and Z. Let $F : X \rightrightarrows Y$ be a closed and convex operator, in the sense of Definition 2.8.1, and $A : Z \to X$ a continuous linear mapping.*

  (i) *If A is surjective then $D(FA)(z, y) = DF(Az, y) \circ A$ for all $(z, y) \in \mathrm{Gph}(F \circ A)$.*

  (ii) *If A^* is surjective then $D(FA)(z, y)^* = A^* \circ DF(Az, y)^*$ for all $(z, y) \in \mathrm{Gph}(F \circ A)$.*

Proof. Let $G = F \circ A$. It is easy to verify that $\mathrm{Gph}(G) = (A \times I)^{-1}(\mathrm{Gph}(F))$, where I is the identity operator in Y. In view of the definition of DF, the statement of item (i) is equivalent to

$$T_{\mathrm{Gph}(G)}(z, y) = (A \times I)^{-1} \left(T_{\mathrm{Gph}(F)}(Az, y) \right), \qquad (3.65)$$

and item (ii) is equivalent to

$$N_{\mathrm{Gph}(G)}(z, y) = (A^* \times I) \left(N_{\mathrm{Gph}(F)}(Az, y) \right). \qquad (3.66)$$

We proceed to prove both inclusions in (3.65). Take $(u, v) \in T_{\mathrm{Gph}(G)}(z, y)$. In view of Proposition 3.6.13, there exist $t_n \downarrow 0$, $u_n \to u$, and $v_n \to v$ such that $y + t_n v_n \in G(z + t_n u_n) = F(Az + t_n Au_n)$. By the same proposition, in order to show that (u, v) belongs to $(A \times I)^{-1} T_{\mathrm{Gph}(F)}(Az, y)$ it suffices to find $s_n \downarrow 0$, $w_n \to Au$, and $\bar{v}_n \to v$ such that $y + s_n \bar{v}_n \in F(Az + s_n w_n)$. It is obvious that this inclusion is achieved by taking $s_n = t_n$, $\bar{v}_n = v_n$, and $w_n = Au_n$ (note that $\lim_n Au_n = Au$ by continuity of A).

For the converse inclusion, we assume the existence of $s_n \downarrow 0$, $w_n \to Au$, and $\bar{v}_n \to v$ such that $y + s_n \bar{v}_n \in F(Az + s_n w_n)$ and we must exhibit $t_n \downarrow 0$, $u_n \to u$, and $v_n \to v$ such that $y + t_n v_n \in G(z + t_n u_n) = F(Az + t_n Au_n)$. In this case we take also $t_n = s_n$ and $v_n = \bar{v}_n$. For choosing the right u_n, we invoke Corollary 2.8.6 with $K = Z$, $x_0 = u$, and $y = w_n$: there exists $L > 0$ and u_n such that $Au_n = w_n$ and $\|u_n - u\| \le L \|w_n - Au\|$. This is the appropriate u_n, because, inasmuch as $\lim_n w_n = Au$, we get that $\lim_n u_n = u$.

The equality in (3.66) is established in a similar fashion, using the surjectivity of A^*.   □

The surjectivity and continuity hypotheses in Proposition 3.7.9 can be somewhat relaxed at the cost of some technical complications; see Theorem 4.2.6 in [17].

A similar result holds for left composition with linear mappings.

Proposition 3.7.10. *Consider reflexive Banach spaces X, Y, and Z. Let $F : X \rightrightarrows Y$ be a closed and convex operator, in the sense of Definition 2.8.1 and a linear mapping $B : Y \to Z$. Then the following formulas hold for all $(x, y) \in \mathrm{Gph}(F)$.*

(i) $D(BF)(x, By) = \overline{B \circ DF(x, y)}$.

(ii) $D(BF)(x, By)^* = DF(x, y)^* \circ B^*$.

Proof. The argument is similar to the one in the proof of Proposition 3.7.9. $\quad\Box$

We combine the results in Propositions 3.7.9 and 3.7.10 in the following corollary.

Corollary 3.7.11. *Consider reflexive Banach spaces W, X, Y, and Z. Let $F : X \rightrightarrows Y$ be a closed and convex operator, in the sense of Definition 2.8.1, $A : W \to X$ a continuous linear mapping, and $B : Y \to Z$ a linear mapping.*

(i) *If A is surjective then $D(BFA)(u, By) = \overline{B \circ DF(Au, y) \circ A}$ for all $u \in W$ and all $y \in F(Au)$.*

(ii) *If A^* is surjective then $D(BFA)(u, By)^* = A^* \circ DF(Au, y)^* \circ B^*$ for all $u \in W$ and all $y \in F(Au)$.*

Proof. The results follow immediately from Propositions 3.7.9 and 3.7.10 $\quad\Box$

The following corollary combines compositions with sums.

Corollary 3.7.12. *Consider reflexive Banach spaces X, Y, and Z. Let $F : X \rightrightarrows Z$ and $G : Y \rightrightarrows Z$ be two closed and convex operators, in the sense of Definition 2.8.1, and such that $D(F) = X$, and $A : X \to Y$ a continuous linear mapping.*

(i) *If A is surjective then $D(F + GA)(x, y + z) = \overline{DF(x, y) + DG(Ax, z) \circ A}$, for all $x \in A^{-1}(D(G))$, all $y \in Fx$, and all $z \in G(Ax)$.*

(ii) *If A^* is surjective then $D(F + GA)(x, y + z)^* = DF(x, y)^* + A^* \circ DG(Ax, z)^*$ for all $x \in A^{-1}(D(G))$, all $y \in Fx$, and all $z \in G(Ax)$.*

Proof. Note that $F + GA = \hat{B} \circ \hat{G} \circ \hat{A}$, where $\hat{A}(x) = (x, Ax)$, $\hat{G}(x, y) = F(x) \times G(y)$, and $\hat{B}(z_1, z_2) = z_1 + z_2$. It suffices to check that $D\hat{G}(x, y, z_1, z_2) = DF(x, z_1) \times DG(y, z_2)$ and to apply Corollary 3.7.11. $\quad\Box$

The case of the derivative of a sum results from Corollary 3.7.12 and is presented next.

Corollary 3.7.13. *Consider reflexive Banach spaces X and Y. Let $F, G : X \rightrightarrows Y$ be two closed and convex point-to-set mappings, in the sense of Definition 2.8.1, and such that $D(F) = X$. Then the following formulas hold for all $x \in D(G)$, all $y \in Fx$, and all $z \in Gx$.*

(i) $D(F + G)(x, y + z) = \overline{DF(x, y) + DG(x, z)}$.

(ii) $D(F + G)(x, y + z)^* = DF(x, y)^* + DG(x, z)^*$.

Proof. The result follows from Corollary 3.7.12 by taking $X = Y$ and the identity operator in X as A.   □

Next we consider the restriction of a mapping to a closed and convex subset of its domain. Note that $F_{/K} = F + \phi_K$, with ϕ_K as in Example 3.7.4.

Corollary 3.7.14. *Consider reflexive Banach spaces X and Y. Let $F : X \rightrightarrows Y$ be two closed and convex point-to-set mappings, in the sense of Definition 2.8.1, with $D(F) = X$, and $K \subset X$ a closed and convex set. Then*

(i) $D\left(F_{/K}\right)(x, y)$ *is the restriction of $DF(x, y)$ to $T_K(x)$.*

(ii) $D\left(F_{/K}\right)(x, y)^*(q) = DF(x, y)^*(q) + N_K(x)$.

Proof. The result follows from Corollary 3.7.13 with $G = \phi_K$, using that $F_{/K} = F + G$.   □

Exercises

3.1. Prove Proposition 3.7.2(ii).

3.2. Prove Proposition 3.7.3.

3.3. Prove the "if" statement of Proposition 3.7.5.

3.4. Prove Proposition 3.7.10.

3.5. Let $f : X \to \mathbb{R} \cup \{\infty\}$ be convex and proper and fix $\bar{x} \in \mathrm{Dom}(f)$. Prove that the following statements are equivalent.

(i) $\bar{x}$ is a minimizer of f.

(ii) $(0, -1)$ belongs to $N_{\mathrm{Epi}(f)}(\bar{x}, f(\bar{x})) = N_{\mathrm{Gph}(E_f)}(\bar{x}, f(\bar{x}))$, with E_f as in Definition 3.1.6(b).

(iii) 0 belongs to $D(E_f)(\bar{x}, f(\bar{x}))^*(1)$.

(iv) $D(E_f)(\bar{x}, f(\bar{x}))(u) \subset \mathbb{R}_+$ for all $u \in X$.

Hint: use Proposition 3.7.6 and Theorem 3.7.8.

3.8   Marginal functions

Definition 3.8.1. *Let X and Y be Hausdorff topological spaces and $F : X \rightrightarrows Y$ a point-to-set mapping. Given $f : \mathrm{Gph}(F) \to \mathbb{R}$, define $g : X \to \mathbb{R} \cup \{\infty\}$ as*

$g(x) := \sup_{y \in F(x)} f(x, y)$, *which is called the* marginal function *associated with F and f.*

The result below relates the continuity properties of F and f with those of g.

Theorem 3.8.2. *Consider F, f, and g as in Definition 3.8.1 and take $x \in D(F)$.*

(a) *Let X be a metric space and Y a topological space. If F is USC at x, $F(x)$ is compact and f is upper-semicontinuous at every $(x, y) \in \{x\} \times F(x)$, then g is upper-semicontinuous at x.*

(b) *Let X and Y be Hausdorff topological spaces. If F is ISC at x and f is lower-semicontinuous at every $(x, y) \in \{x\} \times F(x)$, then g is lower-semicontinuous at x.*

(c) *Assume any of the assumptions below:*

(i) *X is a Banach space, $Y = X^*$, F is (sw^*)-OSC and locally bounded at x, and f is (sw^*)-upper-semicontinuous at any $(x, y) \in \{x\} \times F(x)$.*

(ii) *X and Y are Banach spaces, Y is reflexive, F is (sw)-OSC and locally bounded at x, and f is (sw)-upper-semicontinuous at any $(x, y) \in \{x\} \times F(x)$.*

Then g is upper-semicontinuous at x.

Proof.
(a) In order to prove that g is upper-semicontinuous at x, we must establish that for all $\varepsilon > 0$ there exists $\eta > 0$ such that for all $x' \in B(x, \eta)$ it holds that

$$g(x') < g(x) + \varepsilon. \tag{3.67}$$

Take any $y \in F(x)$. Because f is upper-semicontinuous at (x, y), there exist $r(y) > 0$ and a neighborhood $W_y \in \mathcal{U}_y$ such that, whenever $(x', y') \in B(x, r(y)) \times W_y$ it holds that

$$f(x', y') < f(x, y) + \varepsilon/2. \tag{3.68}$$

When y runs over $F(x)$, the neighborhoods W_y provide an open covering of $F(x)$; that is, $F(x) \subset \cup_{y \in F(x)} W_y$. Because $F(x)$ is compact, there exists a finite subcovering $W_{y_1}, \ldots, W_{y_p}$, such that $F(x) \subset \cup_{i=1}^p W_{y_i} =: W$. Because F is USC at x, there exists $\bar{r} > 0$ such that $F(x') \subset W$ for all $x' \in B(x, \bar{r})$. Set $\eta < \min\{\min\{r(y_i) : i = 1, \ldots, p\}, \bar{r}\}$. Consider the set

$$\bar{W} := B(x, \eta) \times (\cup_{i=1}^p W_{y_i}) \in \mathcal{U}_{(x,y)}.$$

We claim that (3.67) holds for this choice of η. Indeed, if $x' \in B(x, \eta)$ and $z \in F(x')$, we have

$$z \in F(x') \subset F(B(x, \eta)) \subset W = \cup_{i=1}^p W_{y_i}, \tag{3.69}$$

where we used the fact that $\eta < \bar{r}$ and the definition of $\bar{r}$. In view of (3.69), there exists an index i_0 such that $z \in W_{y_{i_0}}$. Thus, $(x', z) \in B(x, r(y_{i_0})) \times W_{y_{i_0}}$. Using now (3.68) for $y' = z$ and $y = y_{i_0} \in F(x)$, and the definition of g, we conclude that

$$f(x', z) < f(x, y_{i_0}) + \varepsilon/2 \leq \sup_{y \in F(x)} f(x, y) + \varepsilon/2 < g(x) + \varepsilon. \qquad (3.70)$$

Inequality (3.67) now follows by taking the supremum with $z \in F(x')$ in the leftmost expression of (3.70), establishing the claim.

(b) Assume g is not lower-semicontinuous at x. So there exist $\alpha \in \mathbb{R}$ such that $g(x) > \alpha$ and a net $\{x_i\}_{i \in I}$ converging to x with

$$g(x_i) \leq \alpha, \qquad (3.71)$$

for all $i \in I$. Because F is ISC at x, we have $F(x) \subset \liminf_{i \in I} F(x_i)$. By definition of g and the fact that $g(x) > \alpha$ there exists $y \in F(x)$ such that $f(x, y) > \alpha$. Using now the fact that f is lower-semicontinuous at (x, y) we can find a neighborhood $U \times W \subset X \times Y$ of (x, y) such that $f(x', y') > \alpha$ for all $(x', y') \in U \times W$. Using the fact that $\{x_i\}_{i \in I}$ converges to x we have that there exists $I_0 \subset I$ a terminal set such that $x_i \in U$ for all $i \in I_0$. Because $y \in W \cap F(x)$, we have by inner-semicontinuity of F that there exists a terminal set $I_1 \subset I$ such that

$$F(x_i) \cap W \neq \emptyset,$$

for all $i \in I_1$. For $i \in I_0 \cap I_1$, choose $y_i \in F(x_i) \cap W$. So $(x_i, y_i) \in U \times W$ and hence $f(x_i, y_i) > \alpha$. Therefore, $g(x_i) \geq f(x_i, y_i) > \alpha$, contradicting (3.71). Hence we must have g lsc at x.

(c) Consider first assumption (i). In order to prove that g is (strongly) upper-semicontinuous at x, we must prove that whenever $g(x) < \gamma$, there exists a strong neighborhood U of x such that $g(x') < \gamma$ for all $x' \in U$. If the latter statement is not true, then there exists a sequence $\{x_n\}$ converging strongly to x such that $g(x_n) \geq \gamma$ for all n. Because $g(x_n) = \sup_{z \in F(x_n)} f(x_n, z) \geq \gamma > \gamma - 1/n$ there exists $z_n \in F(x_n)$ such that $f(x_n, z_n) > \gamma - 1/n$ for all n. We prove the conclusion under the assumption (i). Using the local boundedness of F, choose a neighborhood V of x such that $F(V) \subset B(0, \rho) \subset X^*$. Fix $n_0 \in \mathbb{N}$ such that $x_n \in V$ for all $n \geq n_0$. This gives $z_n \in F(x_n) \subset F(V) \subset B(0, \rho)$ for all $n \geq n_0$. By Theorem 2.5.16(ii) applied to $A := B(0, \rho)$ we have that the *net* $\{z_n\}$ is contained in a weak* compact set. Using now Theorem 2.1.9(a) there exists a weak* convergent subnet of $\{z_n\}$, which we call $\{z_i\}_i$, converging to a weak* limit point z. By (sw*)-outer-semicontinuity of F we get $z \in F(x)$. Call $\{x_i\}_i$ the subnet of $\{x_n\}$ which corresponds to the subnet $\{z_i\}_i$ (in other words, the subnet that uses the same function ϕ in Definition 2.1.7). By Theorem 2.1.9(c), the subnet $\{x_i\}_i$ converges to x. Altogether, the subnet $\{(x_i, z_i)\}_i$ converges (sw*) to (x, z). Using these facts and (sw*)-upper-semicontinuity of f we get

$$f(x, z) \leq g(x) < \gamma \leq \limsup_i f(x_i, y_i) \leq f(x, z),$$

which is a contradiction. This proves that g must be upper-semicontinuous at x. Case (ii) is dealt with in a similar way, quoting Theorem 2.5.19 instead of Theorem 2.5.16(ii), and using a subsequence $\{z_{n_k}\}$ of $\{z_n\}$ instead of a subnet. $\quad\square$

Remark 3.8.3. If F is not USC, or has noncompact values, then g may fail to be upper-semicontinuous. The point-to-set mapping pictured in Figure 3.4(a) has compact values but it is not USC at 0, and the one pictured in 3.4(b) is USC at 0, but has noncompact values. In both cases, g is not upper-semicontinuous at 0. In both (a) and (b), f is given by $f(x,y) := xy$. Indeed, take F as in Figure 3.4(a) and consider the sequence $x_n := 1/n$ for all n. We have

$$g(x_n) = \max_{y \in \{1/n,n\}} y \, x_n = \max\{1/n^2, 1\} = 1,$$

and hence $\limsup_n g(x_n) = 1 > g(0) = 0$. For the case pictured in (b) we can take the same sequence $\{x_n\}$, obtaining

$$g(x_n) = \sup_{y \in [1/n,+\infty)} y/n = +\infty,$$

and henceforth $\limsup_n g(x_n) > g(0) = 0$. In Figure 3.5 we consider $F : \mathbb{R} \rightrightarrows \mathbb{R}$ defined by

$$F(x) := \begin{cases} [0, 1/x] & \text{if } x \neq 0, \\ 0 & \text{if } x = 0. \end{cases}$$

This F is clearly not OSC, and again g is not upper-semicontinuous at 0, because $1 = \limsup_n g(x_n) > g(0) = 0$.

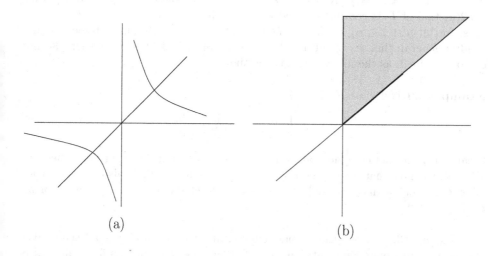

(a)

(b)

Figure 3.4. *Compact-valuedness and upper-semicontinuity*

3.9 Paracontinuous mappings

Let X, Y be Banach spaces, and $F : X \rightrightarrows Y$ a point-to-set mapping. For each $p \in Y^*$, define $\gamma_p : X \to \mathbb{R}$ as $\gamma_p(x) = \sup_{y \in F(x)} \langle p, y \rangle = \sigma_{F(x)}(p)$. If we take now

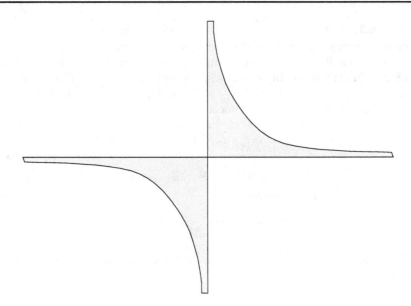

Figure 3.5. *Example with g not upper-semicontinuous at* 0

$\gamma : Y^* \times X \to \mathbb{R} \cup \{\infty\}$ given by $\gamma(p, x) = \gamma_p(x)$, it follows easily from Theorem 3.8.2(a) that if F is compact-valued and USC then γ is Usc on $Y^* \times D(F)$, and thus, for all $p \in Y^*$, γ_p is Usc on $D(F)$. The converse of this statement is not valid in general; that is, γ_p can be Usc in situations in which F is neither USC nor compact-valued, as the following example shows.

Example 3.9.1. Define $F : \mathbb{R} \rightrightarrows \mathbb{R}$ as

$$F(x) = \begin{cases} \{x, 1/x\} & \text{if } x \neq 0, \\ 2\mathbb{Z} + 1 & \text{if } x = 0. \end{cases}$$

Clearly, F is neither compact-valued nor USC at $x = 0$. On the other hand, because $Y^* = \mathbb{R}$, we have that $\gamma_t(0) = \sup_{s \in 2\mathbb{Z}+1} ts = \infty$ for all $t \neq 0$. It also holds that for all $x \neq 0$, $\gamma_t(x) = \max\{tx, t/x\}$, and it is easy to check that γ_t is Usc at 0 for all $t \in \mathbb{R}$.

Example 3.9.1 shows that upper-semicontinuity of γ_p for all $p \in Y^*$ is a weaker condition than upper-semicontinuity of F. Point-to-set mappings for which γ_p is Usc enjoy several interesting properties, thus justifying the following definition.

Definition 3.9.2. *Given Banach spaces* X, Y, *a point-to-set mapping* $F : X \rightrightarrows Y$ *is said to be*

(a) *Paracontinuous (PC) at* $x \in D(F)$ *if and only if for all* $p \in Y^*$ *the function* $\gamma_p : X \to \mathbb{R} \cup \{\infty\}$ *defined as* $\gamma_p(x) := \sigma_{F(x)}(p)$ *(with* σ *as in Definition 3.4.19) is Usc at* x.

(b) Paracontinuous *if it is* PC *at* x *for all* $x \in D(F)$.

Obviously, paracontinuity depends on the topologies of X and Y. We denote with $s-$PC and $w-$PC paracontinuity with respect to the strong and the weak topology in X, respectively. Paracontinuous mappings are called *upper hemicontinuous* in [18, Section 2.6]. Because the expression "hemicontinuity" has been quite frequently used for denoting mappings whose restrictions to segments or lines are continuous, we decided to adopt a different notation. The next result establishes connections between paracontinuity and outer-semicontinuity.

In view of Theorem 2.5.4, $(sw)-$OSC $((ww)-$OSC$)$ means that $\mathrm{Gph}(F)$ is (sw)-closed $((ww)$-closed).

Proposition 3.9.3. *Let X and Y be Banach spaces. Take $F : X \rightrightarrows Y$ such that $F(x)$ is closed and convex for all $x \in X$. Then*

(i) If F is $s-$PC then F is $(sw)-$OSC.

(ii) If F is $w-$PC then F is $(ww)-$OSC.

Proof.
(i) Because $F(x)$ is closed and convex, by (3.26) we have

$$F(x) = \{y \in Y : \langle p, y \rangle \leq \sigma_{F(x)}(p) \,\, \forall \, p \in Y^*\}. \tag{3.72}$$

Take $\{(x_i, y_i)\} \subset \mathrm{Gph}(F)$ a net converging strongly on X and weakly on Y to (x, y). Because $\{y_i\}$ is weakly convergent to y, we have

$$\begin{aligned} \langle p, y \rangle &= \lim_i \langle p, y_i \rangle = \limsup_i \langle p, y_i \rangle \\ &\leq \limsup_i \sigma_{F(x_i)}(p) \leq \gamma_p(x) = \sigma_{F(x)}(p), \end{aligned} \tag{3.73}$$

using (3.72) in the first inequality and the fact that F is $s-$PC at x in the second one. It follows from (3.73), using again (3.72), that y belongs to $F(x)$, and so the result holds.

(ii) Same proof as in (i), using now the fact that F is $w-$PC at x in the second inequality of (3.73). $\square$

The next proposition presents conditions upon which the converse of Proposition 3.9.3 holds; that is, outer-semicontinuity implies paracontinuity.

Proposition 3.9.4. *Let X and Y be Banach spaces and take $F : X \rightrightarrows Y$, $x \in X$. Assume that Y is reflexive and that F is locally bounded at x. If F is $(sw)-$OSC at x then F is $s-$PC at x.*

Proof. The result follows from Theorem 3.8.2(c)(ii) and the definition of paracontinuity. $\square$

3.10 Ky Fan's inequality

We prove in this section Ky Fan's inequality [83] (see also [82]), which is used, together with Brouwer's fixed point theorem, in order to derive two existence results in a point-to-set environment: one deals with equilibrium points and the other is Kakutani's fixed point theorem. We need some topological preliminaries.

Definition 3.10.1. *Given a subset K of a topological space X, and a finite covering $\mathcal{V} = \{V_1, \ldots, V_p\}$ of K with open sets, a* partition of unity *associated with $\mathcal{V}$ is a set $\{\alpha_1, \ldots, \alpha_p\}$ of continuous functions $\alpha_i : X \to \mathbb{R}$ $(1 \leq i \leq p)$ satisfying*

(a) $\sum_{i=1}^{p} \alpha_i(x) = 1$ for all $x \in K$.

(b) $\alpha_i(x) \geq 0$ for all $x \in K$ and all $i \in \{1, \ldots, p\}$.

(c) $\{x \in X : \alpha_i(x) > 0\} \subset V_i$ for all $i \in \{1, \ldots, p\}$.

Proposition 3.10.2. *If K is compact then for every finite covering $\mathcal{V}$ there exists a partition of unity associated with $\mathcal{V}$.*

Proof. See Theorem VIII.4.2 in [73], where the result is proved under a condition on K weaker than compactness, namely paracompactness. □

We also need the following result on lsc functions.

Lemma 3.10.3. *Let K be a compact topological space, S an arbitrary set, and $\psi : K \times S \to \mathbb{R}$ a function such that $\psi(\cdot, s)$ is lsc on K for all $s \in S$. Let $\mathcal{F}$ be the family of finite subsets of S. Then, there exists $\bar{x} \in K$ such that*

$$\sup_{s \in S} \psi(\bar{x}, s) \leq \sup_{M \in \mathcal{F}} \inf_{x \in K} \max_{s \in M} \psi(x, s). \tag{3.74}$$

Proof. Let

$$\beta := \sup_{M \in \mathcal{F}} \inf_{x \in K} \max_{s \in M} \psi(x, s). \tag{3.75}$$

For each $s \in S$, define $T_s = \{x \in K : \psi(x, s) \leq \beta\}$. We claim that T_s is nonempty and closed for all $s \in S$. We proceed to establish the claim. Fix $s \in S$. Because K is compact and $\psi(\cdot, s)$ is lsc, by Proposition 3.1.15 there exists $\tilde{x}$ such that $\psi(\tilde{x}, s) = \inf_{x \in K} \psi(x, s) \leq \beta$, where the last inequality follows from (3.75), taking $M = \{s\}$. It follows that $\tilde{x}$ belongs to T_s, which is henceforth nonempty. The fact that T_s is closed follows from lower-semicontinuity of $\psi(\cdot, s)$ and Proposition 3.1.10(v). The claim holds. Being closed subsets of the compact space K, the sets T_s $(s \in S)$ are compact. We look now at finite intersections of them. Take $M = \{s_1, \ldots, s_p\} \subset S$. By Proposition 3.1.14(iii) and lower-semicontinuity of $\psi(\cdot, s_i)$ $(1 \leq i \leq p)$, we conclude that $\max_{s \in M} \psi(\cdot, s) = \max_{1 \leq i \leq p} \psi(\cdot, s_i)$ is lsc. Using again Proposition 3.1.15 and compactness of K, we conclude that there exists

$x_M \in K$ such that

$$\max_{1 \le i \le p} \psi(x_M, s_i) = \inf_{x \in K} \max_{1 \le i \le p} \psi(x, s_j) \le \beta, \tag{3.76}$$

using (3.75) in the last inequality. It follows from (3.76) that x_M belongs to $\cap_{i=1}^p T_{s_i}$. All finite intersections of the sets T_s are nonempty. It follows from their compactness that the intersection of all of them is also nonempty. In other words, there exists $\bar{x} \in \cap_{s \in S} T_s$. By definition of T_s, we have that $\psi(\bar{x}, s) \le \beta$ for all $s \in S$, and thus $\bar{x}$ satisfies (3.74). □

The following is Brouwer's fixed point theorem (cf. Theorem 4 in [42]), proved for the first time with a statement equivalent to the one given below in [122]. This classical point-to-point result is extended in the remainder of this section to the point-to-set setting.

Theorem 3.10.4. *Let X be a finite-dimensional Euclidean space, K a convex and compact subset of K, and $h : K \to K$ a continuous function. Then there exists $\bar{x} \in K$ such that $h(\bar{x}) = \bar{x}$.*

Proof. See a proof for the case in which K is the unit ball in Theorem XVI.2.1 and Corollary XVI.2.2 of [73]. The extension to a general convex and compact K is elementary. □

Our next result is Ky Fan's inequality.

Theorem 3.10.5. *Let X be a Banach space and K a convex and compact subset of X. Take $\phi : X \times X \to \mathbb{R}$ satisfying:*

(i) $\phi(\cdot, y)$ is lsc on X for all $y \in K$.

(ii) $\phi(x, \cdot)$ is concave in K for all $x \in K$.

(iii) $\phi(x, x) \le 0$ for all $x \in K$.

Then there exists $\bar{x} \in K$ such that $\phi(\bar{x}, y) \le 0$ for all $y \in K$.

Proof. We prove the theorem in two steps. In the first one, we assume that X is finite-dimensional. Suppose, by contradiction, that the result does not hold. Then for all $x \in K$ there exists $y(x) \in K$ such that $\phi(x, y(x)) > 0$, and henceforth, in view of the lower-semicontinuity of $\phi(\cdot, y(x))$ and Definition 3.1.8(a), there exists an open neighborhood $V(x)$ of x such that $\phi(z, y(x)) > 0$ for all $z \in V(x)$. Clearly, the family $\{V(x) : x \in K\}$ is an open covering of K. By compactness of K, there exists a finite subcovering $\mathcal{V} = \{V(x_1), \dots, V(x_p)\}$. We invoke Proposition 3.10.2 to ensure the existence of a partition of unity associated with $\mathcal{V}$, say $\alpha_1, \dots, \alpha_p$. Define $f : X \to X$ as

$$f(x) = \sum_{i=1}^p \alpha_i(x) y(x_i). \tag{3.77}$$

Properties (a) and (b) in Definition 3.10.1 imply that the right-hand side of (3.77) is a convex combination of $\{y(x_1), \ldots, y(x_p)\} \subset K$. Because K is convex, we conclude that $f(x)$ belongs to K for all $x \in K$. The functions $\{\alpha_i(\cdot)\}_{i=1}^p$ are continuous, so f is also continuous. Therefore we can apply Theorem 3.10.4 to the restriction $f_{/K}$ of f to K, concluding that there exists $\bar{y} \in K$ such that $f(\bar{y}) = \bar{y}$. Let $J = \{i : \alpha_i(\bar{y}) > 0\}$. Note that J is nonempty because $\sum_{i=1}^p \alpha_i(\bar{y}) = 1$. Now we use concavity of $\phi(\bar{y}, \cdot)$, getting

$$
\begin{aligned}
\phi(\bar{y}, \bar{y}) = \phi(\bar{y}, f(\bar{y})) &= \phi\left(\bar{y}, \sum_{i=1}^p \alpha_i(\bar{y})y(x_i)\right) \\
&\geq \sum_{i=1}^p \alpha_i(\bar{y})\phi(\bar{y}, y(x_i)) \\
&= \sum_{i \in J} \alpha_i(\bar{y})\phi(\bar{y}, y(x_i)).
\end{aligned}
\tag{3.78}
$$

By property (c) in Definition 3.10.1, $\bar{y}$ belongs to $V(x_i)$ for $i \in J$, and thus, by the definition of $V(x_i)$, it holds that $\phi(\bar{y}, y(x_i)) > 0$ for all $i \in J$. Then, it follows from (3.78) that $\phi(\bar{y}, \bar{y}) > 0$, contradicting (iii). The finite-dimensional case is therefore settled.

We consider now the infinite-dimensional case, which is reduced to the finite-dimensional one in the following way. Let $\mathcal{F}_K$ be the family of finite subsets of K. Take $M = \{y_1, \ldots, y_p\} \in \mathcal{F}_K$ and define $\beta \in \mathbb{R}$ as

$$
\beta = \sup_{M \in \mathcal{F}_K} \inf_{x \in K} \max_{1 \leq i \leq p} \phi(x, y_i).
\tag{3.79}
$$

Calling $\phi_M(x) = \max_{1 \leq i \leq m} \phi(x, y_i)$, it follows from lower-semicontinuity of $\phi(\cdot, y_i)$ and Proposition 3.1.14(iii) that ϕ_M is lsc and thus $\inf_{x \in K} \phi_M(x)$ is attained, by compactness of K and Proposition 3.1.15.

We claim now that $\beta \leq 0$. In order to prove the claim, let Δ^p be the p-dimensional simplex; that is,

$$
\Delta^p = \left\{(\lambda_1, \ldots, \lambda_p) \in \mathbb{R} : \lambda_i \geq 0 \; (1 \leq i \leq p), \sum_{i=1}^p \lambda_i = 1\right\},
$$

and define, for $M \in \mathcal{F}_K$, $\Gamma_M : \Delta^p \times \Delta^p \to \mathbb{R}$ as

$$
\Gamma_M(\lambda, \mu) = \sum_{j=1}^p \mu_j \phi\left(\sum_{i=1}^p \lambda_i y_i, y_j\right).
$$

Now we intend to apply the inequality proved for the finite-dimensional case to the mapping Γ_M. Δ^p is certainly convex and compact. We check next that Γ_M satisfies (i)–(iii). Note that $\Gamma_M(\cdot, \mu) = \sum_{j=1}^p \mu_j \phi(q(\cdot), y_j)$, with $q : \Delta^p \to \Delta^p$ given by $q(\lambda) = \sum_{i=1}^p \lambda_i y_i$. Because q is linear, it is certainly lsc, and then lower-semicontinuity of Γ_M follows from Proposition 3.1.14(i), (iv), and (v). Thus, Γ_M satisfies (i). For (ii), note that $\Gamma_M(\lambda, \cdot)$ is in fact linear, henceforth concave. Finally, because concavity of Γ_M and property (iii) of ϕ imply that

$$
\Gamma_M(\lambda, \lambda) = \sum_{j=1}^p \lambda_j \phi\left(\sum_{i=1}^p \lambda_i y_i, y_j\right) \leq \phi\left(\sum_{i=1}^p \lambda_i y_i, \sum_{j=1}^p \lambda_j y_j\right) \leq 0,
$$

we conclude that Γ_M satisfies (iii). Therefore, we may apply to Γ_M the result proved in the first step for the finite-dimensional problem, and so there exists $\bar{\lambda} \in \Delta^p$ such that $\Gamma_M(\bar{\lambda}, \mu) \leq 0$ for all $\mu \in \Delta^p$, or equivalently, $\sum_{j=1}^p \mu_j \phi\left(\sum_{i=1}^p \bar{\lambda}_i y_i, y_j\right) \leq 0$ for all $\mu \in \Delta^p$. Define $\hat{x} := \sum_{i=1}^p \bar{\lambda}_i y_i$; then $\hat{x} \in K$ and

$$\sum_{j=1}^p \mu_j \phi(\hat{x}, y_j) \leq 0 \tag{3.80}$$

for all $\mu \in \Delta^p$. Note that

$$\phi(\hat{x}, y_j) \leq \sup_{\mu \in \Delta^p} \sum_{i=1}^p \mu_i \phi(\hat{x}, y_i) \tag{3.81}$$

for all $j \in \{1, \ldots, p\}$. Combining (3.80) and (3.81), we get

$$\inf_{x \in K} \max_{1 \leq i \leq p} \phi(x, y_i) \leq \max_{1 \leq i \leq p} \phi(\hat{x}, y_i) \leq \sup_{\mu \in \Delta^p} \sum_{j=1}^p \mu_j \phi(\hat{x}, y_i) \leq 0. \tag{3.82}$$

Because (3.82) holds for all $M \in \mathcal{F}_K$, we conclude from (3.79) that

$$\beta = \sup_{M \in \mathcal{F}_K} \inf_{x \in K} \max_{1 \leq i \leq p} \phi(x, y_i) \leq 0, \tag{3.83}$$

establishing the claim.

In view of (3.83), the result will hold if we prove that there exists $\bar{x} \in K$ such that $\sup_{y \in K} \phi(\bar{x}, y) \leq \beta$. We now invoke Lemma 3.10.3 with $S = K$, $\psi = \phi$. The assumptions of this lemma certainly hold, and the point $\bar{x} \in K$ whose existence it ensures satisfies the required inequality. $\square$

3.11 Kakutani's fixed point theorem

In this section we apply Ky Fan's inequality to obtain fixed points of a point-to-set mapping $F : X \rightrightarrows X$ in a given subset K of a Banach space X. Our basic assumptions are that K is convex and compact, and that F is PC and closed-valued. In the next section we replace the compactness assumption on K by a coercivity hypothesis on F. The next example shows that the former assumptions on K and F (i.e., K convex and compact, and F PC and closed-valued) are not enough to ensure existence of zeroes or fixed points of F in K.

Example 3.11.1. Take $K = [1, 2] \subset \mathbb{R}$ and $F : \mathbb{R} \rightrightarrows \mathbb{R}$ defined as

$$F(x) = \begin{cases} -1 & \text{if } x > 0 \\ [-1, 1] & \text{if } x = 0 \\ 1 & \text{if } x < 0. \end{cases}$$

It can be easily checked that

$$\gamma_t(x) = \begin{cases} -t & \text{if } x > 0 \\ |t| & \text{if } x = 0 \\ t & \text{if } x < 0, \end{cases}$$

and thus γ_t is Usc for all $t \in \mathbb{R}$, so that F is PC. However, F has neither zeroes nor fixed points in K.

Example 3.11.1 shows that an additional assumption, involving both K and F, is needed. The next definition points in this direction.

Definition 3.11.2. *Given $K \subset X$ and $F : X \rightrightarrows X$, K is said to be a* feasibility domain *of F if and only if*

(i) $K \subset D(F)$.

(ii) $F(x) \cap T_K(x) \neq \emptyset$ for all $x \in K$, with T_K as in Definition 3.6.3(c).

Remark 3.11.3. Note that condition (ii) in Definition 3.11.2 is effective only for $x \in \partial K$. Indeed, for $x \in K^\circ$, we have $T_K(x) = X$ by Proposition 3.6.14(a), so that (ii) requires just nonemptiness of $F(x)$, already ensured by (i). Note also that K is not a feasibility domain of F in Example 3.11.1, because $F(1) \cap T_K(1) = \{-1\} \cap [0, \infty) = \emptyset$.

We are now able to state and prove our result on existence of constrained equilibrium points of point-to-set mappings.

Theorem 3.11.4. *Let X be a Banach space, $F : X \rightrightarrows X$ a PC point-to-set mapping such that $F(x)$ is closed and convex for all $x \in X$, and $K \subset X$ a convex and compact feasibility domain for F. Then, there exists $\bar{x} \in K$ such that $0 \in F(\bar{x})$.*

Proof. Let us suppose, for the sake of contradiction, that $0 \notin F(x)$ for all $x \in K$. By Theorem 3.4.14(iii), there exists $H_{p(x), \alpha(x)}$ strictly separating $\{0\}$ from $F(x)$; that is, there exists $(p(x), \alpha(x)) \in X^* \times \mathbb{R}$ such that

$$\langle p(x), y \rangle < \alpha(x) < \langle p(x), 0 \rangle = 0$$

for all $y \in F(x)$, which implies that

$$\gamma_{p(x)}(x) \leq \alpha(x) < 0 \tag{3.84}$$

for all $x \in X$. Define now, for each $p \in X^*$, $V_p = \{x \in X : \gamma_p(x) < 0\}$. Note that, in view of Definition 3.9.2, every γ_p is USc because F is PC. Using Proposition 3.1.10(v) for the lsc function $f := -\gamma_p$, we have that every V_p is open, and they cover K, in view of (3.84). Because K is compact, there exists a finite subcovering $\mathcal{V} =$

$\{V_{p_1}, \ldots, V_{p_m}\}$. By Proposition 3.10.2, there exists a partition of unity associated with $\mathcal{V}$, say $\{\alpha_1, \ldots, \alpha_m\}$. Define $\Phi : X \times X \to \mathbb{R}$ as

$$\Phi(x, y) = \sum_{i=1}^{m} \alpha_i(x) \langle p_i, x - y \rangle. \tag{3.85}$$

We check next that Φ satisfies the assumptions of Theorem 3.10.5: the continuity of each function $\alpha_i(\cdot)$ implies the continuity of $\Phi(\cdot, y)$, and a fortiori, its lower-semicontinuity; concavity of $\Phi(x, \cdot)$ is immediate, because this function is indeed affine, and obviously $\Phi(x, x) = 0$ for all $x \in X$. Thus, Theorem 3.10.5 ensures the existence of $\bar{x} \in K$ such that $\Phi(\bar{x}, y) \le 0$ for all $y \in K$. Let $\bar{p} = \sum_{i=1}^{m} \alpha_i(\bar{x}) p_i$. It is immediate that $\langle -\bar{p}, y - \bar{x} \rangle = \Phi(\bar{x}, y) \le 0$ for all $y \in K$, so that $-\bar{p}$ belongs to $N_K(\bar{x})$. In view of Corollary 3.6.7, $-\bar{p}$ belongs to $T_K^-(\bar{x})$; that is,

$$\langle \bar{p}, x \rangle \ge 0, \tag{3.86}$$

for all $x \in T_K(\bar{x})$. Because K is a feasibility domain for F, there exists $y \in F(\bar{x}) \cap T_K(\bar{x})$, and therefore, using (3.86),

$$\sigma_{F(\bar{x})}(\bar{p}) \ge \langle \bar{p}, y \rangle \ge 0. \tag{3.87}$$

Let $J = \{i \in \{1, \ldots, m\} : \alpha_i(\bar{x}) > 0\}$. Because $\sum_{i=1}^{m} \alpha_i(\bar{x}) = 1$, J is nonempty. Then

$$\begin{aligned}
\sigma_{F(\bar{x})}(\bar{p}) = \sup_{y \in F(\bar{x})} \langle \bar{p}, y \rangle &= \sup_{y \in F(\bar{x})} \left\langle \sum_{j \in J} \alpha_j(\bar{x}) p_j, y \right\rangle \\
&\le \sum_{j \in J} \alpha_j(\bar{x}) \sup_{y \in F(\bar{x})} \langle p_j, y \rangle \\
&= \sum_{j \in J} \alpha_j(\bar{x}) \gamma_{p_j}(\bar{x}).
\end{aligned} \tag{3.88}$$

Note that $\alpha_j(\bar{x}) > 0$ for all $j \in J$, which implies that $\bar{x}$ belongs to V_{p_j}, in view of Definition 3.10.1(c). Therefore $\gamma_{p_j}(\bar{x}) < 0$ for all $j \in J$, and so, by (3.88), $\sigma_{F(\bar{x})}(\bar{p}) < 0$, in contradiction with (3.87). It follows that the result holds. $\square$

Now we get Kakutani's fixed point theorem as an easy consequence of Theorem 3.11.4.

Theorem 3.11.5 (Kakutani's fixed point theorem). *Let X be a Banach space, $K \subset X$ a convex and compact subset of X, and $G : X \rightrightarrows K$ a PC point-to-set mapping such that $G(x)$ is closed and convex for all $x \in X$, and $K \subset D(G)$. Then, there exists $\bar{x} \in K$ such that $\bar{x} \in G(\bar{x})$.*

Proof. We apply Theorem 3.11.4 to $F(x) := G(x) - x$. In view of Definition 3.6.8 and Proposition 3.6.11,

$$K - x = \{y - x : y \in K\} \subset S_K(x) \subset T_K(x) \tag{3.89}$$

for all $x \in K$. Because $G(K) \subset K$, we get from (3.89),

$$F(x) = G(x) - x \subset K - x \subset T_K(x) \tag{3.90}$$

for all $x \in K$. By (3.90), because $K \subset D(G)$, we have that $F(x) \cap T_K(x) = F(x) \neq \emptyset$, and henceforth K is a feasibility domain for F. Therefore, F and K satisfy the hypotheses of Theorem 3.11.4, which implies the existence of $\bar{x} \in K$ such that $0 \in F(\bar{x})$, or equivalently, $\bar{x} \in G(\bar{x})$. $\square$

Remark 3.11.6. Note that the assumption that $G(K) \subset K$ in Theorem 3.11.5 is indeed stronger than needed; it suffices to suppose that K is a feasibility domain for $F = G - I$.

Remark 3.11.6 suggests the following definition.

Definition 3.11.7. *Given $G : X \rightrightarrows X$ and $K \subset X$,*

(a) G is K-internal if and only if $G(x) \cap [T_K(x) + x] \neq \emptyset$ for all $x \in K$.

(b) G is K-external if and only if $G(x) \cap [x - T_K(x)] \neq \emptyset$ for all $x \in K$.

Note that G is K-internal when K is a feasibility domain for $G - I$, and K-external when it is a feasibility domain for $I - G$. In such circumstances the proof of Theorem 3.11.5 still works, therefore without assuming that $G(K) \subset K$, we obtain the following corollary.

Corollary 3.11.8. *Let X be a Banach space, $K \subset X$ a convex and compact subset of X, and $G : X \rightrightarrows X$ a PC point-to-set mapping such that $G(x)$ is closed and convex for all $x \in X$, and either K-internal or K-external. Then, there exists $\bar{x} \in K$ such that $\bar{x} \in G(\bar{x})$.*

3.12   Fixed point theorems with coercivity

In this section we attempt to recover the main results of Section 3.11, but relaxing the hypothesis of compactness of K, and demanding instead some coercivity properties of the point-to-set mapping F. We assume in the remainder of this section that X is finite-dimensional, and thus we identify X with its dual X^*. A formal definition of such coercivity properties follows. We use γ_x as in Definition 3.9.2(a).

Definition 3.12.1. *Given a subset K of X, a point-to-set mapping $F : X \rightrightarrows X$ is said to be*

(a) Anticoercive in K if and only if $\limsup_{\|x\| \to \infty, x \in K} \gamma_x(x) < 0$

(b) Strongly anticoercive in K if and only if

$$\limsup_{\|x\| \to \infty, x \in K} \frac{\gamma_x(x)}{\|x\|} = -\infty$$

Example 3.12.2. Take $X = \mathbb{R}^2$, $K = \mathbb{R}_+^2 = \{x \in \mathbb{R}^2 : x_1 \geq 0,\ x_2 \geq 0\}$, and define $F : \mathbb{R}^2 \rightrightarrows \mathbb{R}^2$ as

$$F(x) = \begin{cases} \{\lambda x : \lambda \in [-2, -1]\} & \text{if } \|x\| \geq 1 \\ x & \text{if } \|x\| < 1. \end{cases}$$

If we take $x = (s, 0) \in \partial K$, with $s > 1$ we get that $T_K(x) = \{(a, t) : t \geq 0, a \in \mathbb{R}\}$, $F(x) = \{(\lambda s, 0) : \lambda \in [-2, -1]\}$, so that $T_K(x) \cap F(x) \neq \emptyset$. It can be easily verified that, for all x such that $\|x\| \geq 1$,

$$\gamma_x(y) = \begin{cases} -2\langle x, y \rangle & \text{if } \langle x, y \rangle < 0, \\ -\langle x, y \rangle & \text{if } \langle x, y \rangle > 0, \\ 0 & \text{if } \langle x, y \rangle = 0. \end{cases}$$

When $\|x\| < 1$ it holds that $\gamma_x(y) = \langle x, y \rangle$, and thus γ_x is Usc in $\mathbb{R}^n$ for all $x \in \mathbb{R}^n$. Also, $F(x)$ is closed for all $x \in K$, and F has both zeroes and fixed points in K, which is closed and convex, but not compact. The results in this section apply to point-to-set mappings like F in this example. We show next that F is strongly anticoercive in K. Take any real $M > 1$. Then

$$\sup_{\|x\| \geq M, x \geq 0} \sup_{y \in F(x)} \left\{ \frac{\langle x, y \rangle}{\|x\|} \right\} = \sup_{\|x\| \geq M, x \geq 0} \sup_{-2 \leq \lambda \leq -1} \{\lambda \|x\|\}$$

$$= \sup_{\|x\| \geq M, x \geq 0} \{-\|x\|\} = -M,$$

and therefore

$$\inf_{M > 0} \sup_{\|x\| \geq M, x \geq 0} \left\{ \frac{\sigma_{F(x)}(x)}{\|x\|} \right\} = -\infty.$$

Theorem 3.12.3. *Let $K \subset \mathbb{R}^n$ be a closed and convex set and $F : \mathbb{R}^n \rightrightarrows \mathbb{R}^n$ a* PC *point-to-set mapping such that $F(x)$ is convex and closed for all $x \in \mathbb{R}^n$ and K is a feasibility domain for F.*

(i) If F is anticoercive in K then there exists $\bar{x} \in K$ such that $0 \in F(\bar{x})$.

(ii) If F is strongly anticoercive in K then for all $y \in K$ there exists $\tilde{x} \in K$ such that $y \in \tilde{x} - F(\tilde{x})$.

Proof.

(i) Because F is anticoercive in K, there exist $\varepsilon > 0$, $M > 0$, such that

$$\sup_{\|x\| \geq M, x \in K} \sigma_{F(x)}(x) \leq -\varepsilon < 0. \tag{3.91}$$

Let $B(0, M) = \{z \in \mathbb{R}^n : \|z\| \leq M\}$. We claim now that $F(x) \subset T_{B(0,M)}(x)$ for all $x \in K \cap B(0, M)$. We proceed to prove the claim. Take $x \in K \cap B(0, M)$. For $x \in B(0, M)^o$, we have $T_{B(0,M)}(x) = \mathbb{R}^n$ by Proposition 3.6.14(a), and the result holds trivially. If $x \in \partial B(0, M)$, Proposition 3.6.19 implies that $T_{B(0,M)}(x) = \{v \in \mathbb{R}^n : \langle v, x \rangle \leq 0\}$; on the other hand we get from (3.91) and the fact that $\|x\| = M$ that $\langle y, x \rangle \leq -\varepsilon < 0$ for all $y \in F(x)$ and thus $F(x) \subset T_{B(0,M)}(x)$ also in this case, so that the claim is established. Take $\bar{M} \geq M$ such that $B(0, \bar{M})^\circ \cap K \neq \emptyset$. By Proposition 3.6.18, $T_{K \cap B(0,\bar{M})}(x) = T_K(x) \cap T_{B(0,\bar{M})}(x)$ and therefore,

$$F(x) \cap T_{K \cap B(0,\bar{M})} = \big(F(x) \cap T_{B(0,\bar{M})}(x)\big) \cap T_K(x) = F(x) \cap T_K(x) \neq \emptyset,$$

so that $K \cap B(0, \bar{M})$ is a domain of feasibility for F. Because finite-dimensionality implies that $K \cap B(\bar{M}, 0)$ is compact, we can apply Theorem 3.11.4 in order to conclude that there exists $\bar{x} \in K \cap B(0, \bar{M})$ such that $0 \in F(\bar{x})$.

(ii) Fix some $y \in K$ and define $G_y : \mathbb{R}^n \rightrightarrows \mathbb{R}^n$ as $G_y(x) = F(x) + y - x$. We claim that G_y is anticoercive in K. Indeed, assume, for the sake of contradiction, that

$$\theta = \limsup_{\|x\| \to \infty, x \in K} \sigma_{G_y(x)}(x) \geq 0.$$

Take a sequence $\{x_n\} \subset K$ such that $\lim_n \|x_n\| = \infty$, $\lim_n \sigma_{G_y(x_n)}(x_n) = \theta$. Because $\theta \geq 0$, for all $\varepsilon > 0$ there exists n_0 such that $\sigma_{G_y(x_n)}(x_n) > -\varepsilon$ for $n \geq n_0$. Take now $z_n \in G_y(x_n)$ such that $\langle z_n, x_n \rangle > \theta - \varepsilon$. If $u_n = z_n + x_n - y$, then we have that $u_n \in F(x_n)$ and

$$\frac{\langle z_n, x_n \rangle}{\|x_n\|} = \langle u_n, \|x_n\|^{-1} x_n \rangle + \langle y, \|x_n\|^{-1} x_n \rangle - \|x_n\| > \frac{\theta - \varepsilon}{\|x_n\|},$$

for all $n \geq n_0$. Hence

$$\langle u_n, \|x_n\|^{-1} x_n \rangle > \frac{\theta - \varepsilon}{\|x_n\|} + \|x_n\| - \langle y, \|x_n\|^{-1} x_n \rangle. \tag{3.92}$$

Taking limits in both sides of (3.92) as n goes to ∞, we conclude that

$$\lim_n \langle u_n, \|x_n\|^{-1} x_n \rangle = \infty. \tag{3.93}$$

On the other hand,

$$\frac{\langle u_n, x_n \rangle}{\|x_n\|} \leq \frac{\sigma_{F(x_n)}(x_n)}{\|x_n\|} \leq \limsup_{\|x\| \to \infty, x \in K} \frac{\sigma_{F(x)}(x)}{\|x\|} = -\infty,$$

contradicting (3.93); thus the claim holds and G_y is anticoercive in K.

We must verify now that K is a feasibility domain for G_y. We claim that $z + (y - x) \in G_y(x) \cap T_K(x)$ for all $z \in F(x) \cap T_K(x)$. Indeed, inasmuch as both z and $y - x$ belong to $T_K(x)$, which is a convex cone, it follows that $z + (y - x)$ belongs to $T_K(x)$, which is enough to establish the claim, in view of the definition of G_y. We are therefore within the hypotheses of item (i) of this theorem, and thus there exists $\bar{x} \in K$ such that $0 \in G_y(\bar{x}) = F(\bar{x}) + y - \bar{x}$, or equivalently, $y \in \bar{x} - F(\bar{x})$.
□

3.13 Duality for variational problems

In this section we develop some duality results in the framework of variational problems, in particular for the problem of finding zeroes of point-to-set operators with a special structure. We follow the approach presented in [12]. We deal specifically with the case in which the point-to-set operator T is written as the sum of two operators; that is, $T = A + B$. This partition is of interest when each term in the sum has a different nature. The prototypical example is the transformation of a constrained optimization problem into an unconstrained one. Take a Banach space X, a convex function $f : X \to \mathbb{R}$, and a convex subset $C \subset X$. The problem of interest is

$$\min f(x)$$

$$\text{s.t. } x \in C.$$

As we show in detail in Section 6.1, this problem is equivalent to the minimization of $\hat{f} := f + \delta_C$, where δ_C is the indicator function of the set C, introduced in Definition 3.1.3. It turns out that $\hat{f}$ is convex, and so its minimizers are precisely the zeroes of its subdifferential $\partial \hat{f}$. Assuming, for example, the constraint qualification (CQ) in Section 3.5.1, it holds that $\partial \hat{f} = \partial f + N_C$, where N_C is the normal cone of C as defined in (3.39). In this setting the term ∂f contains the information on the objective function, whereas N_C encapsulates the information on the feasible set. These two operators enjoy different properties and in many optimization algorithms they are used in a sequential way. For instance, in the case of the projected gradient method (see, e.g., Section 2.3 in [29]) alternate steps are taken first in the direction of a subgradient of f at the current iterate, and then the resulting point is orthogonally projected onto C: the first step ignores the information on C and the second one does not consider at all the objective function f. We are interested thus in developing a duality theory for the following problem, to be considered as the *primal problem*, denoted (P): given vector spaces X and Y, point-to-set mappings $A, B : X \rightrightarrows Y$, and a point $y_0 \in Y$, find $x \in X$ such that

$$y_0 \in Ax + Bx. \tag{3.94}$$

We associate with this problem a so-called *dual problem*, denoted (D), consisting of finding $y \in Y$ such that

$$0 \in A^{-1}y - B^{-1}(y_0 - y). \tag{3.95}$$

The connection between the primal and the dual problem is established in the following result.

Theorem 3.13.1.

(i) *If $x \in X$ solves (P), then there exists $y \in Ax$ which solves (D).*

(ii) *If $y \in Y$ solves (D), then there exists $x \in A^{-1}y$ which solves (P).*

(iii) *The inclusion in (3.95) is equivalent to*

$$y_0 \in y + B(A^{-1}y). \tag{3.96}$$

Proof. Note that x solves (P) if and only if there exists $y \in Ax$ such that $y_0 - y$ belongs to Bx, which happens if and only if x belongs to $A^{-1}y \cap B^{-1}(y_0 - y)$; that is, precisely when y solves (D). We have proved (i) and (ii). The proof of (iii) is immediate. $\square$

We state for future reference the case of "homogeneous" inclusions; that is, (3.94)–(3.96) with $y_0 = 0$.

$$0 \in Ax + Bx \iff \exists y \in Ax \text{ such that } 0 \in A^{-1}y - B^{-1}(-y)$$
$$\iff 0 \in y + B(A^{-1}y). \tag{3.97}$$

Next we rewrite Problem (P) in a geometric way; that is, in terms of the graphs of the operators A and B. We need some notation. Given $A : X \rightrightarrows Y$, we define $A^- : X \rightrightarrows Y$ as $A^-x = A(-x)$.

Lemma 3.13.2. *Problem (P) has solutions if and only if the pair $(0, y_0)$ belongs to $\mathrm{Gph}(A) + \mathrm{Gph}(B^-)$.*

Proof. $(0, y_0) \in \mathrm{Gph}(A) + \mathrm{Gph}(B^-)$ if and only if there exists $y \in Ax$ and $z \in B^-(-x)$ such that $(0, y_0) = (x, y) + (-x, z)$; that is, if and only if $y_0 \in Ax + Bx$ (note that $B^-(-x) = Bx$). $\square$

We recall next an elementary property of point-to-set operators. Given $U \in X \times Y$, we denote U^{-1} the subset of $Y \times X$ defined as $(y, x) \in U^{-1}$ if and only if $(x, y) \in U$. Note that $\mathrm{Gph}(A^{-1}) = [\mathrm{Gph}(A)]^{-1}$ for any operator $A : X \rightrightarrows Y$.

Proposition 3.13.3. $[\mathrm{Gph}(S) + \mathrm{Gph}(T)]^{-1} = [\mathrm{Gph}(S)]^{-1} + [\mathrm{Gph}(T)]^{-1}$ *for all* $S, T : X \rightrightarrows Y$.

Proof.
$$(x, y) \in \mathrm{Gph}(S) + \mathrm{Gph}(T) \iff (x, y)$$
$$= (x_S, y_S) + (x_T, y_T) = (x_S + x_T, y_S + y_T),$$
with $(x_S, y_S) \in \mathrm{Gph}(S)$, and $(x_T, y_T) \in \mathrm{Gph}(T)$. This is equivalent to
$$(y, x) = (y_S + y_T, x_S + x_T) = (y_S, x_S) + (y_T, x_T) \in [\mathrm{Gph}(S)]^{-1} + [\mathrm{Gph}(T)]^{-1},$$
establishing the result. $\square$

We summarize the previous results in the following corollary.

Corollary 3.13.4. *Given $A, B : X \rightrightarrows Y$ and $y_0 \in Y$, the following statements are equivalent.*

(i) $x \in X$ is a solution of (P).

(ii) $(0, y_0) \in \mathrm{Gph}(A) + \mathrm{Gph}(B^-)$, with the decomposition $(0, y_0) = (x, y) + (-x, z)$, $y \in Ax$, $z \in B^-x$.

(iii) $(y_0, 0) \in \mathrm{Gph}(A^{-1}) + \mathrm{Gph}((B^-)^{-1})$ *with the decomposition* $(y_0, 0) = (y, x) + (z, -x), x \in A^{-1}y, -x \in -B^{-1}(z).$

(iv) $0 \in A^{-1}y - B^{-1}z$ *with* $y + z = y_0.$

(v) There exists $y \in Ax$ *which solves (D).*

Proof. The results follow immediately from Theorem 3.13.1 and Lemma 3.13.2.
□

The results above might seem to be of a purely formal nature, with little substance. When applied to more structured frameworks, however, they give rise to more solid constructions. Next we give a few examples.

3.13.1   Application to convex optimization

The first example deals with the Fenchel–Rockafellar duality for convex optimization. Let X be a reflexive Banach space with dual X^*, and take $f, g : X \to \overline{\mathbb{R}}$ convex, proper, and lsc. We apply the results above with $A = \partial f$, $B = \partial g$, in which case problem (P) becomes

$$0 \in \partial f(x) + \partial g(x) \tag{3.98}$$

and (D) becomes

$$0 \in \partial g^*(y) - \partial f^*(-y), \tag{3.99}$$

where f^* and g^* denote the convex conjugates of f and g, respectively, defined in Definition 3.4.5. Corollary 3.5.5 yields the equivalence between (D) and (3.98).

The connection between these two formulations of (P) and (D) is stated in the following corollary.

Corollary 3.13.5. *If x solves (3.98) then there exists $y \in \partial g(x)$ which solves (3.99). Conversely, if y solves (3.99) then there exists $x \in \partial g^*(y)$ which solves (3.98).*

Proof. It follows directly from Theorem 3.13.1.   □

We mention that no constraint qualification is needed for the duality result in Corollary 3.13.5. If we want to move from the (P)–(D) setting to convex optimization, we do need a constraint qualification, for instance, the one called (CQ) in Section 3.5.1, under which we get the following stronger result.

Corollary 3.13.6. *Let X be a reflexive Banach space, and $f, g : X \to \overline{\mathbb{R}}$ convex, proper, and lsc. If (CQ) holds and $\bar{x}$ is a minimizer of $f + g$ then there exists $\bar{y} \in \partial f(\bar{x})$ such that $0 \in \partial f^*(\bar{y}) - \partial g^*(-\bar{y}).$*

Proof. Because $\bar{x}$ minimizes $f + g$, we have that $0 \in \partial(f + g)(\bar{x})$. Under (CQ), we get from Theorem 3.5.7 that $\partial(f + g)(\bar{x}) = \partial f(\bar{x}) + \partial g(\bar{x})$, so that $\bar{x}$ solves (3.98), and the result follows from Corollary 3.13.5.   □

3.13.2   Application to the Clarke–Ekeland least action principle

Next we discuss an application to Hamiltonian systems with convex Hamiltonian. We follow the approach in [39].

We recall next the definitions of adjoint and self-adjoint operators.

Definition 3.13.7. *Let X be a reflexive Banach space and $L : X \to X^*$ a linear operator.*

  (i) *The* adjoint *of L is the linear operator $L^* : X \to X^*$ defined by $\langle Lx, x' \rangle = \langle x, L^*x' \rangle$ for every $x, x' \in X$.*

 (ii) *L is said to be* self-adjoint *if $L(x) = L^*(x)$ for all $x \in X$.*

Let X be a Hilbert space, $V \subset X$ a linear subspace, $L : V \to X$ a self-adjoint linear operator (possibly unbounded, in which case V is not closed), and $F : X \to \overline{\mathbb{R}}$ a convex, proper, and lsc function. Consider the problem

$$0 \in Lx + \partial F(x), \tag{3.100}$$

and the functional $\Phi : X \to \mathbb{R} \cup \{\infty, -\infty\}$, defined as

$$\Phi(x) = \frac{1}{2}\langle Lx, x \rangle + F(x).$$

Note that if either F is differentiable or L is bounded, then the solutions of (3.100) are precisely the stationary points for Φ (in the latter case, if additionally L is positive semidefinite, as defined in Section 5.4, then Φ is convex and such stationary points are indeed minimizers). However, in this setting, we assume neither differentiability of F nor boundedness of L, so that other classical approaches fail to connect (3.100) with optimization of Φ. As is well known, differential operators in $\mathcal{L}^2$ are typically unbounded, and in the examples of interest in Hamiltonian systems, L is of this type. It is in this rather delicate circumstance that the duality principle resulting from Theorem 3.13.1 becomes significant. Without further assumptions, application of Theorem 3.13.1 with $A = \partial F$, $B = L$, and $y_0 = 0$, yields

$$0 \in (\partial F)^{-1}(y) - L^{-1}(-y) = (\partial F^*)(y) + L^{-1}(y) \tag{3.101}$$

for some $y \in \partial F(x)$, if x solves (3.100). Note that $L^{-1}(-y) = -L^{-1}(y)$ by linearity of L, although L^{-1} is in general point-to-set, because $\mathrm{Ker}(L)$ might be a nontrivial subspace, even in the positive semidefinite case. We now make two additional assumptions.

  (i) $R(L)$ is a closed subspace of X.

 (ii) $\mathrm{Dom}(F^*) = X$ (this implies that F^* is continuous, because it is convex).

In view of (i), we can write $X = R(L) \oplus \mathrm{Ker}(L)$, so that the restriction of L to $R(L)$, which we denote $\hat{L}$, is one to one. It follows that $\hat{L}^{-1} : R(L) \to R(L)$ is well defined and has a closed domain, by the assumption of closedness of $R(L)$. Note that $\hat{L}$ is self-adjoint, and hence $\hat{L}^{-1}$ is self-adjoint. It is well-known and easy to prove that the graph of a self-adjoint linear operator with closed domain is closed. Invoking now the closed graph theorem (i.e., Corollary 2.8.9), we conclude that $\hat{L}^{-1}$ is bounded, hence continuous. The definition of $\hat{L}$ yields

$$L^{-1}y = \hat{L}^{-1}y + \mathrm{Ker}(L). \tag{3.102}$$

In view of (3.102), we conclude that (3.101) is equivalent to finding $y \in R(L)$ and $v \in \partial F^*(y)$ such that

$$\hat{L}^{-1}y + v \in \mathrm{Ker}(L). \tag{3.103}$$

We claim that the solutions of (3.103) are precisely the critical points for the problem:

$$\min \Psi(y) := \frac{1}{2}\langle \hat{L}^{-1}y, y \rangle + F^*(y) \tag{3.104}$$

$$\text{s.t. } y \in R(L). \tag{3.105}$$

It is sufficient to verify that

$$\partial(F^* + \delta_{R(L)}) = \partial F^* + N_{R(L)} = \partial F^* + \mathrm{Ker}(L),$$

which follows from Theorem 3.5.7 because (CQ) holds, by assumption (ii).

We summarize the discussion above in the following result.

Theorem 3.13.8. *Under assumptions (i) and (ii) above, if $\bar{y}$ is a critical point for problem (3.104)–(3.105), then there exists $z \in \mathrm{Ker}(L)$ such that $\bar{x} = \hat{L}^{-1}(-\bar{y}) + z$ is a critical point of Φ.*

We describe next an instance of finite-dimensional Hamiltonian systems, for which problem (3.103) is easier than the original problem (3.100).

A basic problem of celestial mechanics consists of studying the periodic solutions of the following Hamiltonian system (see [67]),

$$J\dot{u} + \nabla H(t, u(t)) = 0,$$

where $u = (p, q) \in \mathbb{R}^n \times \mathbb{R}^n$, H is the classical Hamiltonian, and J is the simplectic matrix, given by

$$J = \begin{bmatrix} 0 & I \\ -I & 0 \end{bmatrix},$$

with $I \in \mathbb{R}^{n \times n}$ being the identity matrix. Equivalently, denoting $\nabla H = (H_p, H_q)$,

$$-\dot{p} + H_q = 0, \tag{3.106}$$

$$\dot{q} + H_p = 0. \tag{3.107}$$

We rewrite the Hamiltonian system in our notation. We take $X = \mathcal{L}^2(0, t^*) \times \mathcal{L}^2(0, t^*)$ and $L(u) = J\dot{u}$. In such a case,

$$D(L) = \{u \in X \mid \dot{u} \in X, u(0) = u(t^*)\},$$

$$R(L) = \left\{ v \in X \mid \int_0^{t^*} v(t)dt = 0 \right\},$$

$$F(u) = \int_0^{t^*} H(t, u(t))dt,$$

so that

$$\nabla F(u(t)) = \nabla H(t, u(t)).$$

Observe that with this notation, the system (3.106)–(3.107) becomes precisely $0 \in L(u) + \nabla F(u)$; that is, problem (3.100). The Hamiltonian action is given by

$$\Phi(u) = \int_0^{t^*} \left[\frac{1}{2}\langle J\dot{u}, u \rangle + H(t, u(t)) \right] dt.$$

We describe next the way in which our abstract duality principle leads to the well-known Clarke–Ekeland least action principle (see [67]). First we must check the assumptions required in our framework. Observe that L is self-adjoint, in the sense of Definition 3.13.7(ii) (it follows easily, through integration by parts, from the fact that $u(t^*) - u(0) = v(t^*) - v(0) = 0$). Note also that $R(L)$ is closed, because it consists of the elements in X with zero mean value. Next we identify the dual functional Ψ associated with this problem, as defined in (3.104). An elementary computation gives

$$\left[\hat{L}^{-1}(v) \right](t) = - \int_0^t Jv(s)ds$$

for all $v \in R(L)$. Hence

$$\Psi(v) = \int_0^{t^*} \left[-\frac{1}{2} \left(\int_0^t Jv(s)ds \right) v(t) + H^*(t, v(t)) \right] dt.$$

Thus, the dual problem looks for critical points of Ψ in the subspace of zero mean value functions in X; that is, derivatives of periodic functions. In other words, we must find the critical points of the functional

$$\Psi(\dot{w}) = \int_0^{t^*} \left[-\frac{1}{2} \left(\int_0^t J\dot{w}(s)ds \right) \dot{w}(t) + H^*(t, \dot{w}(t)) \right] dt,$$

in the subspace of periodic functions w with period t^*. An elementary transformation yields the Clarke–Ekeland functional $\hat{\Psi}$ given by

$$\hat{\Psi}(v) = \int_0^{t^*} \left[\frac{1}{2}\langle J\dot{v}, v(t) \rangle + H^*(t, \dot{v}(t)) \right] dt.$$

3.13.3 Singer–Toland duality

Lately, many of the results in convex optimization have been extended to larger classes of functions. One of them is the *DC*-family, consisting of functions that are the difference of two convex ones (see e.g., [19]). We mention that any twice-differentiable function $f : \mathbb{R}^n \to \mathbb{R}$ whose Hessian matrix is uniformly bounded belongs to this class, because the function $g(x) = f(x) + \alpha \|x\|^2$ is convex for large enough α. We consider now functions defined on a reflexive Banach space X. According to Toland [218], x is a *critical point* of $f - g$ if

$$0 \in \partial f(x) - \partial g(x). \tag{3.108}$$

We apply Theorem 3.13.1 (or more precisely, its homogeneous version given by (3.97)), to (3.108), with $A = \partial f$, and $B = -\partial g$. Note that $y \in B^{-1}x$ if and only if $x \in -\partial g(y)$; that is, $-x \in \partial g(y)$, in which case $y \in (\partial g)^{-1}(-x) = \partial g^*(-x)$. Thus, we conclude from Theorem 3.13.1 that the existence of $x \in X$ which satisfies (3.108) is equivalent to the existence of $u \in X^*$ such that

$$0 \in \partial g^*(u) - \partial f^*(u); \tag{3.109}$$

that is, u is a critical point of $g^* - f^*$.

The relevance of this duality approach becomes clear in the following example, presented in [217], taken from the theory of plasma equilibrium. Consider a bounded open set $\Omega \subset \mathbb{R}^n$ with boundary Γ. The problem of interest, which is our primal problem (P), is the following differential equation,

$$-\Delta u = \lambda u^+, \tag{3.110}$$

with boundary conditions $u(x) = \text{constant}$ for all $x \in \Gamma$ and $\int_\Gamma (\partial u / \partial n)(s) ds = \eta$, where $\lambda > 0$ and η are given real parameters, and the constant value of u on Γ is unknown. In (3.110), Δ denotes the Laplacian operator; that is, $\Delta u = \sum_{i=1}^n \partial^2 u / \partial x_i^2$, and $u^+(x) = \max\{u(x), 0\}$.

The space of interest, say X, consists of the differentiable functions u defined on Ω, constant on Γ, and such that both u and its first partial derivatives belong to $\mathcal{L}^2(\Omega)$. X is a Hilbert space, because it can be identified with $H_0^1(\Omega) \oplus \mathbb{R}$, associating with each $u \in X$ the pair $(\tilde{u}, \theta)$, where $\theta = u(x)$ for any $x \in \Gamma$ and $\tilde{u}(x) = u(x) - \theta$.

It has been proved in [217] that u solves (P) if and only if it is a critical point of the functional $E : X \to \mathbb{R}$ defined as

$$E(u) = \frac{1}{2} \|\nabla u\|^2 - \frac{\lambda}{2} |u^+|^2 + \eta u(\Gamma)$$

$$= \frac{1}{2} \sum_{i=1}^n \int_\Omega \left| \frac{\partial u}{\partial x_i}(x) \right|^2 dx - \frac{\lambda}{2} \int_\Omega \max\{u(x), 0\}^2 dx + \eta u(\Gamma),$$

where $u(\Gamma)$ denotes, possibly with a slight abuse of notation, the constant value of u on the set Γ. In general E is not convex, but it is indeed DC-convex, because it can be written as $E = G - F$, with

$$G(u) = \frac{1}{2} \|\nabla u\|^2 + \frac{1}{2} |u(\Gamma)|^2,$$

and
$$F(u) = \frac{\lambda}{2}\left|u^+\right|^2 + \frac{1}{2}\left|u(\Gamma)\right|^2 - \eta u(\Gamma).$$

It can be easily verified that $G = G^*$ because G is the square of the norm in X, with the identification above (see Example 3.4.7). The dual problem becomes rather simple, because both F and G are continuously differentiable on X. In fact, F^* can indeed be computed, in which case (3.109) becomes the Berestycki–Brezis variational formulation of problem (P), consisting of finding the critical points of a certain DC-function (see [25]).

3.13.4 Application to normal mappings

Let $C \subset \mathbb{R}^n$ be a closed and convex set, $P_C : \mathbb{R}^n \to C$ the orthogonal projection onto C, and $T : \mathbb{R}^n \to \mathbb{R}^n$ a continuous mapping. We are interested in the problem of finding $x \in \mathbb{R}^n$ such that

$$T(P_C(x)) + (x - P_C(x)) = 0. \tag{3.111}$$

This problem appears frequently in optimization and equilibrium analysis. As we show, it is indeed equivalent to the variational inequality problem $\mathrm{VIP}(T, C)$, as defined in (5.50), and we establish this equivalence with the help of the duality scheme introduced in this section.

Consider the normal cone N_C of C (see (3.39)). It is easy to check that

$$(I + N_C)^{-1} = P_C. \tag{3.112}$$

It can be verified, using (3.112), that application of Theorem 3.13.1 to (3.111) leads to

$$0 \in T(x) + N_C(x), \tag{3.113}$$

which is precisely $\mathrm{VIP}(T, C)$. An in-depth study of problems of the kind (3.111) can be found in [180]. Here we establish the equivalence between (3.111) and (3.113) for a general variational inequality problem in a Hilbert space X. Now, $C \subset X$ is closed and convex, and $T : X \rightrightarrows X$ is point-to-set. In this setting, $\mathrm{VIP}(T, C)$ consists of finding $\bar{x} \in C$ and $\bar{u} \in T(\bar{x})$ such that $\langle \bar{u}, x - \bar{x} \rangle \geq 0$ for all $x \in C$, which is known to be equivalent to (3.113) (see Section 5.5). Clearly, (3.113) is equivalent to

$$0 \in -x + T(x) + x + N_C(x).$$

Take $A = -I + T$, $B = I + N_C$, where I is the identity operator in X. It follows that (3.113) reduces to finding $x \in X$ such that

$$0 \in A(x) + B(x).$$

In view of (3.97), this problem is equivalent to finding $y \in B(x)$ such that

$$0 \in y + A\left(B^{-1}(y)\right). \tag{3.114}$$

It can be easily verified that $B^{-1} = P_C$, so that (3.114) becomes

$$0 \in (x - P_C(x)) + T(P_C(x)).$$

Other applications of this duality scheme, developed in Section 4.5 of [12], include the study of constraint qualifications for guaranteeing the maximality of the sum of two maximal monotone operators, in the spirit of the results in Section 4.8.

3.13.5   Composition duality

In this section we consider a more general framework, introduced by Robinson in [182], with four vector spaces X, Y, U, and V, and four point-to-set mappings $A : X \rightrightarrows Y$, $P : X \rightrightarrows U$, $B : U \rightrightarrows B$, and $Q : V \rightrightarrows Y$. A similar analysis was performed in [168]. We are interested in the following primal problem (P): find $x \in X$ such that

$$0 \in Ax + QBPx, \tag{3.115}$$

where we use a "multiplicative" notation for compositions; that is, $QBP = Q \circ B \circ P$. We associate a dual problem (D) to (P), namely, finding $v \in V$ such that

$$0 \in (-P)A^{-1}(-Q)(v) + B^{-1}(v). \tag{3.116}$$

Observe that when $X = U$, $Y = B$, and P, Q are the respective identity maps, problems (P) and (D) reduce precisely to those given by (3.94) and (3.95).

The basic duality result is the following.

Theorem 3.13.9. *The problem* (3.115) *has solutions if and only if the problem* (3.116) *has solutions.*

Proof. If (3.115) is solvable, then there exist $x \in X$ and $y \in A(x) \cap (-QBP)(x)$. In other words, there exist $v \in V$ and $u \in U$ such that $-y \in Q(v)$, $v \in B(u)$, and $u \in P(x)$. It follows that $u \in B^{-1}(v)$ and $-u \in (-P)A^{-1}(-Q)(v)$, in which case v solves (3.116). In view of the symmetries in both problems, a similar argument can be used for proving the converse statement. □

We give next an example for which Theorem 3.13.9 entails some interesting computational consequences. The problem of interest, arising in economic equilibrium problems, is the following. Find $p \in \mathbb{R}^k$, $y \in \mathbb{R}^m$ such that

$$0 \in \begin{pmatrix} -f(p) \\ a \end{pmatrix} + \begin{bmatrix} 0 & M \\ -M^t & 0 \end{bmatrix} \begin{pmatrix} p \\ y \end{pmatrix} + N_P(p) \times N_Y(y), \tag{3.117}$$

where $f : \mathbb{R}^k \to \mathbb{R}^k$ is a (possibly highly) nonlinear function, $a \in \mathbb{R}^n$, $M \in \mathbb{R}^{k \times m}$ has transpose M^t, $P \subset \mathbb{R}^k$ is a polyhedron, and $Y \subset \mathbb{R}^m$ is a polyhedral cone. This variational inequality models an economic equilibrium involving quantities given by the vector y, and prices represented by p. The y variables are linear, but the p

variables are not. In general m might be much larger than k, so that it is preferable not to work in the space $\mathbb{R}^{k+m}$.

Consider the linear transformation $S : \mathbb{R}^{k+m} \to \mathbb{R}^k$ defined as $S(x, y) = x$. Note that the adjoint operator S^* (cf. Definition 3.13.7), is given by $S^*(x) = (x, 0)$. We can rewrite (3.117) as

$$0 \in S^*(-f)S(p, y) + A(p, y), \tag{3.118}$$

where $A : \mathbb{R}^{k+m} \rightrightarrows \mathbb{R}^{k+m}$ is given by

$$A(p, y) := \begin{pmatrix} 0 \\ a \end{pmatrix} + \begin{bmatrix} 0 & M \\ -M^t & 0 \end{bmatrix} \begin{pmatrix} p \\ y \end{pmatrix} + N_P(p) \times N_Y(y), \tag{3.119}$$

Observe that (3.118) has the format of (3.115) with $X = Y = \mathbb{R}^{k+m}$, $U = V = \mathbb{R}^k$, $P = S$, $B = -f$, and $Q = S^*$. We mention, parenthetically, that A is a maximal monotone operator (see Chapter 4) which is not, in general, the subdifferential of a convex function, but this is not essential for our current discussion.

Theorem 3.13.9 implies that (3.118), seen as a primal problem, is equivalent to

$$0 \in (-f^{-1})(q) + (-S)A^{-1}(-S^*)(q). \tag{3.120}$$

Now we apply Theorem 3.13.9 to (3.120), obtaining

$$f(p) \in (S^*)^{-1}AS^{-1}(p),$$

where S^{-1} is the inverse of S as a point-to-set operator, not in the sense of linear transformations. It is not difficult to check that

$$(S^*)^{-1}AS^{-1} = N_{P \cap Z},$$

where

$$Z = \{z \in \mathbb{R}^k : M^t z - a \in Y^*\},$$

and Y^* is the positive polar cone of Y, as introduced in Definition 3.7.7. The final form of the dual problem is then

$$0 \in (-f)(p) + N_{P \cap Z}(p),$$

in the space $\mathbb{R}^k$. When m is much larger than k, the dual problem is in principle more manageable than the primal. Additionally, it can be seen that some simple auxiliary computation allows us to get a primal solution starting from a dual one. This scheme, implemented in [181] for a problem with $k = 14$ and $m = 26$ analyzed by Scarf in [198], reduced the computational time by 98%.

Exercises

3.1. Prove the statements in Example 3.12.2. For F as in this example, prove that $K = [-1, 1]$ is a feasibility domain for F, and that F has a zero and a fixed point on K.

3.2. Prove that a self-adjoint operator with a closed domain has a closed graph.

3.3. Prove item (iii) of Theorem 3.13.1.

3.4. Let X be a Hilbert space. Prove that if $C \subset X$ is a closed and convex set, $N_C : X \rightrightarrows X$ is the normality operator associated with C, and $P_C : X \to C$ is the orthogonal projection onto C, then $(I + N_C)^{-1} = P_C$.

3.14 Historical notes

The concept of lower-semicontinuous function, well known in classical analysis, is central in the context of optimization of extended-valued functions. The observation that this is the relevant property to be used in this context is due independently to Moreau [157, 158, 159] and Rockafellar [183]. Both of them, in turn, were inspired by the unpublished notes of Fenchel [85]. Moreau and Rockafellar observed that the notion of "closedness" used by Fenchel could be reduced to lower-semicontinuity of extended-valued functions.

Ekeland's variational principle was announced for the first time in [78] and [79], and published in full form in [80].

Caristi's fixed point theorem in a point-to-point framework appeared for the first time in [58]; its point-to-set version (i.e., Theorem 3.3.1) is due to Siegel [201].

The study of convex sets and functions started with Minkowski in the early twentieth century [148]. The subject became a cornerstone in analysis, and its history up to 1933 can be found in the book by Bonnesen and Fenchel [31]. The use of convexity in the theory of inequalities up to 1948 is described in [23]. This theory grew up to the point that it became an important branch in modern mathematics, with the name of "convex analysis". The basic modern references for convex analysis in finite-dimensional spaces are the comprehensive and authoritative books by Rockafellar [187] and Hiriart-Urruty and Lemaréchal [94].

The infinite-dimensional case is treated in the books by Ekeland and Teman [81], van Tiel [223], and Giles [87]. The lecture notes by Moreau [161] are an early source for the study of convex functions in infinite-dimensional spaces.

Theorem 3.4.22 is due to Roberts and Varberg [178] and its proof was taken from Clarke's book [66].

The "conjugate", or Legendre transform of a function, is a widely used concept in the calculus of variations (it is indeed the basic tool for generating Hamiltonian systems). A study limited to the positive and single-variable case appeared in 1912 in [232] and was extended in 1939 in [139]. The concept in its full generality was introduced by Fenchel in [84] and [85]. The central Theorem 3.4.18 was proved originally by Fenchel in the finite-dimensional case; its modern formulation in infinite-dimensional spaces can be found in [160] and in Brøndsted's dissertation, written under Fenchel's supervision (see [40]).

The notion of subdifferential, as presented in this book, appeared for the first time in Rockafellar's dissertation [183], where the basic calculus rules with subgradients were proved. Independently and simultaneously, Moreau considered individual subgradients (but not the subdifferential as a set) in [158]. The results relating subdifferentials of convex functions and conjugacy can be traced back to

[31], and appeared with a full development in [84].

The concepts of derivatives and coderivatives of point-to-set mappings were introduced independently in [154, 14] and [167]. Further developments were made in [17, 18, 15] and [155].

Notions of tangency related to the tangent cone as presented here were first developed by Bouligand in the 1930s [34, 35]. The use of the tangent cone in optimization goes back to [91]. The normal cone was introduced essentially by Minkowski in [148], and reappeared in [85]. Both cones attained their current central role in convex analysis in Rockafellar's book [187].

Ky Fan's inequality was proved in [83] (see also [82]). Our proof follows [17] and [13], where the equilibrium theorem is derived from Ky Fan's results. This inequality was the object of many further developments (see e.g., [203] and [106]).

The basic fixed point theorem for continuous operators on closed and convex subsets of $\mathbb{R}^n$ was proved by Brouwer [42]. Another proof was published later on in [122]: Knaster established the connection between fixed points and Sperner's lemma, which allowed Mazurkiewicz to provide a proof of the fixed point result, which was in turn completed by Kuratowski. The result was extended to Banach spaces by Schauder in 1930 [199].

In the 1940s the need for a fixed point theory for set-valued maps became clear, as was indeed stated by von Neumann in connection with the proof of general equilibrium theorems in economics. The result was achieved by Kakutani [112] and appears as our Theorem 3.11.5. Our treatment of the fixed point theorem under coercivity assumptions is derived from [18].

The concept of duality for variational problems can be traced back to Mosco [162]. Our treatment of the subject is based upon [12] and [182].

Chapter 4

Maximal Monotone Operators

4.1 Definition and examples

Maximal monotone operators were first introduced in [149] and [234], and can be seen as a two-way generalization: a nonlinear generalization of linear endomorphisms with positive semidefinite matrices, and a multidimensional generalization of nondecreasing functions of a real variable; that is, of derivatives of convex and differentiable functions. Thus, not surprisingly, the main example of this kind of operators in a Banach space is the Fréchet derivative of a smooth convex function, or, in the point-to-set realm, the subdifferential of an arbitrary lower-semicontinuous convex function, in the sense of Definition 3.5.1. As mentioned in Example 1.2.5, monotone operators are the key ingredient of monotone variational inequalities, which extend to the realm of point-to-set mappings the constrained convex minimization problem. In order to study and develop new methods for solving variational inequalities, it is therefore essential to understand the properties of this kind of point-to-set mapping. We are particularly interested in their inner and outer continuity properties. Such properties (or rather, the lack thereof), are one of the main motivations behind the introduction of the enlargements of these operators, dealt with in Chapter 5.

Definition 4.1.1. *Let X be a Banach space and $T : X \rightrightarrows X^*$ a point-to-set mapping.*

(a) *T is* monotone *if and only if $\langle u - v, x - y \rangle \geq 0$ for all $x, y \in X$, all $u \in Tx$, and all $v \in Ty$.*

(b) *T is* maximal monotone *if it is monotone and in addition $T = T'$ for all monotone $T' : X \rightrightarrows X^*$ such that $\mathrm{Gph}(T) \subset \mathrm{Gph}(T')$.*

(c) *A set $M \subset X \times X^*$ is* monotone *if and only if $\langle u - v, x - y \rangle \geq 0$ for all $(x, u), (y, v) \in M$.*

(d) *A monotone set $M \subset X \times X^*$ is* maximal monotone *if and only if $M' = M$*

for every monotone set M' such that $M \subset M'$.

Definition 4.1.2. *Let X be a Banach space and $T : X \rightrightarrows X^*$ a point-to-set mapping. T is* strictly monotone *if and only if $\langle u - v, x - y \rangle > 0$ for all $x, y \in X$ such that $x \neq y$, all $u \in Tx$, and all $v \in Ty$.*

As a consequence of the above definitions, T is maximal monotone if and only if its graph is a maximal monotone set. Thanks to the next result, we can deal mostly with maximal monotone operators, rather than just monotone ones.

Proposition 4.1.3. *If X is a Banach space and $T : X \rightrightarrows X^*$ a monotone point-to-set mapping, then there exists $\bar{T} : X \rightrightarrows X^*$ maximal monotone (not necessarily unique), such that $\mathrm{Gph}(T) \subset \mathrm{Gph}(\bar{T})$.*

Proof. The result follows from a Zorn's lemma argument. Call $M_0 := \mathrm{Gph}(T) \subset X \times X^*$. By hypothesis, the set M_0 is monotone. Consider the set $\mathcal{M}(M_0) := \{L \subset X \times X^* \mid L \supset M_0 \text{ and } L \text{ is monotone}\}$. Because $M_0 \in \mathcal{M}(M_0)$, $\mathcal{M}(M_0)$ is not empty. Take a family $\{L_i\}_{i \in I} \subset \mathcal{M}(M_0)$ such that for all $i, j \in I$ we have $L_i \subset L_j$ or $L_j \subset L_i$. Then $L_0 := \cup_{i \in I} L_i \in \mathcal{M}(M_0)$. Indeed, for every two elements $(x, v), (y, u) \in L_0$, there exist $i, j \in I$ such that $(x, v) \in L_i$ and $(y, u) \in L_j$. The family is totally ordered, thus we can assume without loss of generality that $L_j \subset L_i$. Therefore both elements $(x, v), (y, u) \in L_i$, and hence monotonicity of L_i implies monotonicity of L_0. The fact that $L_0 \supset M_0$ is trivial. By Zorn's lemma, there exists a maximal element in $\mathcal{M}(M_0)$. This maximal element must correspond to the graph of a maximal monotone operator.　□

We list important examples of maximal monotone operators.

Example 4.1.4. *If $X = \mathbb{R}$, then T is maximal monotone if and only if its graph is a closed curve $\gamma \subset \mathbb{R}^2$ satisfying all the conditions below.*

(i) *γ is totally ordered with respect to the order induced by $\mathbb{R}^2_+$; that is, for all $(t_1, s_1), (t_2, s_2) \in \gamma$, either $t_1 \leq t_2$ and $s_1 \leq s_2$ (which is denoted as $(t_1, s_1) \leq (t_2, s_2)$) or $t_1 \geq t_2$ and $s_1 \geq s_2$ (which is denoted $(t_1, s_1) \geq (t_2, s_2)$).*

(ii) *γ is complete with respect to this order; that is, if $(t_1, s_1), (t_2, s_2) \in \gamma$ and $(t_1, s_1) \leq (t_2, s_2)$, then every (t, s) such that $(t_1, s_1) \leq (t, s) \leq (t_2, s_2)$ verifies $(t, s) \in \gamma$.*

(iii) *γ is unbounded with respect to this order; that is, there does not exist $(s, t) \in \mathbb{R}^2$ that is comparable with all elements of γ and is also an upper or lower bound of γ. In other words, there is no $(s, t) \in \mathbb{R}^2$ that is comparable with all elements of γ and such that $(s, t) \geq (\tilde{s}, \tilde{t})$ for all $(\tilde{s}, \tilde{t}) \in \gamma$ or such that $(s, t) \leq (\tilde{s}, \tilde{t})$ for all $(\tilde{s}, \tilde{t}) \in \gamma$.*

An example of such a curve γ is depicted in Figure 4.1. These curves can be seen as the result of "filling up" the discontinuity jumps in the graph of a nondecreasing function.

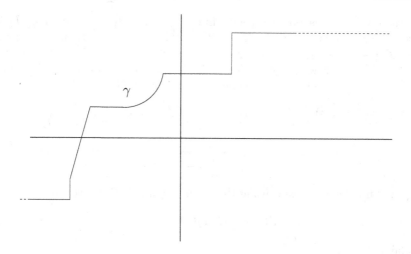

Figure 4.1. *One-dimensional maximal monotone operator*

Example 4.1.5. *A linear mapping $T : X \to X^*$ is maximal monotone if and only if T is* positive semidefinite; *that is, $\langle Tx, x \rangle \geq 0$ for all $x \in X$.*

We recall that a mapping $T : X \to X^*$ is (linearly) Gâteaux differentiable at $x \in X$ if for all $h \in X$ the limit

$$\delta T(x)h := \lim_{t \to 0} \frac{T(x + th) - T(x)}{t}$$

exists and is a linear and continuous function of h. In this situation, we say that T is Gâteaux differentiable at x and the linear mapping $\delta T(x) : X \to X^*$ is the Gâteaux derivative at x. We say that T is Gâteaux differentiable when it is Gâteaux differentiable at every $x \in X$.

Note that when $T : X \to X^*$ is linear, it is Gâteaux differentiable and $\delta T(x) = T$. Thus the previous example is a particular case of the following proposition, whose proof has been taken from [194].

Proposition 4.1.6. *A Gâteaux differentiable mapping $T : X \to X^*$ is monotone if and only if its Gâteaux derivative at x, $\delta T(x)$, is positive semidefinite for every $x \in X$. In this situation, T is maximal monotone.*

Proof. The last assertion is a direct corollary of Theorem 4.2.4 below. Therefore we only need to prove the first part of the statement. Assume that T is monotone, and take $h \in X$. Using monotonicity, we have

$$0 \leq \lim_{t \to 0} \left\langle \frac{T(x + th) - Tx}{t^2}, th \right\rangle = \lim_{t \to 0} \left\langle \frac{T(x + th) - Tx}{t}, h \right\rangle = \langle \delta T(x)h, h \rangle.$$

It follows that $\delta T(x)$ is positive semidefinite. Conversely, assume that $\delta T(\bar{x})$ is positive semidefinite for every $\bar{x} \in X$ and fix arbitrary $x, y \in X$. Define the function $\pi(t) := \langle T(x + t(y - x)) - Tx, y - x \rangle$. Then $\pi(0) = 0$. Our aim is to show that $\pi(1) \geq 0$. This is a consequence of the fact that $\pi'(s) \geq 0$ for all $s \in \mathbb{R}$. Indeed, denote $\bar{x} := x + s(y - x)$, $\alpha := t - s$, and $h := y - x$. We have

$$\frac{\pi(t) - \pi(s)}{t - s} = \left\langle \frac{T(x + t(y - x)) - T(x + s(y - x))}{t - s}, y - x \right\rangle$$

$$= \langle \frac{T(\bar{x} + \alpha h) - T(\bar{x})}{\alpha}, h \rangle.$$

Taking the limit for $t \to s$ and using the assumption on $\delta T(\bar{x})$, we get

$$\pi'(s) = \langle \delta T(\bar{x}) h, h \rangle \geq 0,$$

as we claimed.   □

Example 4.1.7. *Let $f : X \to \mathbb{R} \cup \{\infty\}$ be a proper and convex function. One of the most important examples of a maximal monotone operator is the subdifferential of f, $\partial f : X \rightrightarrows X^*$ (see Definition 3.5.1). The fact that ∂f is monotone is straightforward from the definition. The maximality has been proved by Rockafellar in [189] and we provide a proof in Theorem 4.7.1 below.*

Remark 4.1.8. It follows directly from the definition that a maximal monotone operator is convex-valued.

4.2   Outer semicontinuity

Recall from Theorem 2.5.4 that T is OSC with respect to a given topology in $X \times X^*$ if and only if the $\mathrm{Gph}(T)$ is a closed set with respect to that topology. The natural topology to consider in $X \times X^*$ is the product topology, with the strong topology in X and the weak* topology in X^*. However, unless X is finite-dimensional the graph of T is not necessarily closed with respect to this topology (see [33]).

Nevertheless, there are some cases in which outer-semicontinuity can be established for a maximal monotone operator T. In order to describe those cases, we need some definitions.

Given a topological space Z, we say that a subset $M \subset Z$ is *sequentially closed* if it contains all the limit points of its convergent sequences. Note that every closed set is sequentially closed (using Theorem 2.1.6 and the fact that every sequence is a net), but the converse is not true unless the space is N_1. We say that a point-to-set mapping $F : X \rightrightarrows Y$ is *sequentially* OSC when its graph is sequentially closed.

Proposition 4.2.1. *Let X be a Banach space and $T : X \rightrightarrows X^*$ a maximal monotone operator. Then*

(i) T is sequentially OSC, where $X \times X^*$ is endowed with the (sw)-topology.*

(ii) T is strongly OSC (i.e., $\mathrm{Gph}(T) \subset X \times X^*$ is closed with respect to the strong topology, both in X and X^*). In particular, a maximal monotone operator is OSC when X is finite-dimensional.

Proof. We start with (i). The maximality of T allows us to express $\mathrm{Gph}(T)$ as

$$\mathrm{Gph}(T) = \cap_{(x,v) \in \mathrm{Gph}(T)} \{(y,u) \in X \times X^* : \langle y - x, u - v \rangle \geq 0\}.$$

So it is enough to check sequential (sw*)-closedness of each set

$$C_{(x,v)} := \{(y,u) \in X \times X^* : \langle y - x, u - v \rangle \geq 0\}.$$

Take a sequence $\{(x_n, v_n)\} \subset C_{(x,v)}$ such that $x_n \to y$ strongly and $v_n \to u$ in the weak* topology. Note that

$$0 \leq \langle x_n - x, v_n - v \rangle = \langle x_n - y, v_n - v \rangle + \langle y - x, v_n - v \rangle$$

$$= \langle x_n - y, v_n - v \rangle + \langle y - x, v_n - u \rangle + \langle y - x, u - v \rangle. \tag{4.1}$$

Using the last statement in Theorem 2.5.16, we have that $\{v_n\}$ is bounded. Hence there exists $K > 0$ such that $\|v_n - v\| \leq K$ for all $n \in \mathbb{N}$. Fix $\delta > 0$. Because the sequence $\{(x_n, v_n)\}_{n \in \mathbb{N}}$ converges (sw*) to (y, u), we can choose n_0 such that

$$\|x_n - y\| < \frac{\delta}{2K}, \qquad |\langle v_n - u, y - x \rangle| < \frac{\delta}{2},$$

for all $n \geq n_0$. Using (4.1) for $n \geq n_0$ we obtain

$$0 \leq \delta + \langle y - x, u - v \rangle.$$

Because $\delta > 0$ is arbitrary, we conclude that $(y, u) \in C_{(x,v)}$. Item (ii) follows from (i), because the strong topology is N_1 and hence (sw*)-sequentially closed sets are (ss)-closed. The last assertion follows from the fact that in finite-dimensional spaces the strong and weak topology coincide. □

Remark 4.2.2. We point out that the result above cannot be extended to nets. In [33] an example is given of a maximal monotone operator (in fact, a subdifferential of a lower-semicontinuous convex proper function in a separable Hilbert space), that is not (sw*)-OSC. The proof is based on the fact that a weakly convergent net might be unbounded, as we showed in Example 2.5.17.

We need some notation that is useful for dealing with maximality of monotone operators.

Definition 4.2.3. Given an operator $T : X \rightrightarrows X^*$, a pair $(x, u) \in X \times X^*$ is monotonically related to T if and only if $\langle x - y, u - v \rangle \geq 0$ for all $(y, v) \in \mathrm{Gph}(T)$.

The following result connects maximality and continuity for point-to-point mappings. The proof we present below can be found in [149].

Theorem 4.2.4. *Let X be a Banach space and $T : X \to X^*$ a point-to-point and everywhere defined (i.e., such that $D(T) = X$) monotone operator. If T is (sw^*)-continuous then T is maximal.*

Proof. Take $(x, v) \in X \times X^*$ monotonically related to $\mathrm{Gph}(T)$. We must prove that $v = Tx$. By assumption

$$\langle x - y, v - Ty \rangle \geq 0 \tag{4.2}$$

for all $y \in D(T)$. Take any $h \in X$, $\lambda > 0$. Using (4.2) with $y := x - \lambda h$ and dividing by λ we get

$$\langle h, v - T(x - \lambda h) \rangle \geq 0,$$

(here we use the fact that $D(T) = X$). Taking the limit for $\lambda \downarrow 0$ and using the continuity of T, we conclude that $\langle h, v - Tx \rangle \geq 0$. Because $h \in X$ is arbitrary, we conclude that $Tx = v$. □

We present in Theorem 4.6.4 a sort of converse of Theorem 4.2.4, namely, that maximality and single-valuedness yield continuity in the interior of the domain. The simple remark below shows that the assumption $D(T) = X$ in Theorem 4.2.4 cannot be dropped.

Remark 4.2.5. The above result does not hold if $D(T) \subsetneq X$. For instance, if $T : \mathbb{R} \to \mathbb{R}$ is defined as

$$Tx := \begin{cases} x & \text{if } x < 0, \\ x & \text{if } x > 0, \\ \emptyset & \text{if } x = 0, \end{cases}$$

then T is clearly monotone. It is also continuous at every point of its domain $\mathbb{R} \setminus \{0\}$, but it is not maximal.

Corollary 4.2.6. *If $T : X \rightrightarrows X^*$ is maximal monotone, then Tx is convex and weak* closed.*

Proof. We stated in Remark 4.1.8 the convexity of Tx. In order to prove the weak* closedness of Tx, note that

$$Tx = \cap_{(y,u) \in \mathrm{Gph}(T)} \{ v : \langle v - u, x - y \rangle \geq 0 \} = \cap_{(y,u) \in \mathrm{Gph}(T)} C_{(y,u)},$$

where $C_{(y,u)} := \{ v : \langle v - u, x - y \rangle \geq 0 \}$. The definition of weak* topology readily yields weak* closedness of all sets $C_{(y,u)}$, and hence Tx must be weak* closed. □

Given $T : X \rightrightarrows X^*$, it follows from Definition 2.3.2 that the operator $T^{-1} : X^* \rightrightarrows X$ is given by $x \in T^{-1}(v)$ if and only if $v \in Tx$.

Proposition 4.2.7. *If X is reflexive and $T : X \rightrightarrows X^*$ is maximal monotone then T^{-1} is also maximal monotone.*

Proof. It can be easily seen that $\mathrm{Gph}(T)$ is a maximal monotone set if and only if $\mathrm{Gph}(T^{-1})$ is a maximal monotone set. □

Corollary 4.2.8. *If X is reflexive and $T : X \rightrightarrows X^*$ is maximal monotone, then its set of zeroes $T^{-1}(0)$ is weakly closed and convex.*

Proof. The result follows from Proposition 4.2.7 and Corollary 4.2.6. □

A result stronger than Corollary 4.2.6 can be proved for monotone operators at points in $D(T)^o$. Namely, the set Tx is weak* compact for every $x \in D(T)^o$. This result follows from the local boundedness of T at $D(T)^o$, which is the subject of the next section.

4.2.1 Local boundedness

According to Definition 2.5.15, a point-to-set mapping is said to be *locally bounded* at some point x in its domain if there exists a neighborhood U of x such that the image of U by the point-to-set mapping is bounded. Monotone operators are locally bounded at every point of the interior of their domains. On the other hand, if they are also maximal, they are not locally bounded at any point of the boundary of their domains. These results were established by Rockafellar in [186], and later extended to more general cases by Borwein and Fitzpatrick ([32]). We also see in Theorems 4.6.3 and 4.6.4 how the local boundedness is used for establishing continuity properties of T. Our proof of Theorem 4.2.10(i) has been taken from Phelps [171], whereas the proof of Theorem 4.2.10(ii) is the one given by Rockafellar in [186].

Denote $\mathrm{co}\,A$ the convex hull of the set A, and $\overline{\mathrm{co}}A$ the *closed* convex hull of A; that is, $\overline{\mathrm{co}}A = \overline{\mathrm{co}\,A}$. A subset $A \subset X$ is said to be *absorbing* if for every $x \in X$ there exist $\lambda > 0$ and $z \in A$ such that $\lambda z = x$. If X is a normed space, then any neighborhood of 0 is absorbing. The converse may not be true. The shaded set (including the boundaries) in Figure 4.2 is absorbing, and it is not a neighborhood of zero.

The following classical result (see, e.g., Lemma 12.1 in [206] and [116, page 104]) establishes a case when a given absorbing set is a neighborhood of zero.

Lemma 4.2.9. *Let X be a Banach space and take $C \subset X$ a closed and convex absorbing set. Then C is a neighborhood of 0.*

We use the previous lemma in item (i) of the theorem below.

Theorem 4.2.10. *Let X be a Banach space with dual X^*, and $T : X \rightrightarrows X^*$ a monotone operator.*

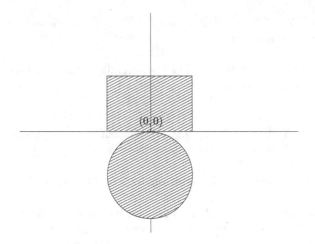

Figure 4.2. *An absorbing set*

(i) *If $x \in D(T)^o$, then T is locally bounded at x.*

(ii) *If T is maximal and $(\overline{co}D(T))^o \neq \emptyset$ then, for all $z \in D(T) \setminus (\overline{co}D(T))^o$,*

   (a) *There exists a nonzero $w \in N_{D(T)}z$.*

   (b) *$Tz + N_{D(T)}z \subset Tz$.*

   (c) *T is not locally bounded at z.*

Proof.
   (i) Note that $x \in D(T)^o$ if and only if $0 \in D(\tilde{T})^o$, where $\tilde{T} := T(x + \cdot)$. Fix $y \in Tx$ and define $\bar{T}(z) := \tilde{T}(z) - y = T(x + z) - y$. Then $0 \in \bar{T}(0)$ and $0 \in D(\bar{T})^o$. Furthermore, T is locally bounded at x if and only if $\bar{T}$ is locally bounded at 0. This translation argument allows us to assume that $0 \in D(T)^o$ and $0 \in T(0)$. Define the function

$$f(x) := \{\sup\langle v, x - y\rangle \mid y \in D(T), \|y\| \leq 1, v \in Ty\},$$

and the set $C := \{x \in X \mid f(x) \leq 1\}$. The function f is convex and lsc, because it is the supremum of a family of affine functions. Hence, the set C is closed and convex. Because $0 \in T(0)$, $f(0) \geq 0$. On the other hand, using again the monotonicity of T and the fact that $0 \in T(0)$, we get $0 \leq \langle v, y\rangle$, for all $v \in Ty$, and so

$$f(0) = \{\sup\langle v, -y\rangle \mid y \in D(T), \|y\| \leq 1, v \in Ty\} \leq 0.$$

Therefore, $f(0) = 0$ and hence $0 \in C$. We claim that $0 \in C^o$. Because C is closed and convex, the claim follows from Lemma 4.2.9 and the fact that C is absorbing. Let us prove next the latter fact. Indeed, because $0 \in D(T)^o$, there exists $r > 0$ such that $B(0, r) \subset D(T)$. Fix $y \in X$. The set $B(0, r)$ is absorbing, so there exists

$t > 0$ such that $x := ty \in B(0, r)$. Because $B(0, r) \subset D(T)$, there exists $u \in Tx$. For every $(y, v) \in \mathrm{Gph}(T)$ we have

$$0 \leq \langle x - y, u - v \rangle,$$

which gives $\langle x - y, u \rangle \geq \langle x - y, v \rangle$. Taking the supremum on $(y, v) \in \mathrm{Gph}(T)$ with $\|y\| \leq 1$, we conclude that

$$f(x) \leq \{ \sup \langle x - y, u \rangle \mid y \in D(T), \|y\| \leq 1 \} \leq \langle x, u \rangle + \|u\| < \infty.$$

This yields the existence of some $\bar{\rho} \in (0, 1)$ such that $\rho f(x) < 1$ for all $0 < \rho \leq \bar{\rho}$. By convexity of f,

$$f(\rho x) = f(\rho x + (1 - \rho)0) \leq \rho f(x) + (1 - \rho)f(0) = \rho f(x) < 1,$$

so $\rho x \in C$ for all $0 < \rho \leq \bar{\rho}$. In particular, $y = (1/t)x = \rho x/(t\rho) \in (1/(t\rho))C$. Because $y \in X$ is arbitrary, we conclude that C is absorbing. Lemma 4.2.9 now implies that $0 \in C^{\circ}$ as claimed. As a consequence, there exists $\eta \in (0, 1)$ such that

$$f(z) \leq 1 \quad \text{for all} \quad z \in B(0, 2\eta). \tag{4.3}$$

We show that $T(B(0, \eta)) \subset B(0, 1/\eta)$, which yields the local boundedness of T at 0. Indeed, take $y \in D(T) \cap B(0, \eta)$ and $z \in B(0, 2\eta)$. By (4.3), we have, for all $v \in Ty$,

$$\langle v, z - y \rangle \leq 1,$$

using the fact that $\|y\| < \eta < 1$. Hence

$$2\eta \|v\| = \sup\{ \langle v, z \rangle \mid \|z\| \leq 2\eta \} \leq \langle v, y \rangle + 1 \leq \eta \|v\| + 1,$$

implying that $\|v\| \leq 1/\eta$. The claim is true and hence T is locally bounded at 0.

(ii) Fix z in $D(T) \setminus (\overline{\mathrm{co}} D(T))^{\circ}$. Because $D(T) \subset \overline{\mathrm{co}} D(T)$, z belongs to the boundary of $\overline{\mathrm{co}} D(T)$ which is closed, convex and with nonempty interior, so that we can use Theorem 3.4.14(ii) to conclude that there exists a supporting hyperplane to $\overline{\mathrm{co}} D(T)$ at z. In other words, there exists $w \neq 0$, $w \in X^{*}$ such that

$$\langle w, z \rangle \geq \langle w, y \rangle \tag{4.4}$$

for all $y \in D(T)$. It is clear that w belongs to $N_{D(T)}z$, establishing (a). Take now $v \in Tz$. For every $\lambda \geq 0, u \in Ty$ we have

$$\langle (v + \lambda w) - u, z - y \rangle = \langle v - u, z - y \rangle + \lambda \langle w, z - y \rangle \geq 0,$$

using the monotonicity of T and the definition of w. By maximality of T, we get

$$v + \lambda w \in Tz \tag{4.5}$$

for all $\lambda \geq 0$. It follows easily that $Tz + N_{D(T)}z \subset Tz$, and hence (b) holds. Also, it follows from (4.5) that the set Tz is unbounded, and hence T is not locally bounded at z, establishing (c). $\quad \square$

Corollary 4.2.11. *If $T : X \rightrightarrows X^*$ is maximal monotone then the set Tx is weak* compact for every $x \in D(T)^o$.*

Proof. By Theorem 4.2.6, we know that Tx is weak* closed and

$$Tx = \cap_{(y,u)\in\mathrm{Gph}(T)}\{v \ : \ \langle v - u, x - y\rangle \geq 0\} = \cap_{(y,u)\in\mathrm{Gph}(T)}C_{(y,u)},$$

where each set $C_{(y,u)} := \{v \ : \ \langle v - u, x - y\rangle \geq 0\}$ is weak* closed. Assume that $x \in D(T)^o$. By Theorem 4.2.10(i), there exists $M > 0$ such that $\|v\| \leq M$ for all $v \in Tx$. By Theorem 2.5.16(ii), the balls in X^* are weak* compact and hence the sets $C_{(y,u)} \cap \{v \in X^* \mid \|v\| \leq M\}$ are also weak* compact. So

$$Tx = Tx \cap B(0, M) = \cap_{(y,u)\in\mathrm{Gph}(T)}C_{(y,u)} \cap \{v \in X^* \mid \|v\| \leq M\},$$

must be weak* compact too.   □

Another important consequence of Theorem 4.2.10(i) is the outer-semicontinuity of a maximal monotone operator T at points in $D(T)^o$.

Corollary 4.2.12. *If $T : X \rightrightarrows X^*$ is maximal monotone then T is (sw*)-OSC at every $x \in D(T)^o$.*

Proof. Take a net $\{x_i\}$ converging strongly to x, with $x \in D(T)^o$. We must prove that $\mathrm{w}^* - \lim\mathrm{ext}_i\, T(x_i) \subset T(x)$, where $\mathrm{w}^* - \lim\mathrm{ext}$ stands for the exterior limit with respect to the weak* topology. Take $v \in \mathrm{w}^* - \lim\mathrm{ext}_i\, T(x_i)$. We claim that (x, v) is monotonically related to T. Indeed, take $(y, u) \in \mathrm{Gph}(T)$; we must show that $\langle v - u, x - y\rangle \geq 0$. Fix $\varepsilon > 0$. Because $v \in \mathrm{w}^* - \lim\mathrm{ext}_i\, T(x_i)$, and the set $W_0 := \{w \in X^* : |\langle w - v, x - y\rangle| < \varepsilon/3\}$ is a weak* neighborhood of v, there exists a cofinal set $J_0 \subset I$ such that

$$F(x_i) \cap W_0 \neq \emptyset \quad \text{for all } i \in J_0.$$

Take $u_i \in F(x_i) \cap W_0$ for every $i \in J_0$. By Theorem 4.2.10(i), there exists $R, M > 0$ such that $T(B(x, R)) \subset B(0, M)$. Because $\{x_i\}$ converges strongly to x, there exists a terminal set I_1 such that

$$\|x_i - x\| < \min\left\{R, \frac{\varepsilon}{3(\|v\| + M)}, \frac{\varepsilon}{3(\|v\| + \|u\|)}\right\}, \qquad \text{for all } i \in I_1.$$

Altogether, the set $J_1 := J_0 \cap I_1$ is cofinal in I and such that for all $i \in J_1$ we have $\|u_i\| \leq M$, $|\langle u_i - v, x - y\rangle| < \varepsilon/3$, $|\langle v - u, x - x_i\rangle| < \varepsilon/3$, and $|\langle v - u_i, x - x_i\rangle| < \varepsilon/3$. So, for all $i \in J_1$ we can write

$$
\begin{aligned}
\langle v - u, x - y\rangle &= \langle v - u, x - x_i\rangle + \langle v - u, x_i - y\rangle \\
&= \langle v - u, x - x_i\rangle + \langle v - u_i, x_i - y\rangle + \langle u_i - u, x_i - y\rangle \\
&\geq \langle v - u, x - x_i\rangle + \langle v - u_i, x_i - y\rangle \\
&\geq \langle v - u, x - x_i\rangle + \langle v - u_i, x_i - x\rangle + \langle v - u_i, x - y\rangle \\
&> -\varepsilon,
\end{aligned}
$$

where we used the fact that $(x_i, u_i), (y, u) \in \mathrm{Gph}(T)$ in the first inequality and the definition of J_1 in the last one. Because $\varepsilon > 0$ is arbitrary, the claim is true, and maximality of T yields $v \in T(x)$. $\quad\square$

The next result establishes upper-semicontinuity of maximal monotone operators in the interior of their domain.

Corollary 4.2.13. *If $T : X \rightrightarrows X^*$ is maximal monotone then T is (sw*)-USC at every $x \in D(T)^o$.*

Proof. The conclusion follows by combining Proposition 2.5.21(ii), local-boundedness of T at x and Corollary 4.2.12 $\quad\square$

We remark that a maximal monotone operator T may be such that each point in $D(T)$ fails to satisfy the assumptions of both (i) and (ii) in Theorem 4.2.10; that is, such that $D(T)^o = \emptyset$, $D(T) \setminus (\overline{\mathrm{co}}D(T))^o = \emptyset$. Consider $V \subset \ell_2$ defined as $V = \{x \in \ell_2 : \sum_{n=1}^{\infty} 2^{2n} x_n^2 < \infty\}$. V is a dense linear subspace of ℓ_2. Define $T : \ell_2 \to \ell_2$ as $(Tx)_n = 2^n x_n$ if $x \in V$; $Tx = \emptyset$ otherwise. It is immediate that $D(T) = V$ and that T is monotone. It can be proved that T is indeed maximal monotone (see [171], page 31). Observe that $D(T)$ has an empty interior, because it is a subspace. On the other hand, inasmuch as V is dense in ℓ_2, we have $\overline{\mathrm{co}}D(T) \supset \overline{D(T)} = \overline{V} = \ell_2$, so that $(\overline{\mathrm{co}}D(T))^o = \ell_2$ and hence $D(T) \setminus (\mathrm{co}\, D(T))^o = \emptyset$. For this example, no assertion of Theorem 4.2.10 is valid for any $x \in D(T)$: it is easy to check that T is an unbounded linear operator, so that it cannot be locally bounded at any point, but it is also point-to-point at all points of $D(T)$, so that no halfline is contained in Tz, for all $z \in D(T)$, at variance with the conclusion of Theorem 4.2.10(ii)-(b) (i.e., the lack of local boundedness is a truly local phenomenon, not a pointwise one, as in Theorem 4.2.10(ii)). It is possible to strengthen Theorem 4.2.10 so that its conclusion refers to all points in $D(T)$, but under a stronger assumption: one must suppose that $(\mathrm{co}\, D(T))^o$ is nonempty, instead of $(\overline{\mathrm{co}}D(T))^o \neq \emptyset$ (note that the new stronger assumption does not hold for T as in the example above, whereas the weaker one does). We present this result in Section 4.6.

4.3 The extension theorem for monotone sets

A very important fact concerning monotone sets is the classical extension theorem of Debrunner and Flor [72]. This theorem is used for establishing some key facts related to domains and ranges of maximal monotone operators. The proofs of all theorems in this section and the two following ones were taken from [172].

Theorem 4.3.1 (Debrunner–Flor). *Let $C \subset X^*$ be a weak* compact and convex set, $\phi : C \to X$ a function that is continuous from the weak* topology to the strong topology, and $M \subset X \times X^*$ a monotone set. Then there exists $v \in C$ such that $\{(\phi(v), v)\} \cup M$ is monotone.*

Proof. For each element $(y, v) \in M$ define the set

$$U(y, v) := \{u \in C \mid \langle u - v, \phi(u) - y \rangle < 0\}.$$

The sets $U(y, v) \subset X^*$ are weak* open because the function $p : C \to \mathbb{R}$, defined as $p(w) = \langle w - v, \phi(w) - y \rangle$, is weak* continuous (Exercise 6). Assume that the conclusion of the theorem is not true. This means that for every $u \in C$, there exists $(y, v) \in M$ such that $u \in U(y, v)$. Thus the family of open sets $\{U(y, v)\}_{(y,v) \in M}$ is an open covering of C. By compactness of C, there exists a finite subcovering $\{U(y_1, v_1), \ldots, U(y_n, v_n)\}$ of C. By Proposition 3.10.2, associated with this finite subcovering there exists a partition of unity as in Definition 3.10.1; that is, functions $\alpha_i : X^* \to \mathbb{R}$ $(1 \leq i \leq n)$ that are weak* continuous and satisfy

(a) $\sum_{i=1}^{n} \alpha_i(u) = 1$ for all $u \in C$.

(b) $\alpha_i(u) \geq 0$ for all $u \in C$ and all $i \in \{1, \ldots, n\}$.

(c) $\{u \in C : \alpha_i(u) > 0\} \subset U_i := U(y_i, v_i)$ for all $i \in \{1, \ldots, n\}$.

Define $K := co\{v_1, \ldots, v_n\} \subset C$. K can be identified with a finite-dimensional convex and compact set. Define also the function $\rho : K \to K$ as $\rho(u) := \sum_{i=1}^{n} \alpha_i(u) v_i$. Clearly, ρ is weak* continuous. Weak* continuity coincides with strong continuity in the finite-dimensional vector space spanned by K, thus we can invoke Theorem 3.10.4 (Brouwer's fixed point theorem) in order to conclude that there exists $w \in K$ such that $\rho(w) = w$. Therefore, we have

$$0 = \langle \rho(w) - w, \sum_j \alpha_j(w) y_j - \phi(w) \rangle$$

$$= \left\langle \sum_i \alpha_i(w)(v_i - w), \sum_j \alpha_j(w) y_j - \phi(w) \right\rangle$$

$$= \sum_{i,j} \alpha_i(w) \alpha_j(w) \langle v_i - w, y_j - \phi(w) \rangle. \tag{4.6}$$

Call $a_{ij} := \langle v_i - w, y_j - \phi(w) \rangle$. Note that

$$a_{ij} + a_{ji} = a_{ii} + a_{jj} + \langle v_i - v_j, y_j - y_i \rangle \leq a_{ii} + a_{jj},$$

using the monotonicity of M. Combining these facts with (4.6) gives

$$0 = \sum_{i,j} \alpha_i(w) \alpha_j(w) a_{ij} = \sum_i \alpha_i(w)^2 a_{ii} + \sum_{i<j} \alpha_i(w) \alpha_j(w)(a_{ij} + a_{ji})$$

$$\leq \sum_i \alpha_i(w)^2 a_{ii} + \sum_{i<j} \alpha_i(w) \alpha_j(w)(a_{ii} + a_{jj}). \tag{4.7}$$

Call $I(w) := \{i \in \{1, \ldots, n\} \mid w \in U_i\}$. Using property (c) above, the inequality in (4.7) can be rewritten as

$$0 \leq \sum_{i \in I(w)} \alpha_i(w)^2 a_{ii} + \sum_{\substack{i<j \\ i,j \in I(w)}} \alpha_i(w) \alpha_j(w)(a_{ii} + a_{jj}). \tag{4.8}$$

It is clear from the definitions that $a_{ii} < 0$ if and only if $w \in U_i$. Because $\alpha_i(w) \geq 0$ by property (b) above, we get that all terms in the right-hand side of (4.8) are nonpositive, and, in view of the inequality in (4.8), they all vanish, implying that $\alpha_i(w) = 0$ for all $i = 1, \ldots, n$, but this contradicts the fact that $\sum_i \alpha_i(w) = 1$, completing the proof. □

We establish next some facts relating domains and ranges of maximal monotone operators in reflexive Banach spaces.

4.4 Domains and ranges in the reflexive case

A central result for the subject of interest in this section is due to Simons. It provides a necessary and sufficient condition for a monotone set to be maximal monotone. We state it below.

Theorem 4.4.1. *Let X be a reflexive Banach space and $M \subset X \times X^*$ a monotone set. Then M is maximal monotone if and only if for all $(x, v) \in X \times X^*$ there exists $(y, u) \in M$ such that*

$$\|y - x\|^2 + \|u - v\|^2 + 2\langle y - x, u - v \rangle = 0.$$

Proof. See Theorem 10.3 in [206].   □

Important properties of a Banach space are studied through the duality map, which we define next.

Definition 4.4.2. *Consider $g : X \to \mathbb{R}$ defined by*

$$g(x) = \frac{1}{2}\|x\|^2.$$

The duality mapping $J : X \rightrightarrows X^*$ is the subdifferential ∂g of g.

Remark 4.4.3. A simple property of J, which we prove in the following proposition, is that $\text{Gph}(J)$ is monotone. Indeed, we show later on (using Theorem 4.7.1) that it is maximal monotone, because it is the subdifferential of a proper convex function. When X is a Hilbert space, then $J = I$, the identity mapping, and hence onto. Moreover, the duality mapping J is onto if and only if X is reflexive (see, e.g., Theorem 3.4 in [65]).

We state next some useful properties of the duality mapping J.

Proposition 4.4.4.

(i) *For all $x \in X$, it holds that $J(x) = \{v \in X^* \mid \langle x, v \rangle = \|x\| \|v\|, \|x\| = \|v\|\}$.*

(ii) *For all $(x, u), (y, v) \in \text{Gph}(J)$, it holds that*

$$\langle x - y, u - v \rangle \geq (\|x\| - \|y\|)^2,$$

and hence J is monotone.

(iii) $J(-x) = -Jx$.

(iv) *Take* $(x, v) \in X \times X^*$. *A pair* (x, v) *belongs to* $\mathrm{Gph}(-J)$ *if and only if*

$$\|x\|^2 + \|v\|^2 + 2\langle x, v \rangle = 0.$$

(v) *If* $U := B(0, r) = \{x \in X \mid \|x\| \leq r\}$ *and* $z \in \partial U = \{x \in X \mid \|x\| = r\}$ *then* $N_U(z) = \{\lambda w \mid w \in Jz \text{ and } \lambda \geq 0\}$.

Proof.

(i) From the definition of J as the subdifferential of $g(x) := \frac{1}{2}\|x\|^2$ we get, for all $v \in Jx$,

$$\langle v, y \rangle - \frac{1}{2}\|y\|^2 \leq \langle v, x \rangle - \frac{1}{2}\|x\|^2.$$

Taking the supremum on $y \in Y$, and recalling that $g^*(v) = \frac{1}{2}\|v\|^2$ (see Example 3.4.7), we get

$$\frac{1}{2}\|v\|^2 \leq \langle v, x \rangle - \frac{1}{2}\|x\|^2 \leq \|v\|\,\|x\| - \frac{1}{2}\|x\|^2,$$

which yields $0 \geq (1/2)(\|v\| - \|x\|)^2$ and hence $\|v\| = \|x\|$. Using this fact in the inequality above we get

$$\|x\|^2 = \|x\|\,\|v\| \geq \langle v, x \rangle \geq \|x\|^2,$$

as required.

(ii) By (i), $\|v\| = \|y\|$ for all $y \in X$, $v \in Jy$. Hence, $-\langle x, v \rangle \geq -\|x\|\,\|v\| = -\|x\|\,\|y\|$ for every $x \in X$, $v \in Jy$. Thus, for all $u \in Jx$, $v \in Jy$, we have

$$\langle x - y, u - v \rangle = \langle x, u \rangle + \langle y, v \rangle - \langle x, v \rangle - \langle y, u \rangle$$

$$\geq \|x\|^2 + \|y\|^2 - 2\|x\|\,\|y\| = (\|x\| - \|y\|)^2 \geq 0.$$

The proof of (iii) follows directly from (i), and (iv) is a direct consequence of the definitions. Let us establish (v). Call $S(z)$ the rightmost set in the statement of (v). Fix z such that $\|z\| = r$. Because $S(z)$ and $N_U(z)$ are cones, it is enough to prove that $Jz \subset N_U(z)$, in order to conclude that $S(z) \subset N_U(z)$. Take $w \in Jz$. For all $x \in U$, we have

$$\langle w, x - z \rangle = \langle w, x \rangle - \langle w, z \rangle \leq \|w\|\,\|x\| - \langle w, z \rangle \leq r^2 - r^2 = 0,$$

using (i) and the Cauchy–Schwartz inequality. The above expression yields $w \in N_U(z)$ and hence $S(z) \subset N_U(z)$.

Let us prove the converse inclusion. Take $w \in N_U(z)$ and assume that $\|w\| = r$. Using the same argument as above, if we prove that every such w belongs to Jz, we get that $N_U(z) \subset S(z)$, because both sets are cones. Because $w \in N_U(z)$, we get

$$\langle w, x \rangle \leq \langle w, z \rangle,$$

for all x such that $\|x\| \leq r$. Taking the supremum on $x \in \{x' \in X \;\|x'\| \leq r\}$, we conclude that

$$r \|w\| = r^2 \leq \langle w, z \rangle \leq \|w\| \|z\| = r^2,$$

so that $\langle w, z \rangle = r^2$. Because we also have $\|w\| = \|z\| = r$ by (i), we obtain that $w \in Jz$. $\square$

Theorem 4.4.5. *Let X be reflexive and $M \subset X \times X^*$ a monotone set. Then M is maximal monotone if and only if $X \times X^* = M+ \mathrm{Gph}(-J)$.*

Proof. It follows from Proposition 4.4.4(iii) that, for all $(z, w) \in X \times X^*$, (z, w) belongs to $\mathrm{Gph}(-J)$ if and only if $(-z, -w) \in \mathrm{Gph}(-J)$.

By Proposition 4.4.4(iv) and Theorem 4.4.1, we have that M is maximal monotone if and only if for all $(x, v) \in X \times X^*$ there exists $(y, u) \in M$ such that $(y - x, u - v) \in \mathrm{Gph}(-J)$, which is equivalent, in view of the observation just made, to $(-y + x, -u + v) \in \mathrm{Gph}(-J)$, which in turn is equivalent to $(x, v) \in M + \mathrm{Gph}(-J)$, because (y, u) belongs to M. $\square$

The result above can be also stated in the following way.

Corollary 4.4.6. *Let X be reflexive and $M \subset X \times X^*$ a monotone set. M is maximal monotone if and only if $X \times X^* = M + \mathrm{Gph}(-\lambda J)$ for all $\lambda > 0$.*

Proof. The "if" statement follows by taking $\lambda = 1$ and using Theorem 4.4.5. For the "only if" statement, take any $(z, w) \in X \times X^*$. Apply Theorem 4.4.5 to the maximal monotone set $M_\lambda := \{(x, v) \in X \times X^* \,|\, (x, \lambda v) \in M\}$, and use the fact that $(z, \lambda^{-1} w) \in X \times X^* = M_\lambda + \mathrm{Gph}(-J)$. This readily gives $(z, w) \in M + \mathrm{Gph}(-\lambda J)$. $\square$

A key result, proved by Minty in [153] for Hilbert spaces, is the fact that T is maximal monotone if and only if $R(T + J) = X^*$. Rockafellar extended this result to reflexive Banach spaces for which both J and J^{-1} are single-valued, in which case the square of the norm is differentiable. This result is known as Rockafellar's characterization of maximal monotone operators. We emphasize that the requirement on J is not a restrictive one. Indeed, it has been proved by Asplund in [8] that every reflexive Banach space X can be renormed in such a way that both J and J^{-1} result single-valued, preserving the monotonicity and/or the maximality properties of operators defined on X. However, this renorming result is quite involved, and thus it is worthwhile to present alternative proofs of the above-mentioned result that do not require a renorming of the space. The proof of the "only if" statement of Rockafellar's characterization of maximal monotone operators without renorming the space is due to Simons, and is given in [206, Theorem 10.7]. We present this proof below.

Theorem 4.4.7. *Assume that X is reflexive and $T : X \rightrightarrows X^*$ a point-to-set mapping.*

(i) *If T is maximal monotone, then $R(T + J) = X^*$. Conversely, if T is monotone, $R(T + J) = X^*$ and J and J^{-1} are single-valued, then T is maximal monotone.*

(ii) *If J and J^{-1} are single-valued, and either T is maximal monotone or T is monotone and $R(T + J) = X^*$, then the operator $(T + J)^{-1} : X^* \rightrightarrows X$ is single-valued, maximal monotone, and continuous from the strong topology to the weak topology.*

Proof.

(i) Assume that T is maximal monotone. We must show that $R(T+J) = X^*$. Take $w \in X^*$. From Theorem 4.4.5, we know that $(0, w) \in \mathrm{Gph}(T) + \mathrm{Gph}(-J)$. This implies that there exists $(x, v) \in \mathrm{Gph}(T)$, $z \in X$, and $u \in Jz$ such that $(0, w) = (x, v) + (z, -u)$. In other words, $x = -z$ and $w = v - u$. Because $J(-x) = -Jx$ by Proposition 4.4.4(iii), we get that

$$w \in v - J(z) = v + J(-z) \subset T(-z) + J(-z) \subset R(T + J).$$

Hence $R(T + J) = X^*$.

Suppose now that T is monotone, $R(T + J) = X^*$, and J and J^{-1} are single-valued. This means that Jx is a point for all $x \in X$ and that the equality $Jz = Jz'$ implies $z = z'$. Take now (z, w) monotonically related to $\mathrm{Gph}(T)$, in the sense of Definition 4.2.3. In order to establish maximality of T, it suffices to check that w belongs to Tz. Because $R(T + J) = X^*$, we have that $w + Jz \in R(T + J)$. This implies that there exists $x \in D(T)$ such that $w + Jz \in (T + J)x$, and hence there exists $u \in Tx$ such that

$$w + Jz = u + Jx. \tag{4.9}$$

We show that $(z, w) = (x, u) \in \mathrm{Gph}(T)$. Note that (4.9) implies that

$$0 = \langle z - x, (w + Jz) - (u + Jx) \rangle = \langle z - x, w - u \rangle + \langle z - x, Jz - Jx \rangle. \tag{4.10}$$

By monotonicity of T and J (see Proposition 4.4.4(ii)), we conclude that both terms on the rightmost expression in (4.10) vanish, which implies, in view of Proposition 4.4.4(ii), $\|z\| = \|x\|$. Because $\|Js\| = \|s\|$ for all $s \in X$, by Proposition 4.4.4(i), we conclude that $\|z\| = \|x\| = \|Jz\| = \|Jx\|$. Using this fact in (4.10) yields

$$0 = (\|z\| \, \|Jx\| - \langle z, Jx \rangle) + (\|x\| \, \|Jz\| - \langle x, Jz \rangle), \tag{4.11}$$

where both terms are nonnegative. So both terms in (4.11) vanish, implying $\|Jz\|^2 = \|x\|^2 = \langle Jz, x \rangle$, and $\|Jx\|^2 = \|z\|^2 = \langle Jx, z \rangle$. Because J is single-valued, it follows from Proposition 4.4.4(i) that

$$Jx = Jz. \tag{4.12}$$

Using (4.12) in (4.9) we obtain $w = u$. Applying now J^{-1} to the equality $Jx = Jz$ and using the fact that J^{-1} is single-valued we get $x = z$, but then (z, w) belongs to $\mathrm{Gph}(T)$ as claimed, and hence T is maximal.

(ii) We begin by establishing that $(T + J)^{-1}$ is maximal monotone. Indeed, by the first statement of item (i), applied to the maximal monotone operator $(1/2)T$, we get $R((1/2)T + J) = X^*$, which gives $R(T + 2J) = R((T + J) + J) = X^*$. The assumption on J and J^{-1} allows us to use the second statement of item (i), applied to the monotone operator $T + J$, in order to conclude that it is maximal monotone, in which case $(T + J)^{-1}$ is maximal monotone. Under any of the assumptions on T in this item, we get $R(T + J) = D((T + J)^{-1}) = X^*$. We claim now that $(T + J)^{-1}$ is single-valued. We proceed to prove this claim. Fix any $u \in X^*$ and take $x, y \in (T + J)^{-1}(u)$, so that $u \in (T + J)x$ and $u \in (T + J)y$. Assume that $x \neq y$. Because J^{-1} is single-valued, we get $Jx \neq Jy$. By the definition of u, there exists $v_1 \in Tx$ and $v_2 \in Ty$ such that $u = v_1 + Jx = v_2 + Jy$. Thus, using the monotonicity of T,

$$0 \leq \langle x - y, Jx - Jy \rangle = \langle x - y, v_2 - v_1 \rangle \leq 0.$$

Using now the same argument as in (4.11), we get $Jx = Jy$, which is a contradiction. Hence $(T + J)^{-1}$ is single-valued. The mapping $(T + J)^{-1}$ is also maximal monotone and everywhere defined, therefore we conclude from Theorem 4.6.4 that it is strong-weak continuous. □

The assumption on J of the previous result cannot be dropped. Indeed, it can be proved that when either J or J^{-1} is not single-valued, the "if" statement of Theorem 4.4.7 is not true. See [206, page 39] for more details.

Remark 4.4.8. In the statement of Theorem 4.4.7, $T + J$ can be everywhere replaced by $T + \lambda J$, with $\lambda > 0$. Indeed, let S be a maximal monotone operator and fix $\lambda > 0$. By Theorem 4.4.7, $R(\lambda^{-1}S + J) = X^*$. It is elementary to check that $R(S + \lambda J) = X^*$. It is also elementary that $(\lambda F)^{-1}(u) = F^{-1}(\lambda^{-1}u)$ for every point-to-set mapping F and for every $\lambda \neq 0$. Therefore $(S + \lambda J)^{-1}$ is single-valued, maximal monotone, and continuous if and only if $(\lambda^{-1}S + J)^{-1}$ is single-valued, maximal monotone, and continuous.

The following result is due to Rockafellar [191]; we present here the proof given in [206].

Theorem 4.4.9. *Let X be a reflexive Banach space and $T : X \rightrightarrows X^*$ a maximal monotone operator. Then $\overline{R(T)}$ is convex.*

Proof. Because $\overline{R(T)}$ is closed, it is enough to prove that $1/2(v_1 + v_2) \in \overline{R(T)}$ for all $v_1, v_2 \in R(T)$. Take $v_1, v_2 \in R(T)$, and pick up $x_1, x_2 \in D(T)$ such that $(x_1, v_1), (x_2, v_2) \in \mathrm{Gph}(T)$. Call $(x_0, v_0) := ((x_1 + x_2)/2, (v_1 + v_2)/2)$. Fix $\lambda > 0$ and define the operator $S : X \rightrightarrows X^*$ as $Sx := (1/\lambda)T(x + x_0) - (v_0/\lambda)$. It is clear that S is maximal monotone (see Exercise 1), $(x_1 - x_2)/2, (x_2 - x_1)/2 \in D(S)$, and

$$\left(\frac{x_1 - x_2}{2}, \frac{v_1 - v_2}{2\lambda}\right), \left(\frac{x_2 - x_1}{2}, \frac{v_2 - v_1}{2\lambda}\right) \in \mathrm{Gph}(S). \qquad (4.13)$$

On the other hand, by Theorem 4.4.7, we have that $0 \in R(S + J)$ and hence there exists $x_\lambda \in D(S)$ such that $0 \in Sx_\lambda + Jx_\lambda$. This implies the existence of some $w_\lambda \in Jx_\lambda$ such that $-w_\lambda \in Sx_\lambda$ and $\langle w_\lambda, x_\lambda \rangle = \|w_\lambda\|^2$. Rearranging the definition of S and using the fact that $-\lambda w_\lambda \in \lambda Sx_\lambda$, we get

$$-\lambda w_\lambda + v_0 \in \lambda S(x_\lambda) + v_0 \in T(x_\lambda + x_0) \subset R(T). \qquad (4.14)$$

Let us prove that $\lim_{\lambda \to 0} \|\lambda w_\lambda\| = 0$. Indeed, using (4.13) and the fact that $-w_\lambda \in Sx_\lambda$, we get

$$\left\langle -w_\lambda - \left(\frac{v_2 - v_1}{2\lambda}\right), x_\lambda - \left(\frac{x_2 - x_1}{2}\right)\right\rangle \geq 0$$

and

$$\left\langle -w_\lambda - \left(\frac{v_1 - v_2}{2\lambda}\right), x_\lambda - \left(\frac{x_1 - x_2}{2}\right)\right\rangle \geq 0.$$

Summing up both inequalities and using the definition of w_λ, we get $\|w_\lambda\|^2 = \langle w_\lambda, x_\lambda \rangle \leq \langle (v_1 - v_2)/2\lambda, (x_1 - x_2)/2 \rangle$, yielding

$$\|\lambda w_\lambda\| \leq \frac{1}{2}\sqrt{\lambda \langle v_1 - v_2, x_1 - x_2 \rangle},$$

proving that $\lim_{\lambda \to 0} \|\lambda w_\lambda\| = 0$. In view of (4.14), $v_0 \in \mathrm{cl}R(T)$. This completes the proof. $\square$

The reflexivity of X allows us to obtain the same result for $D(T)$.

Theorem 4.4.10. *If X is a reflexive Banach space and $T : X \rightrightarrows X^*$ is maximal monotone, then $\overline{D(T)}$ is convex.*

Proof. As commented on above, maximal monotonicity of T^{-1} follows from maximal monotonicity of T. Clearly, if X is reflexive, then X^* is also reflexive. So Theorem 4.4.9 applies, and hence $\overline{R(T^{-1})}$ is convex. The result follows from the fact that $R(T^{-1}) = D(T)$. $\square$

We finish this section with a result, proved in [49], that extends one of the consequences of the surjectivity property in Theorem 4.4.7 to operators other than J. We first need a definition.

Definition 4.4.11. *An operator $T : X \rightrightarrows X^*$ is regular when $\sup_{(y,u) \in \mathrm{Gph}(T)} \langle x - y, u - v \rangle < 0$ for all $(x, v) \in D(T) \times R(T)$.*

Theorem 4.4.12. *Assume that X is a reflexive Banach space and J its normalized duality mapping. Let $C, T_0 : X \rightrightarrows X^*$ be maximal monotone operators such that*

(a) C is regular.

(b) $D(T_0) \cap D(C) \neq \emptyset$ and $R(C) = X^*$.

(c) $C + T_0$ is maximal monotone.

Then $R(C + T_0) = X^*$.

Proof. The proof of this theorem consists of two steps:
Step 1.
In this step we prove the following statement: Let T_1 be a maximal monotone operator. If there exists a convex set $F \subset X^*$ satisfying that for all $u \in F$ there exists $y \in X$ such that

$$\sup_{(z,v)\in \mathrm{Gph}(T_1)} \langle v - u, y - z \rangle < \infty, \tag{4.15}$$

then $F^o \subset R(T_1)$.

If $F^o = \emptyset$ then the conclusion holds trivially. Assume that $F^o \neq \emptyset$. By Theorem 4.4.7(i) and Remark 4.4.8, for each $f \in F$ and each $\varepsilon > 0$, there exists x_ε such that

$$f \in (T_1 + \varepsilon J)(x_\varepsilon), \tag{4.16}$$

where J is the normalized duality mapping in X. Take $v_\varepsilon \in Jx_\varepsilon$ such that $f - \varepsilon v_\varepsilon \in T_1 x_\varepsilon$. Let $a \in X$ be such that (4.15) holds for $y = a$; that is, there exists $\theta \in \mathbb{R}$ with

$$\langle v - f, a - z \rangle \leq \theta \tag{4.17}$$

for all $(z, v) \in \mathrm{Gph}(T_1)$. For the choice $(z, v) := (x_\varepsilon, f - \varepsilon v_\varepsilon)$, (4.17) becomes

$$\varepsilon \|x_\varepsilon\|^2 \leq \varepsilon \langle v_\varepsilon, a \rangle + \theta. \tag{4.18}$$

By Proposition 4.4.4(i), $\|p\|^2 - 2\langle w, q \rangle + \|q\|^2 \geq 0$, for any $p, q \in X$, $w \in J(p)$. Thus, (4.18) becomes

$$\frac{\varepsilon}{2} \|x_\varepsilon\|^2 \leq \frac{\varepsilon}{2} \|a\|^2 + \theta,$$

which gives $\sqrt{\varepsilon} \|x_\varepsilon\| \leq r$, for some $r > 0$ and $\varepsilon > 0$ small enough.

Now, take any $\tilde{f} \in F^o$, and let $\rho > 0$ be such that $\tilde{f} + B^*(0, \rho) := \{\tilde{f} + \varphi \in X^* : \|\varphi\| \leq \rho\} \subset F$. By (4.15), for any $\varphi \in B^*(0, \rho)$ there exists $a(\varphi) \in X$ and $\theta(\varphi) \in \mathbb{R}$ such that

$$\langle v - (\tilde{f} + \varphi), a(\varphi) - z \rangle \leq \theta(\varphi) \tag{4.19}$$

for all $(z, v) \in \mathrm{Gph}(T_1)$. Take $f = \tilde{f}$, $x_\varepsilon = \tilde{x}_\varepsilon$, and $\tilde{v}_\varepsilon \in J\tilde{x}_\varepsilon$ in (4.16) and choose $(z, v) := (\tilde{x}_\varepsilon, \tilde{f} - \varepsilon \tilde{v}_\varepsilon)$ in (4.19), to get

$$\langle -\varepsilon \tilde{v}_\varepsilon - \varphi, a(\varphi) - \tilde{x}_\varepsilon \rangle \leq \theta(\varphi). \tag{4.20}$$

Rearranging (4.20) yields

$$\langle \varphi, \tilde{x}_\varepsilon \rangle \leq \theta(\varphi) + \langle \varphi, a(\varphi) \rangle + \varepsilon \langle \tilde{v}_\varepsilon, a(\varphi) \rangle - \varepsilon \|\tilde{x}_\varepsilon\|^2,$$

which implies

$$\langle \varphi, \tilde{x}_\varepsilon \rangle \le \theta(\varphi) + \langle \varphi, a(\varphi) \rangle + \frac{\varepsilon}{2}(\|a(\varphi)\|^2 - \|\tilde{x}_\varepsilon\|^2).$$

Altogether, we conclude that

$$\langle \varphi, \tilde{x}_\varepsilon \rangle \le \theta(\varphi) + \langle \varphi, a(\varphi) \rangle + \frac{\varepsilon}{2}\|a(\varphi)\|^2, \tag{4.21}$$

where we look at $\tilde{x}_\varepsilon$ as a functional defined in X^*. Taking $\varphi = -\psi \in B^*(0, \rho)$ in (4.21), we obtain a bound $K(\varphi)$ such that

$$|\tilde{x}_\varepsilon(\varphi)| \le K(\varphi) \tag{4.22}$$

for all $\varphi \in B^*(0, \rho)$. Define now $x^k := \tilde{x}_\varepsilon$ for $\varepsilon = 1/k$. By (4.22), $|x^k(\varphi)| \le K(\varphi)$ for all k. By the Banach–Steinhaus theorem (cf. Theorem 15-2 in [22]), there exists a constant $\bar{K}$ such that $|x^k(\varphi)| \le \bar{K}$ for all k, and all $\varphi \in B^*(0, \rho)$. This implies that the sequence $\{x^k\}$ is bounded, and hence it has a subsequence, say $\{x^{k_j}\}$ with a weak limit, say $\bar{x}$. Take $v^{k_j} \in Jx^{k_j}$ such that (4.16) holds for $f = \tilde{f}$ and $x_\varepsilon = x^{k_j}$. Observe that J maps bounded sets on bounded sets. So, there exists a subsequence of $\{x^{k_j}\}$ (which we still call $\{x^{k_j}\}$ for simplicity) such that

$$\begin{aligned}
&\text{w} - \lim_{j \to \infty} x^{k_j} = \bar{x}, \\
&\tilde{f} - \tfrac{1}{k_j} v^{k_j} \in T_1(x^{k_j}), \\
&\text{w} - \lim_{j \to \infty} \tilde{f} - \tfrac{1}{k_j} v^{k_j} = \tilde{f}.
\end{aligned} \tag{4.23}$$

By Proposition 4.2.1(i) and maximal monotonicity of T_1, we have that $\tilde{f} \in T_1(\bar{x})$. Because $\tilde{f}$ is an arbitrary element of F^o, the result of step 1 holds.

Step 2.

We prove now that the set $F = R(C) + R(T_0)$ satisfies (4.15) for the operator $C + T_0$. Take $u \in R(C) + R(T_0)$, $x \in D(C) \cap D(T_0)$ and $w \in T_0(x)$. Write $u = w + (u - w)$. Because $R(C) = X^*$, we can find $y \in X$ such that $(u - w) \in C(y)$. Using now the regularity of C, we know that given $(u - w) \in R(C)$ and $x \in D(C)$, there exists some $\theta \in \mathbb{R}$ with $\sup_{(z,s) \in \text{Gph}(C)} \langle s - (u - w), x - z \rangle \le \theta$, which implies that

$$\langle s - (u - w), x - z \rangle \le \theta \tag{4.24}$$

for any $(z, s) \in \text{Gph}(C)$. Take $z \in D(T_0) \cap D(C)$ and $v \in T_0(z)$. By monotonicity of T_0 we get

$$\langle v - w, x - z \rangle \le 0. \tag{4.25}$$

Adding (4.24) and (4.25) we obtain

$$\langle (s + v) - u, x - z \rangle \le \theta,$$

for any $s \in C(z)$, $v \in T_0(z)$; that is, for any $s + v =: t \in (C + T_0)z$. Therefore,

$$\sup_{(z,t) \in \text{Gph}(C + T_0)} \langle t - u, x - z \rangle < \infty,$$

which establishes (4.15) for $F = R(C) + R(T_0)$ and $T_1 = C + T_0$. Now by step 1 and the fact that $F = X^*$, we obtain that $C + T_0$ is onto. $\square$

4.5 Domains and ranges without reflexivity

In this section we present some results concerning the domain and range of a maximal monotone operator in an arbitrary Banach space. We start with a direct consequence of the extension theorem of Debrunner and Flor, which states that a maximal monotone operator with bounded range must be defined everywhere.

Corollary 4.5.1. *Let $T : X \rightrightarrows X^*$ be maximal monotone and suppose that $R(T)$ is bounded. Then $D(T) = X$.*

Proof. Let C be a weak* compact set containing $R(T)$ and fix $x \in X$. Define $\psi : C \to X$ as $\psi(v) = x$. Then ψ is weak* strong continuous and by Theorem 4.3.1 there exists $v \in C$ such that the set $\{(\psi(v), v)\} \cup \mathrm{Gph}(T) = \{(x, v)\} \cup \mathrm{Gph}(T)$ is monotone. Because T is maximal, we have $v \in Tx$, yielding $x \in D(T)$. $\square$

We need the following technical lemma for proving Theorem 4.5.3.

Lemma 4.5.2. *Let D be a subset of a Banach space X. Then*

(i) *If $(\mathrm{co}\, D)^\circ \subset D$, then D° is convex.*

(ii) *If $\emptyset \neq (\mathrm{co}\, D)^\circ \subset D$, then $\overline{D} = \overline{[(\mathrm{co}\, D)^\circ]}$.*

(iii) *If $\{C_n\}$ are closed and convex sets with nonempty interiors such that $C_n \subset C_{n+1}$ for all n, then*
$$\left(\cup_n C_n\right)^\circ \subset \cup_n C_n^\circ.$$

(iv) *Let $A, B \subset X$. If A is open and $\overline{A \cap B} \subset B$, then $A \cap \overline{B} \subset A \cap B$.*

(v) *If D° is convex and nonempty, then $D^\circ = \left(\overline{D}\right)^\circ$.*

Proof.

(i) Combining the assumption with the fact that $D \subset \mathrm{co}\, D$, we get $(\mathrm{co}\, D)^\circ \subset D \subset \mathrm{co}\, D$, and taking interiors in the previous chain of inclusions we obtain $(\mathrm{co}\, D)^\circ \subset D^\circ \subset (\mathrm{co}\, D)^\circ$, so that $D^\circ = (\mathrm{co}\, D)^\circ$, implying that D° is the interior of a convex set, hence convex.

(ii) It is well-known that if a convex set A has a nonempty interior, then $\overline{A} = \overline{(A^\circ)}$ (see, e.g., Proposition 2.1.8 in Volume I of [94]), so that $A \subset \overline{(A^\circ)}$. Taking $A = \mathrm{co}\, D$, we get that
$$D \subset \overline{D} \subset \overline{[(\mathrm{co}\, D)^\circ]}. \tag{4.26}$$

On the other hand, because $(\mathrm{co}\, D)^\circ \subset D$ by assumption, we have that
$$\overline{[(\mathrm{co}\, D)^\circ]} \subset \overline{D}. \tag{4.27}$$

By (4.26) and (4.27), $\overline{[(\mathrm{co}\, D)^0]}$ is a closed set containing D and contained in $\overline{D}$. The result follows.

(iii) Take $x \in (\cup_n C_n)^o$ and suppose that $x \notin \cup_n (C_n)^o$. In particular, $x \in \cup_n C_n$ and hence there exists m such that $x \in C_m = \overline{C_m^o} \subset \overline{\cup_n C_n^o} =: T$. The set T is a closed and convex set, and $T = T^o \cup \partial T$. Because $x \notin \cup_n (C_n)^o$, we have $x \in \partial T$. Because $T^o \neq \emptyset$, by Theorem 3.4.14(ii) there exists a supporting hyperplane H of T at x. In other words, $x \in H$ and $T^o = \cup_n C_n^o$ is contained in one of the open half-spaces H_{++} defined by H. Thus, for all n we have that $C_n = \overline{C_n^o}$ is contained in the closure H_+ of H_{++}, and hence $\cup_n C_n \subset H_+$. Taking interiors we get $(\cup_n C_n)^o \subset H_{++}$, but $x \in (\cup_n C_n)^o$ and hence $x \in H_{++}$, a contradiction.

(iv) This property follows at once from the definitions.

(v) Assume that $z \in (\overline{D})^o$, so that $z \in \overline{D} = D^o \cup \partial D$. If $z \notin D^o$, then $z \in \partial D$. Because $D^o \neq \emptyset$, there exists a supporting hyperplane to D^o at z. In other words, there exists a nonzero $v \in X^*$ such that $\langle z, v \rangle =: \alpha \geq \langle x, v \rangle$ for all $x \in \overline{D}$. Because $z \in (\overline{D})^o$, there exists $r > 0$ such that $B(z, r) \subset \overline{D}$. Take now $x := z + \frac{rv}{2\|v\|} \in B(z, r) \subset \overline{D}$. So $\langle x, v \rangle = \alpha + (r/2) > \alpha$, contradicting the separating property of v. $\Box$

Now we are ready to prove the main theorem of this section, which states that $D(T)^o$ is convex, with closure $\overline{D(T)}$. This is a result due to Rockafellar [186]. The proof of the fact that (a) implies (i) was taken from [172], and the remainder of the proof is based upon the original paper [186].

Theorem 4.5.3. *Assume that one of the conditions below holds:*

(a) The set $(\mathrm{co}(D(T)))^o \neq \emptyset$.

(b) X is reflexive and there exists a point $x_0 \in D(T)$ at which T is locally bounded.

Then

(i) $D(T)^o$ is a nonempty convex set whose closure is $\overline{D(T)}$.

(ii) If $\overline{D(T)} = X$, then $D(T) = X$.

Proof. We prove first that (a) implies (i). Call $B_n := \{x \in X \mid \|x\| \leq n\}$ and $B_n^* := \{v \in X^* \mid \|v\| \leq n\}$. For every fixed n, define the set

$$S_n := \{x \in B_n \mid Tx \cap B_n^* \neq \emptyset\}.$$

We claim that $C := (\mathrm{co}\, D(T))^o \subset D(T)$. We prove the claim in two steps.

Step 1. We prove that there exists n_0 such that $C \subset \cup_{n \geq n_0} (\overline{\mathrm{co}}\, S_n)^o$, where $(\overline{\mathrm{co}}\, S_n)^o \neq \emptyset$ for all $n \geq n_0$. It is easy to check that $S_n \subset S_{n+1}$ and

$$D(T) = \cup_n S_n \subset \cup_n \overline{\mathrm{co}}\, S_n.$$

Because $S_n \subset S_{n+1}$, we have $\overline{\mathrm{co}}\, S_n \subset \overline{\mathrm{co}}\, S_{n+1}$, and hence $\cup_n \overline{\mathrm{co}}\, S_n$ is convex. Thus, $\mathrm{co}\, D(T) \subset \cup_n \overline{\mathrm{co}}\, S_n$, and by definition of C, we get $C \subset \mathrm{co}\, D(T) \subset \cup_n \overline{\mathrm{co}}\, S_n$. Defining $C_n := C \cap \overline{\mathrm{co}}\, S_n$ we can write

$$C = \cup_n C_n,$$

where each C_n is closed relative to C. Inasmuch as C is an open subset of the Banach space X, we can apply Corollary 2.7.5 and conclude that there exists n_0 such that $(C_{n_0})^o \neq \emptyset$. Because the family of sets $\{C_n\}$ is increasing, it follows that $(C_n)^o \neq \emptyset$ for all $n \geq n_0$. Because $C_{n_0} \supset C_p$ for $p < n_0$, we obtain

$$C = \cup_{n \geq n_0} C_n \subset \cup_{n \geq n_0} \overline{co} S_n,$$

where $(\overline{co} S_n)^o \supset (C_n)^o \neq \emptyset$ for all $n \geq n_0$. Taking interiors and applying Lemma 4.5.2(iii) we conclude that

$$C \subset \cup_{n \geq n_0} (\overline{co} S_n)^o,$$

which completes the proof of the step.

Step 2. We prove that $(\overline{co} S_n)^o \subset D(T)$ for all $n \geq n_0$, with n_0 as in step 1. Fix $n \geq n_0$. Note that, because $(\overline{co} S_n)^o \neq \emptyset$ by step 1, we get that $S_n \neq \emptyset$. Then, the definition of S_n yields $R(T) \cap B_n{}^* \neq \emptyset$. Define

$$M_n = \{(x,v) \in \mathrm{Gph}(T) \mid \|v\| \leq n\} = \{(x,v) \in \mathrm{Gph}(T) \mid v \in B_n{}^*\}.$$

Note that $M_n \subset X \times B_n{}^*$ is a nonempty monotone subset of $X \times B_n{}^*$. Take now $x_0 \in (\overline{co} S_n)^o$ and define the family of subsets of X^*:

$$A_n(x_0) := \{w \in X^* \mid \langle w - u, x_0 - y \rangle \geq 0 \ \forall \ (y,u) \in \mathrm{Gph}(T), \|u\| \leq n\}$$

$$= \{w \in X^* \mid \langle w - u, x_0 - y \rangle \geq 0 \ \forall \ (y,u) \in M_n\}.$$

It is clear from this definition that $A_{n+1}(x_0) \subset A_n(x_0)$ for all $n \geq n_0$, and that $T(x_0) \subset A_n(x_0)$. Moreover, $A_n(x_0)$ is weak* closed for all $n \geq n_0$, because it is an intersection of closed half-spaces in X^*. In order to prove that each $A_n(x_0)$ is not empty, consider the function $\phi_n : B_n{}^* \to X$ defined as $\phi_n(v) = x_0$ for all $v \in B_n{}^*$. Clearly, ϕ_n is weak* strong continuous. By Theorem 4.3.1, we conclude that there exists $v_n \in B_n{}^*$ such that $\{(\phi_n(v_n), v_n)\} \cup M_n = \{(x_0, v_n)\} \cup M_n$ is monotone. Hence there exists $v_n \in A_n(x_0)$. Now we claim that each $A_n(x_0)$ is weak* compact. Because $x_0 \in (\overline{co} S_n)^o$, there exists $\varepsilon > 0$ such that $B(x_0, \varepsilon) \subset \overline{co} S_n$. Fix $w \in A_n(x_0)$. Then, for all $(y,u) \in \mathrm{Gph}(T)$,

$$\langle w, y - x_0 \rangle \leq \langle u, y - x_0 \rangle \leq \langle u, y \rangle - \langle u, x_0 \rangle \leq \langle u, y \rangle + n^2, \tag{4.28}$$

using the facts that $\|u\| \leq n$ and $x_0 \in S_n \subset B_n$. Taking $y \in S_n$, we get from (4.28)

$$\langle w, y - x_0 \rangle \leq 2n^2. \tag{4.29}$$

The inequality in (4.29) yields $S_n \subset \{z \in X \mid \langle w, z - x_0 \rangle \leq 2n^2\}$. Because $\{z \in X \mid \langle w, z - x_0 \rangle \leq 2n^2\}$ is convex, we have that

$$\overline{co} S_n \subset \{z \in X \mid \langle w, z - x_0 \rangle \leq 2n^2\}. \tag{4.30}$$

Now take $h \in X$ with $\|h\| \leq \varepsilon$, so that $x_0 + h \in B(x_0, \varepsilon) \subset \overline{co} S_n$. Using (4.30), we get

$$\langle w, h \rangle = \langle w, (x_0 + h) - x_0 \rangle \leq 2n^2.$$

Therefore, $\varepsilon\|w\| = \sup_{\|h\|\leq\varepsilon}\langle w, h\rangle \leq 2n^2$. This gives $\|w\| \leq (2n^2/\varepsilon)$, and hence $A_n(x_0) \subset B(0, 2n^2/\varepsilon)$, which implies that for each $n \geq n_0$, the set $A_n(x_0)$ is weak* compact, establishing the claim. Because $A_{n+1}(x_0) \subset A_n(x_0)$, we invoke the finite intersection property of decreasing families of compact sets, for concluding that there exists some $v_0 \in \cap_{n\geq n_0}A_n(x_0)$. We claim that (x_0, v_0) is monotonically related to $\mathrm{Gph}(T)$, in the sense of Definition 4.2.3. We proceed to prove the claim. Take $(y, u) \in \mathrm{Gph}(T)$. There exists $p \geq n_0$ such that $\|u\| \leq p$. Because $v_0 \in A_p(x_0)$, we have $\langle v_0 - u, x_0 - y\rangle \geq 0$, and the claim holds. By maximality of T, $v_0 \in T(x_0)$, which yields $x_0 \in D(T)$, completing the proof of step 2.

It follows from steps 1 and 2 that $C \subset D(T)$, and so C is convex by Lemma 4.5.2(i). The fact that the closure of C is the closure of $D(T)$ follows from Lemma 4.5.2(ii) with $D = D(T)$, and hence we have established that (a) implies (i). Next we prove that (b) implies (i). Take $x_0 \in D(T)$ at which T is locally bounded, and an open and convex neighborhood U of x_0 such that $T(U)$ is bounded. Because X is reflexive we can use weak instead of weak* topology. Call B the weak closure of $T(U)$. Then B is weakly compact by Theorem 2.5.16(i). The latter result also implies that B is bounded. By [186, Lemma 2], $T^{-1}(B) = \{x \in X \mid Tx \cap B \neq \emptyset\}$ is weakly closed. We claim that $U \subset D(T)$. Otherwise, the definitions of B and U yield

$$U \cap D(T) \subset T^{-1}(B) \subset D(T).$$

Because $T^{-1}(B)$ is weakly closed, we get

$$\overline{U \cap D(T)} \subset T^{-1}(B) \subset D(T).$$

By Lemma 4.5.2(iv) applied to $A := U$ and $B := D(T)$ we obtain

$$U \cap \overline{D(T)} \subset U \cap D(T). \tag{4.31}$$

Thus, every boundary point of $D(T)$ that is also in U must belong to $D(T)$. By Theorem 4.4.10, the set $\overline{D(T)}$ is convex. By Corollary 5.3.13, the set of points of $\overline{D(T)}$ at which $\overline{D(T)}$ has a supporting hyperplane is dense in $\partial D(T)$, and hence some of those points belong to U. Using (4.31), we conclude that those supporting points are in $D(T)$. Now the same argument used in the last part of the proof of Theorem 4.2.10(ii) can be used to prove that Tx_0 is unbounded, which contradicts the fact that $T(U)$ is bounded. The statement (ii) of the theorem follows from the facts that $\emptyset \neq U \subset D(T)^o$ and that (a) implies (i).

For proving that (a) or (b) imply (ii), we show that (i) implies (ii). We proved already that $(D(T))^o$ is a nonempty convex set, so that $(\mathrm{co}\, D(T))^o = (D(T))^o$, and hence, by Lemma 4.5.2(ii), we have

$$\overline{D(T)} = \overline{(D(T))^o}. \tag{4.32}$$

Applying now Lemma 4.5.2(v) to the set $A := (D(T))^o$, we get from (4.32),

$$(D(T))^o = ((\overline{(D(T))^o})^o = (\overline{D(T)})^o.$$

Because $\overline{D(T)} = X$ by the assumption of conclusion (ii), we get $X = (D(T))^o$ from which $X = D(T)$ follows immediately. $\square$

4.6 Inner semicontinuity

In this section we study inner-semicontinuity of a maximal monotone operator. We follow the approach of [171]. The next result is a stronger version of Theorem 4.2.10.

Theorem 4.6.1. *Let X be a Banach space with dual X^*, and $T : X \rightrightarrows X^*$ a maximal monotone operator such that $(\operatorname{co} D(T))^o \neq \emptyset$. Then*

(i) $D(T)^o$ is convex and its closure is $\overline{D(T)}$.

(ii) T is locally bounded at any $x \in D(T)^o$.

(iii) For all $z \in D(T) \setminus (D(T))^o$:

    (a) There exists a nonzero $w \in N_{D(T)}z$.

    (b) $Tz + N_{D(T)}z \subset Tz$.

    (c) T is not locally bounded at z.

Proof. Item (i) is just a restatement of Theorem 4.5.3(i) and item (ii) follows from Theorem 4.2.10(i). In view of Theorem 4.2.10(ii), for establishing (iii) it suffices to prove that under the new assumption we have $D(T)^o = (\overline{\operatorname{co}}D(T))^o$. Because $D(T)^o \subset D(T) \subset \overline{D(T)}$, we have $\operatorname{co}(D(T)^o) \subset \operatorname{co} D(T) \subset \operatorname{co} \overline{D(T)}$. By (i), $D(T)^o$ is convex, and $\overline{D(T)}$ is the closure of the convex set $D(T)^o$, hence convex by Proposition 3.4.4. Thus,

$$D(T)^o \subset \operatorname{co} D(T) \subset \overline{D(T)}. \tag{4.33}$$

Let $A = D(T)^o$. Taking closures throughout (4.33) and using again the fact that $\overline{A} = \overline{D(T)}$, we get $\overline{\operatorname{co}}D(T) = \overline{A}$, so that $(\overline{\operatorname{co}}D(T))^o = \overline{A}^o = A$, using Proposition 3.4.15 in the last equality. In view of the definition of A, the claim holds. $\square$

Next we recall a well-known fact on convex sets.

Lemma 4.6.2. *Let C be a convex set. If $x \in C^o$ and $y \in C$, then $\alpha x + (1 - \alpha)y \in C^o$ for all $\alpha \in (0, 1]$.*

Proof. See, for instance, Theorem 2.23(b) in [223]. $\square$

Theorem 4.6.3. *Let $T : X \rightrightarrows X^*$ be maximal monotone and suppose that $(\operatorname{co}(D(T)))^o \neq \emptyset$. Then*

(i) T is not (sw^*)-ISC at any boundary point of $D(T)$.

(ii) Tx is a singleton if and only if T is (sw^*)-ISC at x.

Proof.

(i) Let $x \in D(T)$ be a boundary point of $D(T)$. Theorem 4.6.1(iii-a) implies that there exists a nonzero element $w \in N_{D(T)}(x)$. Therefore,

$$D(T) \subset \{z \in X \mid \langle w, z \rangle \leq \langle w, x \rangle\}. \tag{4.34}$$

By Theorem 4.5.3(i), $(D(T))^o$ is a nonempty convex set and, taking interiors in both sides of (4.34), we get

$$(D(T))^o \subset \{z \in X \mid \langle w, z \rangle < \langle w, x \rangle\}. \tag{4.35}$$

Fix $y \in (D(T))^o$. By (4.35), there exists $\alpha < 0$ such that $\langle w, y - x \rangle < \alpha$. Because $N_{D(T)}(x)$ is a cone, we can assume that $w \in N_{D(T)}(x)$ is such that $\langle w, y - x \rangle < -1$. Suppose now that T is (sw*)-ISC at x and fix $u \in Tx$. Because $Tx + N_{D(T)}x \subset Tx$ by Theorem 4.2.10(ii-b), we conclude that $u + w \in Tx$. Define the sequence $x_n := ((n-1)/n)x + (1/n)y$, which strongly converges to x for $n \to \infty$. By Lemma 4.6.2, $\{x_n\} \subset (D(T))^o$. By (sw*)-inner-semicontinuity of T at x, we have

$$Tx \subset \limint_{n \in \mathbb{N}} Tx_n. \tag{4.36}$$

Combining the above inclusion with the fact that $u + w \in Tx$, we conclude that every weak* neighborhood of $u + w$ must meet the net of sets $\{Tx_n\}_n$ eventually. Consider the weak* neighborhood of $u + w$ (see (2.19)) given by

$$W := \{z \in X^* : \ |\langle z - (u + w), y - x \rangle| < \tfrac{1}{2}\}.$$

By (4.36) we have that there exists $n_0 \in \mathbb{N}$ such that $W \cap Tx_j \neq \emptyset$ for all $j \geq n_0$. Define now the net $\{v_j\}_{j \geq n_0}$ such that $v_j \in W \cap Tx_j$ for all $j \geq n_0$. By monotonicity and definition of x_j we can write

$$
\begin{aligned}
0 \ &\leq \ \langle v_j - u, x_j - x \rangle = \langle v_j - (u + w), x_j - x \rangle + \langle w, x_j - x \rangle \\
&= \ \tfrac{1}{j}\langle v_j - (u + w), y - x \rangle + \tfrac{1}{j}\langle w, y - x \rangle \\
&< \ \tfrac{1}{2j} - \tfrac{1}{j} < 0,
\end{aligned}
$$

where we also used the definition of W in the inequality. Having thus arrived at a contradiction, we conclude that T cannot be (sw*)-ISC at a point x in the boundary of $D(T)$.

(ii) Assume first that Tx is a singleton. Then we must have $x \in D(T)^o$. Because T is maximal monotone and $x \in D(T)^o$, we have by Corollary 4.2.12 that T is (sw*)-outer-semicontinuous at x. So we are in the conditions of Proposition 2.5.21(ii), which gives T (sw*)-upper-semicontinuous at x. Using now Proposition 2.5.21(iii) we get (sw*)-inner-semicontinuity of T at x. Conversely, assume that T is (sw*)-ISC at x. Suppose that there exist $v_1, v_2 \in Tx$, with $w := v_1 - v_2 \neq 0$. Thus, there exists $z \in X$ such that $\langle w, z \rangle = a > 0$. Define the sequence $x_n := x + (z/n)$. Then $\{x_n\}$ converges strongly to x. Because T is (sw*)-ISC at x, we must have

$T(x) \subset \mathrm{w}^* - \lim\mathrm{int}_n T(x_n)$. Because the set $W := \{u \in X^* : |\langle v_2 - u, z\rangle| < a/2\}$ is a weak* neighborhood of $v_2 \in T(x)$, there exists $n_0 \in \mathbb{N}$ such that

$$W \cap T(x_n) \neq \emptyset, \quad \text{for all } n \geq n_0.$$

Choose $u_n \in W \cap T(x_n)$ for all $n \geq n_0$. Then we have

$$
\begin{aligned}
0 &\leq \langle u_n - v_1, x_n - x\rangle = \tfrac{1}{n}\langle u_n - v_1, z\rangle \\
&= \tfrac{1}{n}\langle u_n - v_2, z\rangle + \tfrac{1}{n}\langle v_2 - v_1, z\rangle \\
&\leq \tfrac{a}{2n} - \tfrac{a}{n} < 0,
\end{aligned}
$$

a contradiction, which implies that Tx is a singleton. $\quad\square$

The following result can be seen as a converse of Theorem 4.2.4.

Theorem 4.6.4. *Let X be a Banach space and $T : X \to X^*$ a point-to-point monotone operator such that $(\mathrm{co}\, D(T))^0 \neq \emptyset$. If T is maximal, then $D(T)$ is open and T is continuous with respect to the strong topology in X and the weak* topology in X^* at every point of $D(T)$.*

Proof. Assume that there exists a point $x \in D(T)$ which is at the boundary of $D(T)$. Then by Theorem 4.6.1(iii-c) Tx is unbounded, contradicting the fact that T is point-to-point. Therefore, $D(T) \subset (D(T))^\circ$ and hence $D(T)$ is open. Fix $x \in D(T)$. From Theorem 4.6.3(ii) and the fact that Tx is a singleton we know that T is (sw*)-ISC. Because T is maximal and $x \in D(T) = D(T)^\circ$, it is (sw*)-OSC at x by Corollary 4.2.12, and the conclusion follows. $\quad\square$

A remarkable property of maximal monotone operators is that, for a large class of Banach spaces, they are point-to-point except in a "small" subset of their domains. The required assumption on the Banach space X is the *separability* of its dual X^*, meaning the existence of a countable dense subset of X^* (recall that in a metric space separability is equivalent to the second countability axiom, i.e., the existence of a countable base of open sets). Among the Banach spaces with separable dual, we mention the spaces ℓ_p, and $\mathcal{L}^p[\alpha, \beta]$ with $1 < p < \infty$ ($\alpha < \beta \in \mathbb{R}$), and more generally, all reflexive separable Banach spaces. An example of a nonreflexive Banach space that also enjoys this property is the space c_0 of sequences with a finite set of nonnull terms, with the norm of the supremum, whose dual is the separable space ℓ_1.

We prove in the next theorem that the subset where a maximal monotone operator is single-valued is, generally speaking, much "larger" than a dense set. We need some preliminary definitions in order to introduce the appropriate class of "large" sets.

In the following three definitions, X is a Banach space with dual X^*.

Definition 4.6.5. *Given $0 \neq z \in X^*$ and $\alpha \in (0,1)$, the revolution cone $K(z,\alpha)$ is defined as $K(z,\alpha) = \{x \in X : \langle z, x\rangle \geq \alpha \|x\| \|z\|\}$.*

When X is a Hilbert space, we can geometrically describe $K(z, x)$ as the set of points in X that make an angle with z not bigger that $\arccos \alpha$. In general, we say that z is the *axis* and $\arccos \alpha$ is the *angle* of $K(z, \alpha)$. It is easy to check that revolution cones are closed and convex and that $K(z, \alpha)^o = \{x \in X : \langle z, x \rangle > \alpha \|x\| \|z\|\}$.

Definition 4.6.6. *Given $\alpha \in (0, 1)$, a subset C of X is α-cone meager if for all $x \in C$ and all $\varepsilon > 0$ there exists $y \in B(x, \varepsilon)$ and $0 \neq z \in X^*$ such that $C \cap [y + K(z, \alpha)^o] = \emptyset$.*

In words, C is α-cone meager if arbitrarily close to any point in C lies the vertex of a shifted revolution cone with angle $\arccos \alpha$ whose interior does not meet C. Clearly, α-cone meager sets have an empty interior for all $\alpha \in (0, 1)$. It is also easy to check that the closure of an α-cone meager set has an empty interior. Note that, because in the real line cones are halflines, for $X = \mathbb{R}$ an α-cone meager set can contain at most two points.

Definition 4.6.7. *A subset C of X is said to be* angle-small *if for any $\alpha \in (0, 1)$ it can be expressed as a countable union of α-cone meager sets.*

The closure of an α-cone meager set has an empty interior, therefore any angle-small set is contained in a countable union of closed and nowhere dense sets, hence it is of first category, but the converse inclusion does not hold: a first category set might fail to be angle-small: in view of the observation above, angle-small subsets of $\mathbb{R}$ are countable, although there are first category sets that are uncountable, for example, the Cantor set. In this sense, angle-small sets are generally "smaller" than first category ones, and so their complements are generally "larger" than residual ones. We prove next that a monotone operator defined on a Banach space with separable dual is point-to-point except in an angle-small set.

Theorem 4.6.8. *Let X be a Banach space with separable dual X^* and $T : X \rightrightarrows X^*$ a monotone operator. Then there exists an angle-small subset A of $D(T)$ such that T is point-to-point at all $x \in D(T) \setminus A$. If in addition T is maximal monotone, then it is (ss)-continuous at all $x \in D(T) \setminus A$.*

Proof. Let $A = \{x \in D(T) : \lim_{\delta \to 0^+} \operatorname{diam}[T(B(x, \delta))] > 0\}$. We claim that A is angle-small. Write $A = \cup_{n \in \mathbb{N}} A_n$ with

$$A_n = \left\{ x \in D(T) : \lim_{\delta \to 0^+} \operatorname{diam}[T(B(x, \delta))] > \frac{1}{n} \right\}. \tag{4.37}$$

Let $\{z_k\}_{k \in \mathbb{N}} \subset X^*$ be a dense subset of X^* and take $\alpha \in (0, 1)$. Define

$$A_{n,k} = \left\{ x \in A_n : d(z_k, T(x)) < \frac{\alpha}{4n} \right\}. \tag{4.38}$$

Because $\{z_k\}$ is dense, $A_n = \cup_{k \in \mathbb{N}} A_{n,k}$, so that $A = \cup_{(n,k) \in \mathbb{N} \times \mathbb{N}} A_{n,k}$, and hence for establishing the claim it suffices to prove that $A_{n,k}$ is α-cone meager for all $(n, k) \in \mathbb{N} \times \mathbb{N}$, which we proceed to do.

Take $x \in A_{n,k}$ and $\varepsilon > 0$. We must exhibit a cone with angle $\arccos \alpha$ and vertex belonging to $B(x, \varepsilon)$ whose interior does not intersect $A_{n,k}$. Because x belongs to A_n, there exist $\delta \in (0, \varepsilon)$, points $x^1, x^2 \in B(x, \delta)$, and points $u^1 \in T(x^1), u^2 \in T(x^2)$ such that $\|u^1 - u^2\| > 1/n$. Thus, for any $u \in T(x)$ either $\|u^1 - u\| > 1/(2n)$ or $\|u^2 - u\| > 1/(2n)$. Because $d(z_k, T(x)) < \alpha/(4n)$, we can choose $\bar{z} \in T(x)$ such that $\|z_k - \bar{z}\| < \alpha/(4n)$ and hence there exist $y \in B(x, \varepsilon)$ and $\hat{z} \in T(y)$ such that

$$\|\hat{z} - z_k\| \geq \|\hat{z} - \bar{z}\| - \|z_k - \bar{z}\| > \frac{1}{2n} - \frac{\alpha}{4n} > \frac{1}{4n}, \qquad (4.39)$$

using the fact that $\alpha \in (0, 1)$, together with (4.37) and (4.38). Let now $z = \hat{z} - \bar{z}$. We claim that $y + K(z, \alpha)$ is the shifted revolution cone whose interior does not meet $A_{n,k}$. We must show that $A_{n,k} \cap [y + K(z, \alpha)^o] = \emptyset$. Note first that

$$y + K(z, \alpha)^o = \{v \in X : \langle \hat{z} - \bar{z}, v - y \rangle > \alpha \|\hat{z} - \bar{z}\| \|v - y\| \}. \qquad (4.40)$$

Take $v \in D(T) \cap [y + K(z, \alpha)^o]$ and $u \in T(v)$. Then

$$\|u - z_k\| \|v - y\| \geq \langle u - z_k, v - y \rangle = \langle u - \hat{z}, v - y \rangle + \langle \hat{z} - z_k, v - y \rangle$$

$$\geq \langle \hat{z} - z_k, v - y \rangle > \alpha \|\hat{z} - z_k\| \|v - y\| \geq \frac{\alpha}{4n} \|v - y\|, \qquad (4.41)$$

using monotonicity of T in the second inequality, (4.40) together with the fact that v belongs to $y + K(z, \alpha)^o$ in the third one, and (4.39) in the fourth one. It follows from (4.41) that $\|u - z_k\| \geq \alpha/(4n)$. Because u belongs to $T(v)$, we get from (4.38) that v does not belong to $A_{n,k}$. Hence, $A_{n,k} \cap [y + K(z, \alpha)^o] = \emptyset$, as required, $A_{n,k}$ is α-cone meager, and A is angle-small, establishing the claim. We claim now that $T(x)$ is a singleton for all $x \in D(T) \setminus A$. Take $x \in D(T) \setminus A$. Because $\mathrm{diam}[T(B(x,t))]$ is non increasing as a function of t, if $x \notin A$ then the definition of A implies that $\lim_{\delta \to 0^+} \mathrm{diam}[T(B(x, \delta))] = 0$. Because $T(x) \subset T(B(x, \delta))$ for all $\delta > 0$, we get $\mathrm{diam}[T(x)] = 0$, so that $T(x)$ is a singleton. We have proved single-valuedness of T outside an angle-small set.

We establish now the statement on continuity of T in $D(T) \setminus A$. In view of Theorem 4.6.3(ii), T is (sw*)-ISC, and a fortiori (ss)-ISC, in $D(T) \setminus A$, where it is single-valued. By Proposition 4.2.1(ii), T is (ss)-OSC. We conclude that T is (ss)-continuous at all $x \in D(T) \setminus A$. $\quad \square$

The subdifferential of a convex function f defined on an open convex subset of a Banach X space is monotone, as follows easily from its definition, therefore it reduces to a singleton outside an angle-small subset of its domain when X^* is separable, as a consequence of Theorem 4.6.8. It can be seen that a convex function is Fréchet differentiable at any point where its subdifferential is a singleton (see Proposition 2.8 and Theorem 2.12 in [171]). A Banach space X is said to be an *Asplund* space if every continuous function defined on an open convex subset U of

X is Fréchet differentiable on a residual subset of U. It follows from Theorem 4.6.8 and the above-discussed relation between angle-small and residual sets that Banach spaces with separable duals are Asplund spaces. In fact the converse result also holds: a space is Asplund if and only if its dual is separable (see Theorem 2.19 in [171]).

4.7   Maximality of subdifferentials

We already mentioned at the end of Section 3.5 that the subdifferential of a proper, convex, and lsc function is maximal monotone. This result, due to Rockafellar [189], is one of the most important facts relating maximal monotonicity and convexity. In fact, it allows us to use subdifferentials as a testing tool for all maximal monotone operators. The proof of the following theorem given below is due to Simons [204, Theorem 2.2].

Theorem 4.7.1. *If X is a Banach space and $f : X \to \mathbb{R} \cup \{\infty\}$ a proper and lower-semicontinuous convex function, then the subdifferential of f is maximal monotone.*

Proof. It follows easily from Definition 3.5.1 that $\langle x - y, u - v \rangle \geq 0$ for all $x, y \in \mathrm{Dom}(f)$, all $u \in \partial f(x)$, and all $v \in \partial f(y)$, establishing monotonicity of ∂f. We proceed to prove maximality. We claim that for every $x_0 \in X$, and every $v_0 \notin \partial f(x_0)$, there exists $(z, u) \in \mathrm{Gph}(T)$ such that $\langle x_0 - z, v_0 - u \rangle < 0$. We proceed to prove the claim. Fix $x_0 \in X$ and $v_0 \notin \partial f(x_0)$. We must find $(z, u) \in \mathrm{Gph}(T)$ which is not monotonically related to (x_0, v_0). Define $g : X \to \overline{\mathbb{R}}$ as $g(y) := f(y) - \langle v_0, y \rangle$. Then $v_0 \notin \partial f(x_0)$ if and only if $0 \notin \partial g(x_0)$, in which case x_0 is not a global minimizer of g; that is,
$$\inf_{y \in X} g(x) < g(x_0).$$
Using now Corollary 5.3.14, applied to the function g, we conclude that there exists $z \in \mathrm{Dom}(g) = \mathrm{Dom}(f)$ and $w \in \partial g(z) = \partial f(z) - v_0$ such that
$$g(z) < g(x_0) \quad \text{and} \quad \langle w, z - x_0 \rangle < 0. \tag{4.42}$$
Because $w \in \partial g(z) = \partial f(z) - v_0$, there exists $u \in \partial f(z)$ such that $w = u - v_0$. In view of the rightmost inequality in (4.42), we get
$$\langle x_0 - z, v_0 - u \rangle = \langle x_0 - z, (-w) \rangle < 0,$$
proving that $(z, u) \in \mathrm{Gph}(T)$ is not monotonically related to (x_0, v_0). We have proved the claim, showing that no pair (x_0, v_0) can be added to $\mathrm{Gph}(T)$ while preserving monotonicity. In other words, T is maximal monotone. This completes the proof.   □

Corollary 4.7.2.

    (i) If X is a Banach space and $f : X \to \mathbb{R} \cup \{\infty\}$ a proper and lower-semicontinuous strictly convex function, then the subdifferential of f is maximal monotone and strictly monotone.

(ii) *A strictly monotone operator* $T : X \rightrightarrows X^*$ *can have at most one zero.*

Proof. Item (i) follows easily from Definitions 3.4.2(c), 4.1.2, and Theorem 4.7.1. For (ii), assume that $0 \in Tx \cap Ty$ with $x \neq y$. It follows from Definition 4.1.2 that $0 = \langle x - y, 0 - 0 \rangle > 0$, a contradiction. □

Next we establish maximal monotonicity of saddle point operators, induced by Lagrangians associated with cone constrained vector-valued convex optimization problems.

Let X_1 and X_2 be real reflexive Banach spaces and $K \subset X_2$ a closed and convex cone.

Definition 4.7.3. *A function* $H : X_1 \rightarrow X_2$ *is* K-*convex if and only if* $\alpha H(x) + (1 - \alpha)H(x') - H(\alpha x + (1 - \alpha)x') \succsim 0$ *for all* $x, x' \in X_1$ *and all* $\alpha \in [0, 1]$.

An easy consequence of this definition is the following generalization of the gradient inequality.

Proposition 4.7.4. *If* $H : X_1 \rightarrow X_2$ *is* K-*convex and Gâteaux differentiable, then* $H(x) - H(x') - H'(x')(x - x') \in K$ *for all* $x, x' \in X_1$, *where the linear operator* $H'(x') : X_1 \rightarrow X_2$ *is the Gâteaux derivative of* H *at* x'.

Proof. Elementary from the definition of K-convexity (see, e.g., Theorem 6.1 in [57]). □

The problem of interest is

$$\min g(x) \tag{4.43}$$

$$\text{s.t. } - G(x) \in K, \tag{4.44}$$

with $g : X_1 \rightarrow \mathbb{R}$, $G : X_1 \rightarrow X_2$, satisfying

(A1) g is convex and G is K-convex.

(A2) g and G are continuously Fréchet differentiable functions with Gâteaux derivatives denoted g' and G', respectively.

We recall that the positive polar cone $K^* \subset X_2^*$ of K is defined as $K^* = \{y \in X_2^* : \langle y, z \rangle \geq 0 \; \forall z \in K\}$ (cf. Definition 3.7.7). We define the Lagrangian $L : X_1 \times X_2^* \rightarrow \mathbb{R}$, as

$$L(x, y) = g(x) + \langle y, G(x) \rangle, \tag{4.45}$$

where $\langle \cdot, \cdot \rangle$ denotes the duality pairing in $X_2^* \times X_2$, and the saddle point operator $T_L : X_1 \times X_2^* \rightrightarrows X_1^* \times X_2$ as

$$T_L(x, y) = (g'(x) + [G'(x)]^*(y), -G(x) + N_{K^*}(y)), \tag{4.46}$$

where $[G'(x)]^* : X_2^* \rightarrow X_1^*$ is the adjoint of the linear operator $G'(x) : X_1 \rightarrow X_2$ (cf. Definition 3.13.7), and $N_{K^*} : X_2^* \rightrightarrows X_2$ denotes the normality operator of the cone K^*. Note that $T_L = (\partial_x L, \partial_y(-L) + N_{K^*})$, where $\partial_x L$, $\partial_y(-L)$ denote the subdifferentials of $L(\cdot, y)$, $-L(x, \cdot)$, respectively.

In the finite-dimensional case (say $X_1 = \mathbb{R}^n$, $X_2 = \mathbb{R}^m$), if $G(x) = (g_1(x), \ldots, g_m(x))$ and $K = \mathbb{R}^m_+$, then L reduces to the standard Lagrangian $L : \mathbb{R}^n \times \mathbb{R}^m_+ \to \mathbb{R}$, given by $L(x, y) = g(x) + \sum_{i=1}^m y_i g_i(x)$. See a complete development of this topic in Sections 6.5, 6.6, and 6.8. In Chapter 6 we need the following result.

Theorem 4.7.5. *The operator T_L defined by (4.46) is maximal monotone.*

Proof. In order to establish monotonicity of T_L, we must check that

$$0 \leq \langle (u, w) - (u', w'), (x, y) - (x', y') \rangle \tag{4.47}$$

for all (x, y), $(x', y') \in D(T_L)$, all $(u, w) \in T_L(x, y)$, and all $(u', w') \in T_L(x', y')$. In view of (4.46), (4.47) can be rewritten as

$$
\begin{aligned}
0 \;\leq\; & \langle g'(x) - g'(x'), x - x' \rangle + \langle G'(x)^* y - G'(x')^* y', x - x' \rangle \\
& + \langle y - y', G(x') - G(x) \rangle + \langle y - y', v - v' \rangle \\
=\; & \langle g'(x) - g'(x'), x - x' \rangle + \langle y, G(x') - G(x) - G'(x)(x' - x) \rangle \\
& + \langle y', G(x) - G(x') - G'(x')(x - x') \rangle + \langle y - y', v - v' \rangle,
\end{aligned} \tag{4.48}
$$

for all (x, y), $(x', y') \in D(T_L)$, all $v \in N_{K^*}(y)$, and all $v' \in N_{K^*}(y')$, using the definition of the adjoint operators $G'(x)^*$, $G'(x')^*$ in the equality.

We look now at the four terms in the rightmost expression of (4.48). The first one and the last one are nonnegative by Theorem 4.7.1, because g is convex, and N_{K^*} is the subdifferential of the convex function δ_{K^*}. By definition of T_L and N_{K^*}, $D(T_L) = X_1 \times K^*$, so that we must check nonnegativity of the remaining two terms only for $y, y' \in K^*$. By K-convexity of G and Proposition 4.7.4, we have $G(x) - G(x') - G'(x')(x - x') \in K$, $G(x') - G(x) - G'(x)(x' - x) \in K$, and thus the two middle terms in the rightmost expression of (4.48) are nonnegative by definition of the positive polar cone K^*.

We have established monotonicity of T_L and we proceed to prove maximality. Take a monotone operator $\widehat{T}$ such that $T_L \subset \widehat{T}$. We need to show that $\widehat{T} = T_L$, for which it suffices to establish that given $(x, y) \in X_1 \times X_2^*$ and $(u, w) \in \widehat{T}(x, y)$, we have in fact $(u, w) \in T_L(x, y)$; that is, $u = g'(x) + G'(x)^* y$, $w = -G(x) + v$ with $v \in N_{K^*}(y)$. Defining $a := u - g'(x) - G'(x)^* y$, $b := w + G(x)$, it is enough to prove that $a = 0$, $b \in N_{K^*}(y)$. Because (u, w) belongs to $\widehat{T}(x, y)$ and $T_L \subset \widehat{T}$, we have, by monotonicity of $\widehat{T}$,

$$0 \leq \langle (u, w) - (u', w'), (x, y) - (x', y') \rangle$$

for all $(x', y') \in X \times X_2^*$ and all $(u', w') \in T_L(x', y')$; that is, by definition of T_L, taking into account the definitions of a and b given above,

$$
\begin{aligned}
0 \;\leq\; & \langle g'(x) - g'(x') + G'(x)^* y - G'(x')^* y' + a, x - x' \rangle \\
& + \langle y - y', G(x') - G(x) + b - v' \rangle,
\end{aligned} \tag{4.49}
$$

for all $(x', y') \in X \times X_2^*$ and all $v' \in N_{K^*}(y')$. Take first $x' = x$, so that (4.49) reduces to $0 \leq \langle y - y', b - v' \rangle$ for all $v' \in N_{K^*}(y')$. Because $N_{K^*} = \partial \delta_{K^*}$ is maximal monotone by Theorem 4.7.1, we conclude that b belongs to $N_{K^*}(y)$, because otherwise the operator $\widehat{N}$, defined as $\widehat{N}(z) = N_{K^*}(z)$ for $z \neq y$, and $\widehat{N}(y) = N_{K^*}(y) \cup \{b\}$, would be a monotone operator strictly bigger than N_{K^*}.

It remains to prove that $a = 0$. Take now $y' = y$, $x' = x + ta$, with $t > 0$. Then, (4.49) becomes

$$0 \leq t \langle g'(x + ta) - g'(x), a \rangle + t \langle y, [G'(x + ta) - G'(x)]a \rangle - t \|a\|^2, \qquad (4.50)$$

using again the definition of the adjoint operators $G'(x)^*$, $G'(x')^*$. Dividing both sides of (4.50) by $t > 0$, and taking the limit with $t \to 0^+$, we obtain, using A2, $0 \leq -\|a\|^2$, so that $a = 0$. We have proved that (u, w) belongs indeed to $T_L(x, y)$, and thus T_L is maximal monotone. $\quad\square$

4.8 Sum of maximal monotone operators

The subdifferential of the sum of two convex functions may not be equal to the sum of the subdifferentials, unless some "constraint qualification" holds (see Theorem 3.5.7). This happens because the sum of the subdifferentials may fail to be maximal. In the same way, if we sum two maximal monotone operators, it is clear that the resulting operator will be monotone, but it may not be maximal. For instance, if the domains of the operators have an empty intersection, then the domain of the sum is empty, and hence the sum will never be maximal. Determining conditions under which the sum of maximal monotone operators remains maximal is one of the central problems in the theory of point-to-set mappings. Most of the results regarding this issue hold in a reflexive Banach space, and very little can be said for nonreflexive Banach spaces. Therefore, we restrict our attention to the reflexive case. The most important result regarding this issue is due to Rockafellar [190], and establishes the maximality of the sum of maximal monotone operators when the interior of the domain of one of the operators intersects the domain of the other one. There are weaker constraint qualifications that ensure (always in reflexive spaces) maximality of the sum (see, e.g., [11, 64]). However, they are in nature very similar to the original one given by Rockafellar. Before proving Theorem 4.8.3, we need some preliminary results, also due to Rockafellar. All the proofs of this section were taken from [190]. We have already mentioned that every reflexive Banach space can be renormed in such a way that J and J^{-1} are single-valued (see [8]). Throughout this section we always assume that the norm of X already has these special properties.

Proposition 4.8.1. *Let X be reflexive and $T : X \rightrightarrows X^*$ a maximal monotone operator. If there exists $r > 0$ such that $\langle v, x \rangle \geq 0$ whenever $\|x\| > r$ and $(x, v) \in \text{Gph}(T)$, then there exists $x \in X$ such that $0 \in Tx$.*

Proof. Call $B_r := \{x \in X \mid \|x\| \leq r\}$. The conclusion of the theorem will follow from the fact that $0 \in T(B_r)$. The set $T(B_r)$ is weakly closed in X (see [186, Lemma 2]). So for proving that $0 \in T(B_r)$ it is enough to show that $0 \in \overline{T(B_r)}$. In other words, we prove that for every $\varepsilon > 0$ there exists $u \in T(B_r)$ such that $\|u\| \leq \varepsilon$. By Theorem 4.4.7 and Remark 4.4.8, we know that $(T + (\varepsilon/r)\,J)$ is onto. This means that there exists $x \in X$ such that

$$0 \in \left(T + \frac{\varepsilon}{r}\,J\right)x. \tag{4.51}$$

By (4.51) and the definition of J we can write $0 = u + (\varepsilon/r)\,Jx$ with $u \in Tx$, which gives

$$\langle x, u \rangle = \frac{-\varepsilon}{r}\langle x, Jx \rangle = \frac{-\varepsilon}{r}\|x\|^2 < 0. \tag{4.52}$$

It follows from (4.52) and the hypothesis of the theorem that $\|x\| \leq r$, implying that $u \in Tx \subset T(B_r)$ and also

$$\|u\| = \frac{\varepsilon}{r}\|Jx\| = \frac{\varepsilon}{r}\|x\| \leq \varepsilon,$$

yielding $\|u\| \leq \varepsilon$. Because $\varepsilon > 0$ is arbitrary, we proved that $0 \in \overline{T(B_r)} = T(B_r)$ and hence the conclusion of the theorem holds. $\square$

Theorem 4.4.7 and Proposition 4.8.1 show that the "perturbations" $T + \lambda J$ with $\lambda > 0$ are a very important device for analyzing and understanding T. Another important way of perturbing T is by "truncating" it. More precisely, for each $r > 0$, denote N_r the normality operator associated with the set $B_r := \{x \in X \mid \|x\| \leq r\}$, where the norm has the special property that both J and J^{-1} are single-valued. As observed before, N_r is the subdifferential of the indicator function of B_r and hence N_r is maximal monotone by Theorem 4.7.1. Recall that $N_r(x) = \{0\}$ whenever $\|x\| < r$, $N_r(x) = \emptyset$ when $\|x\| > r$, and also

$$N_r(x) := \{\lambda\,Jx \mid \lambda \geq 0\} \quad \text{when } \|x\| = r, \tag{4.53}$$

by Proposition 4.4.4(v), using the fact that J is single-valued.

Proposition 4.8.2. *Let X be reflexive, and $T : X \rightrightarrows X^*$ a monotone operator such that $0 \in D(T)$. Let $N_r : X \rightrightarrows X^*$ be the normality operator of the set $B_r = \{x \in X : \|x\| \leq r\}$. If there exists $r_0 > 0$ such that $T + N_{r_0}$ is maximal for every $r \geq r_0$, then T is maximal.*

Proof. Replacing T by $T(\cdot) - v$, where $v \in T(0)$, we can assume that $0 \in T(0)$. Let J be the duality mapping with the properties described above. In order to prove that T is maximal, we show that $R(T + J) = X^*$, and hence the conclusion follows from Theorem 4.4.7. Take any $u \in X^*$. We must find $x \in X$ such that $u \in (T+J)x$. Take $r \geq r_0$ such that $\|u\| < r$. By assumption, $T + N_r$ is maximal monotone, and hence there exists $x \in X$ such that

$$u \in (T + N_r + J)x = (T + N_r)x + Jx. \tag{4.54}$$

Because $x \in D(N_r)$, $\|x\| \leq r$. We claim that $\|x\| < r$. Indeed, if $\|x\| = r$ then by (4.53) we get $u \in (T + N_r + J)x = Tx + (1 + \lambda)Jx$. So there exists $v \in Tx$ such that $u = v + (1 + \lambda)Jx$, in which case

$$\langle u, x \rangle = \langle v + (1 + \lambda)Jx, x \rangle = \langle v, x \rangle + (1 + \lambda)\langle Jx, x \rangle \geq (1 + \lambda)\|x\|^2,$$

using the definition of J and the fact that $0 \in T(0)$. Therefore

$$(1 + \lambda)\|x\|^2 = (1 + \lambda)\langle Jx, x \rangle \leq \langle u, x \rangle \leq \|u\|\,\|x\| < r\|x\|, \qquad (4.55)$$

using the fact that $\|u\| < r$. Rearranging (4.55), we obtain $\|x\| < (1 + \lambda)^{-1}r < r$. Because $\|x\| < r$, $N_r(x) = 0$, which in combination with (4.54) gives $u \in (T + J)x$ as required. $\square$

We are now ready to present the main theorem of this section.

Theorem 4.8.3. *Assume that X is reflexive and let $T_1, T_2 : X \rightrightarrows X^*$ be maximal monotone operators. If $D(T_1) \cap D(T_2)^\circ \neq \emptyset$, then $T_1 + T_2$ is maximal monotone.*

Proof. We consider first the case in which $D(T_2)$ is bounded. Recall that the norm on X is such that J and J^{-1} are single-valued. By translating the sets $D(T_1)$ and $D(T_2)$ if necessary, we can assume that $0 \in D(T_1) \cap D(T_2)^\circ$, and there is also no loss of generality in requiring that $0 \in T_1(0)$. We claim that $R(T_1 + T_2 + J) = X^*$. We proceed to prove the claim. Given an arbitrary $u \in X^*$, we must find $x \in X$ such that $u \in (T_1 + T_2 + J)x$. Replacing T_2 by $T_2(\cdot) - u$, we can reduce the argument to the case in which $u = 0$; that is, we must find $x \in X$ such that $0 \in (T_1 + T_2 + J)x$, which happens if and only if there exists a point $v \in X^*$ such that

$$-v \in (T_1 + (1/2)J)x \quad \text{and} \quad v \in (T_2 + (1/2)J)x. \qquad (4.56)$$

Define the mappings $S_1, S_2 : X^* \rightrightarrows X$ as

$$\begin{aligned} S_1(w) &:= -(T_1 + (1/2)J)^{-1}(-w), \\ S_2(w) &:= (T_2 + (1/2)J)^{-1}(w). \end{aligned}$$

By Theorem 4.4.7(ii), S_1 and S_2 are single-valued maximal monotone operators, such that $D(S_1) = D(S_2) = X^*$. By Theorem 4.6.4, S_1 and S_2 are continuous from the strong topology in X^* to the weak topology in X, and hence $S_1 + S_2$ is monotone, single-valued, and continuous from the strong topology in X^* to the weak topology in X, and such that $D(S_1 + S_2) = X^*$. Thus, we can apply Theorem 4.2.4 to the operator $T := S_1 + S_2$, and conclude that $S_1 + S_2$ is maximal. It is easy to check that the existence of $x \in X$ and $v \in X^*$ verifying (4.56) is equivalent to the existence of some $v \in X^*$ verifying

$$0 \in (S_1 + S_2)v. \qquad (4.57)$$

Because $0 \in T_1(0)$ and $0 = J(0)$, we have that $0 \in (T_1 + (1/2)J)(0)$ and hence $0 \in S_1(0)$. Thus, by monotonicity of S_1 we have

$$\langle S_1(u), u \rangle \geq 0, \quad \text{for all} \quad u \in X^*. \qquad (4.58)$$

On the other hand, $R(S_2) = D(T_2 + (1/2)J) = D(T_2)$ is bounded by assumption. Because $0 \in (D(T_2))^o$, we also have that $0 \in (R(S_2))^o$. We claim that these properties imply the existence of some $r > 0$ such that

$$\langle S_2(u), u \rangle \geq 0 \tag{4.59}$$

for all $\|u\| > r$. We prove the claim. S_2 is monotone, thus we have that

$$\langle S_2(v) - S_2(u), v - u \rangle \geq 0,$$

which gives

$$\langle S_2(v), v \rangle \geq \langle S_2(u), v \rangle + \langle S_2(v) - S_2(u), u \rangle. \tag{4.60}$$

Because $R(S_2)$ is bounded, there exists $r_1 > 0$ such that $R(S_2) \subset B_{r_1}$, and hence

$$\langle S_2(v) - S_2(u), u \rangle \leq 2r_1 \|u\|. \tag{4.61}$$

We noted above that $0 \in (R(S_2))^o = (D(S_2^{-1}))^o$. By Theorem 4.2.10(i) this implies that S_2^{-1} is locally bounded at 0. Therefore there exists $\varepsilon > 0$ and $r_2 > 0$ such that $S_2^{-1}(B_\varepsilon) \subset B_{r_2}$; that is, $B_\varepsilon \subset S_2(B_{r_2})$. By (4.61) and (4.60) we have

$$\langle S_2(v), v \rangle \geq \langle S_2(u), v \rangle - 2r_1 r_2, \tag{4.62}$$

for all $u \in B_{r_2}$. Combining this with the fact that $B_\varepsilon \subset S_2(B_{r_2})$, we get

$$\langle S_2(v), v \rangle \geq \sup_{u \in B_{r_2}} \langle S_2(u), v \rangle - 2r_1 r_2$$

$$= \sup_{\eta \in S_2(B_{r_2})} \langle \eta, v \rangle - 2r_1 r_2 \geq \sup_{w \in B_\varepsilon} \langle w, v \rangle - 2r_1 r_2 = \varepsilon \|v\| - 2r_1 r_2,$$

and the rightmost expression is nonnegative when $\|v\| \geq (2r_1 r_2)/\varepsilon$. Therefore the claim is true for $r := (2r_1 r_2)/\varepsilon$; that is, (4.59) holds for this value of r. Now we combine (4.59) with (4.58), getting

$$\langle (S_1 + S_2)(u), u \rangle \geq 0, \quad \text{for all} \quad \|u\| > r,$$

and then by Proposition 4.8.1 applied to $T := (S_1 + S_2)$ we conclude the existence of some $v \in X^*$ such that (4.57) holds, completing the proof for the case in which $D(T_2)$ is bounded.

Now we consider the general case. Again, we assume without loss of generality that $0 \in D(T_1) \cap D(T_2)^o$ and that $0 \in T_1(0)$. For $r > 0$ consider the mapping N_r defined as the normality operator of $B_r = \{x \in X : \|x\| \leq r\}$. Note that $0 \in (D(N_r))^o$ for all $r > 0$. Because $0 \in D(T_2)$, we get $D(T_2) \cap (D(N_r))^o \neq \emptyset$ for all $r > 0$. Clearly $D(N_r) = B_r$, which is a bounded set. Applying the result of the theorem, already established for the case of bounded $D(T_2)$, we conclude that $T_2 + N_r$ is maximal for all $r > 0$. Note that

$$0 \in (D(T_2))^o \cap (D(N_r))^o \subset (D(T_2) \cap D(N_r))^o = (D(T_2 + N_r))^o.$$

Also $0 \in D(T_1)$, thus we get $0 \in D(T_1) \cap (D(T_2 + N_r))^o$. Because $D(T_2 + N_r) \subset D(N_r)$, which is bounded, we can use again the result for the case of bounded $D(T_2)$, and obtain that $T_1 + (T_2 + N_r)$ is maximal monotone for all $r > 0$. Finally, we apply Proposition 4.8.2 for concluding that $T_1 + T_2$ is maximal monotone. $\quad\square$

When X is finite-dimensional, the requirement on the domains can be weakened to $\operatorname{ri} D(T_1) \cap \operatorname{ri} D(T_2)^o \neq \emptyset$. For proving this result, we need the following tools. Recall that, given a proper subspace S of a finite-dimensional vector space X, the quotient $X/S^\perp$ is defined as

$$\pi(z) \in X/S^\perp \quad \text{if and only if} \quad \pi(z) = \{z' \in X \mid z - z' \in S^\perp\}.$$

The quotient induces a partition of X as the disjoint union of the sets $\{z+S^\perp\}_{z\in S} = \{\pi(z)\}_{z\in S}$. Thus we can see X as the disjoint union of all the parallel translations of $S^\perp$. It is a basic algebraic fact that $X/S^\perp$ can be identified with S, which in turn can be identified with the dual of S. Let $T : X \rightrightarrows X$ be a point-to-set mapping, and assume, without loss of generality, that $0 \in D(T)$. Let S be the affine hull of $D(T)$, which in this case is a linear subspace. We can associate with T the mapping $\tilde{T} : S \rightrightarrows X/S^\perp$ defined as

$$\tilde{T}(x) := \{\pi(y) \mid y \in Tx\}. \tag{4.63}$$

We abuse notation by writing $\pi(z) = z + S^\perp$, so that (4.63) also reads $\tilde{T}(x) = Tx + S^\perp$. Consider the duality product in $S \times X/S^\perp$ given by

$$\langle x, \pi(y) \rangle := \langle x, y \rangle = \langle x, P_S(y) \rangle,$$

where $P_S : X \to S$ is the orthogonal projection on S. The following lemma shows that maximality is preserved when passing from T to $\tilde{T}$.

Lemma 4.8.4. *Assume that X is finite-dimensional and let S be a proper subspace of X. Let $T : X \rightrightarrows X$ be a monotone operator such that S is the affine hull of $D(T)$. If $\tilde{T}$ is the operator defined by (4.63), then $\tilde{T}$ is also monotone. In addition, T is maximal if and only if $\tilde{T}$ is maximal.*

Proof. We leave the proof of monotonicity as an exercise for the reader. Let us prove the second statement. Suppose that T is maximal, and take $(x, \pi(z)) \in S \times X/S^\perp$ monotonically related to $\operatorname{Gph}(\tilde{T})$, in the sense of Definition 4.2.3. Thus, for all $y \in S$ and all $\pi(t) \in \tilde{T}(y)$ we have

$$0 \leq \langle y - x, \pi(t) - \pi(z) \rangle = \langle y - x, t - z \rangle.$$

Note that $\pi(t) \in \tilde{T}(y)$ if and only if $t \in Ty$. Because $(y, t) \in \operatorname{Gph}(T)$ is arbitrary and T is maximal, we conclude that $z \in Tx$, implying that $\pi(z) \in \tilde{T}(x)$, as required. The proof of the converse statement is analogous. $\quad\square$

Theorem 4.8.5. *If X is finite-dimensional and $T_1, T_2 : X \rightrightarrows X^*$ are maximal monotone operators such that $\operatorname{ri} D(T_1) \cap \operatorname{ri} D(T_2) \neq \emptyset$, then $T_1 + T_2$ is maximal monotone.*

Proof. Again, we can assume without loss of generality that $0 \in \text{ri}\, D(T_1) \cap \text{ri}\, D(T_2)$, so that the affine hulls of the domains $D(T_1)$ and $D(T_2)$ are subspaces L_1 and L_2, respectively. Assume that both subspaces are proper (otherwise we are within the hypotheses of Theorem 4.8.3, and the result follows without further ado). Define $N_1 := N_{L_1}$, $N_2 := N_{L_2}$, and $N_0 := N_{L_1 \cap L_2}$. These definitions yield $\mathcal{T}_1 = T_1 + N_1$ and $\mathcal{T}_2 = T_2 + N_2$, so that

$$\mathcal{T}_1 + \mathcal{T}_2 = T_1 + T_2 + N_1 + N_2 = T_1 + T_2 + N_0. \tag{4.64}$$

Consider the maximal monotone operators $\tilde{T}_1 : L_1 \rightrightarrows X/L_1^{\perp}$ and $\tilde{T}_2 : L_2 \rightrightarrows X/L_2^{\perp}$ as in Lemma 4.8.4. Define also $\tilde{N}_0 : L_1 \rightrightarrows X/L_1^{\perp}$. We claim that $\tilde{T}_1 + \tilde{N}_0$ is maximal monotone. Indeed, note that $0 \in D(T_1) = D(\tilde{T}_1) = \text{int}_{L_1}(D(\tilde{T}_1))$, where $\text{int}_{L_1}(A)$ denotes the interior relative to the subspace L_1. Because $0 \in D(\tilde{N}_0) = L_1 \cap L_2$, we have that $0 \in \text{int}_{L_1}(D(\tilde{T}_1)) \cap D(\tilde{N}_0)$. So $\tilde{T}_1 + \tilde{N}_0$ is maximal. By Lemma 4.8.4, $T_1 + N_0$ is maximal. Now we can use the same argument for proving that $(T_1 + N_0) + T_2$ is maximal. Indeed, $0 \in \text{int}_{L_2}(D(\tilde{T}_2))$ and $0 \in \text{int}_{L_1}(D(\tilde{T}_1)) \cap D(\tilde{N}_0) \subset D(\tilde{T}_1 + \tilde{N}_0)$, so that

$$0 \in \text{int}_{L_2}(D(\tilde{T}_2)) \cap D(\tilde{T}_1 + \tilde{N}_0),$$

which implies that $(\tilde{T}_1 + \tilde{N}_0 + \tilde{T}_2)$ is maximal. Using again Lemma 4.8.4, we conclude that $(T_1 + N_0 + T_2)$ is maximal. The maximality of $T_1 + T_2$ follows from (4.64). □

Exercises

4.1. Let $T : X \rightrightarrows X^*$ maximal monotone. Then for all $b \in X, b' \in X^*$, and all $\alpha > 0$ the operators $T_1(x) := T(\alpha x + b)$ and $T_2(x) := \alpha T x + b'$ are maximal monotone.

4.2. Let X, Y be Banach spaces and let $T_1 : X \rightrightarrows X^*$ and $T_2 : Y \rightrightarrows Y^*$ be maximal monotone operators. Prove that $T_1 \times T_2 : X \times Y \rightrightarrows X^* \times Y^*$, defined as $T_1 \times T_2(x, y) := \{(v_1, v_2) \in X^* \times Y^* \mid v_1 \in T_1 x, v_2 \in T_2 y\}$, is also maximal monotone.

4.3. Prove the statement of Example 4.1.4.

4.4. Let $T : X \rightrightarrows X^*$ be a monotone operator. Consider the following statements.

   (a) If $\inf_{(y,u) \in \text{Gph}(T)} \langle y - x, u - v \rangle = 0$ then $(x, v) \in \text{Gph}(T)$.

   (b) If $(x, v) \notin \text{Gph}(T)$ then $\inf_{(y,u) \in \text{Gph}(T)} \langle y - x, u - v \rangle < 0$.

   (c) If $(x, v) \in X \times X^*$, then $\inf_{(y,u) \in \text{Gph}(T)} \langle y - x, u - v \rangle \leq 0$.

   (d) T is maximal.

   Prove that $(a) \iff (b) \iff (d)$ and $(d) \implies (c)$. Note that the converse of (b) holds by monotonicity of T.

4.5. Prove Remark 4.1.8.

4.6. Prove that the function $p : C \to \mathbb{R}$ defined as $p(w) := \langle w - v, \phi(w) - y \rangle$ and used in Theorem 4.3.1 is weak* continuous.

4.7. Prove Proposition 4.4.4, items (iii) and (iv).

4.8. Assume that X is reflexive and $T : X \rightrightarrows X^*$ is maximal monotone, with $(D(T))^o \neq \emptyset$. Prove that T is s–PC at all $x \in (D(T))^o$. Hint: see Definition 3.9.2, Proposition 3.9.4, and Theorem 4.5.3.

4.9. Prove that if $A \subset X$ is α-cone meager for some $\alpha \in (0,1)$ then $\overline{A}$ has an empty interior.

4.10. Prove Proposition 4.7.4.

4.9 Historical notes

The earliest known extension of positive semidefiniteness to nonlinear mappings appeared in 1935 in [88]. This work did not achieve recognition, and the effective introduction of monotone maps in its current formulation is due to Zarantonello [234], who used them for developing iterative methods for functional equations. The observation of the fact that gradients of convex functions are monotone was made by Kachurovskii [110], where also the term "monotonicity" was introduced. The first significant contributions to the theory of monotone operators appeared in the works by Minty [150, 151]. Most of the early research on monotone mappings dealt with applications to integral and differential equations (see [111, 37] and [43]).

Maximality of continuous point-to-point operators was established in [149]. Local boundedness of monotone mappings was proved by Rockafellar in [186], and later extended to more general cases by Borwein and Fitzpatrick [32]. The extension theorem by Debrunner and Flor was proved in [72]; the proof in our book comes from [172]. Theorem 4.4.1 appeared in [206].

The connection between maximal monotonicity of an operator T and surjectivity of $T + J$ (Theorem 4.4.7) was established by Minty in [153] in the setting of Hilbert spaces, and extended to Banach spaces by Rockafellar [190]. Theorem 4.5.3 appeared in [186].

The issue of conditions on a Banach space ensuring that Fréchet differentiability of continuous convex functions is generic, started with Mazur (see [145]), who proved in 1933 that such is the case when the space is separable. The result saying that a Banach space is Asplund if and only if its dual is separable was established in [9]. The theorem stating that monotone operators are single-valued outside an angle-small set was first proved in [174].

The theorem which says that the subdifferential of a possibly nondifferentiable convex function is monotone is due to Minty [152]. Maximal monotonicity of these operators in Hilbert spaces was established by Moreau [160]. The case of Banach spaces (i.e., our Theorem 4.7.1) is due to Rockafellar (see [184, 189]). Theorem 4.8.3 on the maximality of sums of maximal monotone operators was proved in [190].

Chapter 5

Enlargements of Monotone Operators

5.1 Motivation

Let X be a real Banach space with dual X^* and $T : X \rightrightarrows X^*$ a maximal monotone operator. As we have seen in Chapter 4, many important problems can be formulated as

$$\text{Find } x \in X \text{ such that } 0 \in Tx,$$

or equivalently, find $x \in X$ such that $(x, 0) \in G(T)$. Hence, solving such problems requires knowledge of the graph of T. If T is not point-to-point, then it will not be inner-semicontinuous (see Theorem 4.6.3). This fact makes the problem ill behaved and the involved computations become hard. In such a situation, it is convenient to work with a set $G' \supset G(T)$, with better continuity properties. In other words, we look for a point-to-set mapping E which, while staying "close" to T (in a sense which becomes clear later on), has a better-behaved graph. Such an E allows us to define perturbations of the problem above, without losing information on the original problem.

This is a reason for considering point-to-set mappings that extend (i.e., with a bigger graph than) T. Namely, we say that a point-to-set mapping $E : \mathbb{R}_+ \times X \rightrightarrows X^*$ is an *enlargement* of T when

$$E(\varepsilon, x) \supseteq T(x) \quad \text{for all } x \in X, \; \varepsilon \geq 0. \tag{5.1}$$

We must identify which extra properties on E will allow us to

(i) Acquire a better understanding of T itself.

(ii) Define suitable perturbations of problems involving T.

Property (i) indicates that we aim to use E as a theoretical tool, whereas (ii) aims to apply the abstract concept of enlargement for solving concrete problems associated with T. Whether E has the advantages above depends on how the graph of E approaches the graph of T, or more precisely, how the set $\mathrm{Gph}(E) \subseteq \mathbb{R}_+ \times X \times X^*$ approaches the set $\{0\} \times \mathrm{Gph}(T) \subseteq \mathbb{R}_+ \times X \times X^*$.

A well-known example of such an enlargement appears when $T = \partial f$, where f is a proper, convex, and lower-semicontinuous function. In this case, the ε-*subdifferential of f* [41] has been useful both from the theoretical and the practical points of view. This enlargement is defined, for all $x \in X$ and all $\varepsilon \geq 0$, as

$$\partial_\varepsilon f(x) := \{w \in X^* \ : \ f(y) \geq f(x) + \langle w, y - x \rangle - \varepsilon \text{ for all } y \in X\}. \tag{5.2}$$

This notion has been used, for instance, in the proof of maximality of the subdifferential of a proper, convex, and lsc function in [189] (see Theorem 4.7.1). Another important application of this enlargement is in the computation of sub-differentials of sums of convex and lsc functions (see [95]). Relevant examples of practical applications of this enlargement are the development of inexact algorithms for nonsmooth convex optimization, for example, ε-subgradient algorithms (see, e.g., [3, 119]) and bundle methods (see, e.g., [120, 132, 200]).

Inspired by the notion above, we define in Section 5.2 an enlargement, denoted T^e, of an arbitrary maximal monotone operator T. Most of the theoretical properties verified by the ε-subdifferential can be recovered in this more general case, and hence applications of the above-discussed kinds can be obtained also for this enlargement. Practical applications of enlargements are the subject of Section 5.5 and theoretical applications are presented in Section 5.6.

After performing this step in our quest for enlargements of maximal monotone operators, we identify which are the "good" properties in T^e; that is, the ones that allow us to get results of the kinds (i) and (ii). Having identified these properties, we can go one step further and define a whole family of "good" enlargements, denoted $\mathbb{E}(T)$, containing those enlargements that share with T^e these properties. This family contains T^e as a special member (in fact the largest one, as we show later on). We study the family $\mathbb{E}(T)$ in Section 5.4.

5.2   The T^e-enlargement

Given $T : X \rightrightarrows X^*$, define $T^e : \mathbb{R}_+ \times X \rightrightarrows X^*$ as

$$T^e(\varepsilon, x) := \{v \in X^* \mid \langle u - v, y - x \rangle \geq -\varepsilon \text{ for all } y \in X, u \in T(y)\}. \tag{5.3}$$

This enlargement was introduced in [46] for the finite-dimensional case, and extended first to Hilbert spaces in [47, 48] and then to Banach spaces in [50].

We start with some elementary properties that require no assumptions on T. The proofs are left as an exercise.

Lemma 5.2.1.

(a) T^e *is nondecreasing; that is,*

$$0 \leq \varepsilon_1 \leq \varepsilon_2 \ \Rightarrow \ T^e(\varepsilon_1, x) \subset T^e(\varepsilon_2, x)$$

for all $x \in X$.

(b) If $T_1, T_2 : X \rightrightarrows X^*$, and $\varepsilon_1, \varepsilon_2 \geq 0$, then

$$T_1^e(\varepsilon_1, x) + T_2^e(\varepsilon_2, x) \subset (T_1 + T_2)^e(\varepsilon_1 + \varepsilon_2, x)$$

for all $x \in X$.

(c) T^e is convex-valued.

(d) $\left(T^{-1}\right)^e (\varepsilon, \cdot) = (T^e(\varepsilon, \cdot))^{-1}$.

(e) If $A \subset \mathbb{R}_+$ with $\bar{\varepsilon} = \inf A$, then $\cap_{\varepsilon \in A} T^e(\varepsilon, \cdot) = T^e(\bar{\varepsilon}, \cdot)$.

(f) T^e is an enlargement of T; that is, $T^e(\varepsilon, x) \supset T(x)$, for all $x \in X$ and $\varepsilon \geq 0$. If T is maximal monotone, then $T^e(0, \cdot) = T(\cdot)$

Remark 5.2.2. When T is a maximal monotone operator, conditions (a) and (f) above assert that T^e is a *nondecreasing enlargement of* T. We return to this concept in Section 5.4.

We proved before (see Proposition 4.2.1) that a maximal monotone operator is sequentially (sw*)-OSC. The same is true for T^e.

Theorem 5.2.3. Let T^e be defined as in (5.3).

(i) T^e is sequentially OSC, where $X \times X^*$ is endowed with the (sw*)-topology.

(ii) T^e is strongly OSC. In particular, T^e is OSC when X is finite-dimensional.

Proof. The proof follows the same steps as in Proposition 4.2.1 and is left as an exercise. $\square$

We mentioned in Section 5.1 that the ε-subdifferential is an enlargement of $T = \partial f$. The proposition below establishes a relation between $\partial_\varepsilon f$ and ∂f^e.

Proposition 5.2.4. If $T = \partial f$, where $f : X \to \mathbb{R} \cup \{\infty\}$ is a lsc and convex function, then $\partial f^e(\varepsilon, x) \supset \partial_\varepsilon f(x)$ for all $x \in X$ and all $\varepsilon \geq 0$.

Proof. Take $v \in \partial_\varepsilon f(x)$. It holds that

$$\varepsilon + \langle v, x - y \rangle \geq f(x) - f(y) \geq \langle u, x - y \rangle,$$

for all $y \in X$, $u \in T(y) = \partial f(y)$, using (5.2) in the first inequality and the definition of subgradient in the second one. By (5.3), it follows that $v \in \partial f^e(\varepsilon, x)$. $\square$

The next set of examples shows that the inclusion in Proposition 5.2.4 can be strict.

Example 5.2.5. Let $X = \mathbb{R}$ and $f : \mathbb{R} \to \mathbb{R}$ given by $f(x) = (1/p)|x|^p$, with $p \geq 1$. Take $T = \partial f$.

(i) *If $p = 1$ then $\partial f^e(\varepsilon, x) = \partial_\varepsilon f(x)$ for all $x \in X$. Namely,*

$$\partial f^e(\varepsilon, x) = \partial_\varepsilon f(x) = \begin{cases} [1 - (\varepsilon/x), 1] & \text{if } x > \varepsilon/2, \\ [-1, 1] & \text{if } |x| \le \varepsilon/2, \\ [-1, -1 - (\varepsilon/x)] & \text{if } x < -\varepsilon/2. \end{cases}$$

(ii) *If $p = 2$, then*

$$\partial_\varepsilon f(x) = [x - \sqrt{2\varepsilon}, x + \sqrt{2\varepsilon}] \subset [x - 2\sqrt{\varepsilon}, x + 2\sqrt{\varepsilon}] = \partial f^e(\varepsilon, x).$$

(iii) *If $p > 1$ and $x = 0$, then*

$$\partial_\varepsilon f(0) = [-(q\varepsilon)^{1/q}, (q\varepsilon)^{1/q}],$$

$$\partial f^e(\varepsilon, 0) = [-p^{1/p}(q\varepsilon)^{1/q}, p^{1/p}(q\varepsilon)^{1/q}],$$

where $(1/p) + (1/q) = 1$.

(iv) *Assume now that $f(x) = -\log x$. For $x > 0$*

$$\partial_\varepsilon f(x) = [-(1/x)s_1(\varepsilon), -(1/x)s_2(\varepsilon)],$$

$$\partial f^e(\varepsilon, x) = \begin{cases} [-(1/x)(1 + \sqrt{\varepsilon})^2, -(1/x)(1 - \sqrt{\varepsilon})^2] & \text{if } \varepsilon \le 1, \\ [-(1/x)(1 + \sqrt{\varepsilon})^2, 0] & \text{if } \varepsilon \ge 1, \end{cases}$$

where $s_1(\varepsilon) \ge 1 \ge s_2(\varepsilon) > 0$ are the two roots of $-\log s = 1 - s + \varepsilon$ (see Figure 5.1). Note that in this case $\partial f^e(\varepsilon, x)$ is easier to compute than $\partial_\varepsilon f(x)$.

(v) *Take $f(x) = (1/2)\langle Ax, x \rangle$, with $A \in \mathbb{R}^{n \times n}$ symmetric and positive definite. Then*

$$\partial_\varepsilon f(x) = \{Ax + w \ : \ \langle A^{-1}w, w \rangle \le 2\varepsilon\},$$

$$\partial f^e(\varepsilon, x) = \{Ax + w \ : \ \langle A^{-1}w, w \rangle \le 4\varepsilon\},$$

so that $\partial f^e(\varepsilon, x) = \partial_{2\varepsilon} f(x)$.

Example 5.2.5(iv) shows that, for all $\varepsilon \ge 1$ and all $\bar{\varepsilon} > 0$, $\partial f^e(\varepsilon, x)$ is not contained in $\partial_{\bar{\varepsilon}} f(x)$, because $0 \in \partial f^e(\varepsilon, x)$ for all $\varepsilon \ge 1$, whereas $0 \notin \partial_{\bar{\varepsilon}} f(x)$ for all $\bar{\varepsilon} > 0$.

5.3   Theoretical properties of T^e

We have seen already in Theorem 4.2.10 that a maximal monotone operator is locally bounded in the interior of its domain. Inasmuch as $T^e(\varepsilon, x) \supset T(x)$ for all x, it is natural to ask whether this enlargement preserves the local boundedness. This result holds, meaning that T^e does not enlarge T too much. Moreover, we show below that the bound can be taken as an affine function of the parameter

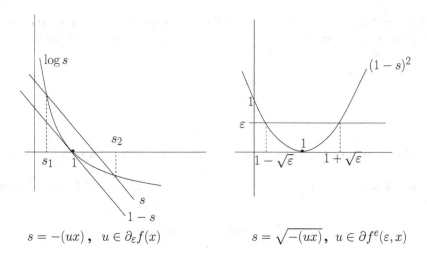

$$s = -(ux), \quad u \in \partial_\varepsilon f(x) \qquad\qquad s = \sqrt{-(ux)}, \quad u \in \partial f^e(\varepsilon, x)$$

Figure 5.1. *Examples of ε-subdifferentials and ε-enlargements*

ε. This claim is made precise in Definition 5.3.2. Another key matter is whether and how we may obtain an element of $T^e(\varepsilon, x)$. This question is solved by means of the so-called "transportation formula." This formula gives a specific procedure for obtaining an element in the graph of T^e, by means of known elements of the graph of T. We can interpret the transportation formula as a way of going from the graph of T to the graph of T^e. In order to have a better understanding of how these graphs approach each other, we must also know how to go the opposite way. For the ε-subdifferential, Brøndsted and Rockafellar [41] addressed this question proving that every element $v \in \partial_\varepsilon f(x)$ can be approximated by an element in the graph of ∂f in the following way: for all $\eta > 0$, there exists $v' \in \partial f(x')$ such that $\|x - x'\| \leq \varepsilon/\eta$ and $\|v - v'\| \leq \eta$. This fact is known as *Brøndsted–Rockafellar's lemma*. We end this section by proving that the graph of T^e can be approximated by the graph of T in exactly the same way. In this section we follow [50].

We start with the formal definition of an enlargement of T.

Definition 5.3.1. *Let $T : X \rightrightarrows X^*$be a multifunction. We say that a point-to-set mapping $E : \mathbb{R}_+ \times X \rightrightarrows X^*$ is an* enlargement *of T when*

$$E(\varepsilon, x) \supseteq T(x) \tag{5.4}$$

for all $x \in X$ and all $\varepsilon \geq 0$.

5.3.1 Affine local boundedness

Next we introduce a special notion of local boundedness for operators defined on $\mathbb{R}_+ \times X$. We require the usual local boundedness of $E(0, \cdot)$, and an additional upper bound proportional to the first argument, when the latter moves away from 0.

Definition 5.3.2. *Let* $E : \mathbb{R}_+ \times X \rightrightarrows X^*$ *be a point-to-set mapping. We say that* E *is* affine locally bounded *at* $x \in X$ *when there exists a neighborhood* V *of* x *and positive constants* L, M *such that*

$$\sup_{\substack{y \in V \\ v \in E(\varepsilon, y)}} \|v\| \leq L\varepsilon + M.$$

For proving the affine local boundedness of T^e, we need a technical lemma.

Lemma 5.3.3. *Let* $T : X \rightrightarrows X^*$ *be a point-to-set mapping and* E *an enlargement of* T. *For a set* $U \subset D(T)$, *define*

$$M := \sup_{u \in T(U)} \|u\| \in (-\infty, \infty].$$

Assume that $V \subset X$ *and* $\rho > 0$ *are such that* $\{y \in X \,:\, d(y, V) < \rho\} \subset U$ *(in other words,* $d(\overline{V}, U^c) \geq \rho$*). Then*

$$\sup_{\substack{y \in V \\ v \in T^e(\varepsilon, y)}} \|v\| \leq (\varepsilon/\rho) + M, \tag{5.5}$$

for all $\varepsilon \geq 0$.

Proof. Take $y \in V$, $v \in T^e(\varepsilon, y)$. Recall from (2.17) that the norm of $v \in X^*$ is given by $\|v\| := \sup\{\langle v, z \rangle \,|\, z \in X, \|z\| = 1\}$. Thus, we can find a sequence $\{z^k\}$ such that $\lim_{k \to \infty} \langle v, z^k \rangle = \|v\|$ with $\|z^k\| = 1$ for all k. For a fixed $\sigma \in (0, \rho)$, define $y^k := y + \sigma z^k$. Then $d(y^k, V) \leq d(y^k, y) = \sigma < \rho$. The assumption on V and ρ implies that $\{y^k\} \subset U$. Because $U \subset D(T)$, we can take a sequence $\{w^k\}$ with $w^k \in T(y^k)$ for all k. By the definition of M, we have $\|w^k\| \leq M$ for all k. Because $v \in T^e(\varepsilon, y)$ and $w^k \in T(y^k)$, we can write

$$-\varepsilon \leq \langle v - w^k, y - y^k \rangle = \langle v - w^k, -\sigma z^k \rangle,$$

using also the definition of $\{y^k\}$. Dividing the expression above by σ and using the definitions of $\{z^k\}$ and M, we get

$$-\frac{\varepsilon}{\sigma} \leq -\langle v, z^k \rangle + \langle w^k, z^k \rangle \leq -\langle v, z^k \rangle + \|w^k\| \|z^k\|$$

$$= -\langle v, z^k \rangle + \|w^k\| \leq -\langle v, z^k \rangle + M.$$

Rearranging the last inequality yields

$$\langle v, z^k \rangle \leq \varepsilon/\sigma + M.$$

Letting now k go to infinity, and using the assumption on $\{z^k\}$, we obtain

$$\|v\| \leq \varepsilon/\sigma + M.$$

Taking the infimum of the rightmost expression for $\sigma \in (0, \rho)$, we conclude that $\|v\| \leq \varepsilon/\rho + M$, establishing (5.5). $\square$

Now we can prove the affine local boundedness of T^e in the set $D(T)^o$. This property is used later on for proving Lipschitz continuity of T^e.

Theorem 5.3.4 (Affine local boundedness). *If $T : X \rightrightarrows X^*$ is monotone, then T^e is affine locally bounded in $D(T)^o$. In other words, for all $x \in D(T)^o$ there exist a neighborhood V of x and positive constants L, M such that*

$$\sup\{\|v\| \; : \; v \in T^e(\varepsilon, y), \; y \in V\} \leq L\varepsilon + M$$

for all $\varepsilon \geq 0$.

Proof. Take $x \in D(T)^o$. By Theorem 4.2.10(i), there exists $R > 0$ such that $T(B(x, R))$ is bounded. Choose $U := B(x, R)$, $\rho := R/2$, and $V := B(x, \rho)$ in Lemma 5.3.3. Because $x \in D(T)^o$, $R > 0$ can be chosen so that $T(B(x, R))$ is bounded. Hence $\{y \in X \; : \; d(y, V) < \rho\} \subset U$ and we can apply the lemma in order to conclude that

$$\sup\{\|v\| \; : \; v \in T^e(\varepsilon, y), \; y \in V\} \leq \frac{2\varepsilon}{R} + M,$$

where M is as in Lemma 5.3.3. Note that $M < \infty$ because $T(B(x, R))$ is bounded. $\square$

Enlargements are functions defined in $\mathbb{R}_+ \times X$, and the ε-subdifferential is an important example of enlargement. Inasmuch as ε is now a variable of the enlargement, the name ε-subdifferential does not fit any more and we need a different terminology. We denote from now on $\breve{\partial}f(\varepsilon, x)$ the set $\partial_\varepsilon f(x)$. In other words, we define $\breve{\partial}f : \mathbb{R}_+ \times X \rightrightarrows X^*$ as

$$\breve{\partial}f(\varepsilon, x) = \partial_\varepsilon f(x).$$

We call this enlargement the *BR-enlargement* of ∂f, (for Brøndsted and Rockafellar).

Combining Proposition 5.2.4 with Theorem 5.3.4 for $T := \partial f$, we conclude that $\breve{\partial}f$ is also affine locally bounded. However, this fact can be proved directly.

Theorem 5.3.5 (Affine local boundedness of $\breve{\partial}f$). *Let $f : X \to \overline{\mathbb{R}}$ be a convex function. Assume that $U \subset X$ is such that f is bounded above on a neighborhood of some point of U. Then, for all $x \in U$, there exists a neighborhood V of x and positive constants L, M such that*

$$\sup\{\|v\| \; : \; v \in \breve{\partial}f(\varepsilon, y), \; y \in V\} \leq L\varepsilon + M$$

for all $\varepsilon \geq 0$.

Proof. For $x \in U$, Theorem 3.4.22 implies that there exists $\rho > 0$ such that f is Lipschitz continuous on $B(x, \rho)$. We claim that the statement of the theorem holds for $V = B(x, \rho/2)$. Indeed, take $y \in B(x, \rho/2)$, and $v \in \breve{\partial} f(\varepsilon, y)$. As in Lemma 5.3.3, we can choose a sequence $\{z^k\}$ such that $\lim_{k \to \infty} \langle v, z^k \rangle = \|v\|$ with $\|z^k\| = 1$ for all k. The vectors $y^k := y + (\rho/4) z^k$ belong to $B(x, \rho)$, and by Lipschitz continuity of f, there exists a constant $C_1 \geq 0$ such that

$$|f(y^k) - f(y)| \leq C_1 \|y^k - y\| = \rho C_1 / 4 . \tag{5.6}$$

On the other hand, using (5.6) and the fact that $v \in \breve{\partial} f(\varepsilon, y)$, we can write

$$\langle v, y^k - y \rangle \leq f(y^k) - f(y) + \varepsilon \leq |f(y^k) - f(y)| + \varepsilon \leq \rho C_1 / 4 + \varepsilon.$$

Using now the definitions of $\{z^k\}$, $\{y^k\}$, and letting k go to infinity, we get $\|v\| \leq 4\varepsilon/\rho + C_1$, and hence the result holds for $L := 4/\rho$ and $M = C_1$. $\square$

5.3.2   Transportation formula

The set $T^e(\varepsilon, x)$ enlarges the set $T(x)$, but this property is of no use if there is no way of computing elements of $T^e(\varepsilon, x)$. A constructive way of doing this is given in the following result.

Theorem 5.3.6.   *Take* $v^1 \in T^e(\varepsilon_1, x^1)$ *and* $v^2 \in T^e(\varepsilon_2, x^2)$. *For any* $\alpha \in [0, 1]$ *define*

$$\hat{x} := \alpha x^1 + (1 - \alpha) x^2, \tag{5.7}$$

$$\hat{v} := \alpha v^1 + (1 - \alpha) v^2, \tag{5.8}$$

$$\hat{\varepsilon} := \alpha \varepsilon_1 + (1 - \alpha) \varepsilon_2 + \alpha \langle x^1 - \hat{x}, v^1 - \hat{v} \rangle + (1 - \alpha) \langle x^2 - \hat{x}, v^2 - \hat{v} \rangle. \tag{5.9}$$

Then $\hat{\varepsilon} \geq 0$ *and* $\hat{v} \in T^e(\hat{\varepsilon}, \hat{x})$.

Proof. By (5.3), we have to prove that

$$\langle u - \hat{v}, y - \hat{x} \rangle \geq -\hat{\varepsilon},$$

for all $y \in X$, $u \in T(y)$. Using (5.7) and (5.8), we can write

$$\langle u - \hat{v}, y - \hat{x} \rangle = \alpha \langle u - \hat{v}, y - x^1 \rangle + (1 - \alpha) \langle u - \hat{v}, y - x^2 \rangle$$

$$= \alpha \big[\langle u - v^1, y - x^1 \rangle + \langle v^1 - \hat{v}, y - x^1 \rangle \big] + (1 - \alpha) \big[\langle u - v^2, y - x^2 \rangle + \langle v^2 - \hat{v}, y - x^2 \rangle \big]$$

$$\geq -\alpha \varepsilon_1 - (1 - \alpha) \varepsilon_2 + \alpha \langle v^1 - \hat{v}, y - x^1 \rangle + (1 - \alpha) \langle v^2 - \hat{v}, y - x^2 \rangle. \tag{5.10}$$

On the other hand,

$$\alpha \langle v^1 - \hat{v}, y - x^1 \rangle + (1 - \alpha) \langle v^2 - \hat{v}, y - x^2 \rangle$$

$$= \alpha \big[\langle v^1 - \hat{v}, y - \hat{x} \rangle + \langle v^1 - \hat{v}, \hat{x} - x^1 \rangle \big] + (1 - \alpha) \big[\langle v^2 - \hat{v}, y - \hat{x} \rangle + \langle v^2 - \hat{v}, \hat{x} - x^2 \rangle \big]$$

$$= \alpha\langle v^1 - \hat{v}, \hat{x} - x^1 \rangle + (1 - \alpha)\langle v^2 - \hat{v}, \hat{x} - x^2 \rangle, \tag{5.11}$$

using the fact that $\alpha\langle v^1 - \hat{v}, y - \hat{x} \rangle + (1 - \alpha)\langle v^2 - \hat{v}, y - \hat{x} \rangle = 0$. Combining (5.11) with (5.10) and (5.9), we get

$$\langle u - \hat{v}, y - \hat{x} \rangle \geq -\hat{\varepsilon}. \tag{5.12}$$

It remains to prove that $\hat{\varepsilon} \geq 0$. If this is not true, then (5.12) yields $\langle u - \hat{v}, y - \hat{x} \rangle > 0$ for all $(y, v) \in \mathrm{Gph}(T)$. Because T is maximal monotone, this implies that $(\hat{x}, \hat{v}) \in \mathrm{Gph}(T)$, in which case we can take $(y, v) = (\hat{x}, \hat{v})$ in (5.12), and a contradiction results. Therefore $\hat{\varepsilon} \geq 0$, and we conclude that $\hat{v} \in T^e(\hat{\varepsilon}, \hat{x})$. $\square$

Remark 5.3.7. Using the definition of $\breve{\partial}f$, it can be proved that a transportation formula like the one given above also holds for this enlargement (Exercise 8).

An alternative transportation formula for the case $T = \partial f$ is given by the next result.

Theorem 5.3.8. *Let $f : X \to \overline{\mathbb{R}}$ be a convex lsc function.*

(i) *If $v \in \breve{\partial}f(\varepsilon_1, x)$, then $v \in \breve{\partial}f(\varepsilon, y)$, for every $\varepsilon \geq \varepsilon_1 + f(y) - f(x) - \langle v, y - x \rangle$. In particular, when $\varepsilon_1 = 0$, v belongs to $\breve{\partial}f(\varepsilon, y)$, for every $\varepsilon \geq f(y) - f(x) - \langle v, y - x \rangle$.*

(ii) *Assume that f is Lipschitz continuous in U and that $V \subset X$, $\rho > 0$ verify $\{y \in X : d(y, V) < \rho\} \subset U$. Take $\beta > 0$. Fix arbitrary $x, y \in V$, $\varepsilon_1 \in (0, \beta]$, $v \in \breve{\partial}f(\varepsilon_1, x)$. Then there exists $\varepsilon \geq 0$ such that $v \in \breve{\partial}f(\varepsilon, y)$. Moreover, $|\varepsilon - \varepsilon_1| \leq K\|x - y\|$ for some nonnegative constant K.*

Proof.
(i) Let $\varepsilon_0 := \varepsilon_1 + f(y) - f(x) - \langle v, y - x \rangle$. Because $v \in \breve{\partial}f(\varepsilon_1, x)$, it holds that $\varepsilon_0 \geq 0$. Take $\varepsilon \geq \varepsilon_0$. We must show that $f(z) - f(y) - \langle v, z - y \rangle \geq -\varepsilon$ for all $z \in X$. Indeed,

$$f(z) - f(y) - \langle v, z - y \rangle = [f(z) - f(x) - \langle v, z - x \rangle] + f(x) - f(y) - \langle v, x - y \rangle$$

$$\geq -\varepsilon_1 + f(x) - f(y) - \langle v, x - y \rangle = -\varepsilon_0 \geq -\varepsilon,$$

using the assumption on ε and the fact that $v \in \breve{\partial}f(\varepsilon_1, x)$.
(ii) Take $\varepsilon := \varepsilon_0$. By item (i), $v \in \breve{\partial}f(\varepsilon, y)$. We have

$$|\varepsilon - \varepsilon_1| \leq |f(y) - f(x)| + \|v\| \, \|y - x\| \leq (C + \|v\|) \, \|y - x\|, \tag{5.13}$$

using the definition of ε_0, the Lipschitz continuity of f, and the Cauchy–Schwartz inequality. Define $y^k := x + (\rho/2)z^k$, where $\{z^k\}$ is such that $\lim_{k \to \infty} \langle v, z^k \rangle = \|v\|$, and $\|z^k\| = 1$ for all k. Using the fact that $v \in \breve{\partial}f(\varepsilon_1, x)$, we get

$$\langle v, y^k - x \rangle \leq f(y^k) - f(x) + \varepsilon_1 \leq |f(y^k) - f(x)| + \varepsilon_1.$$

Because $\|y^k - x\| = \rho/2 < \rho$ we have that $y^k \in U$, and hence the above inequality yields $(\rho/2)\langle v, z^k \rangle \leq C\rho/2 + \varepsilon_1$. Taking now the limit as $k \to \infty$, we get $\|v\| \leq C + (2/\rho)\beta =: C_0$. Combine this fact with (5.13) and conclude that $|\varepsilon - \varepsilon_1| \leq (C + C_0)\|y - x\|$ as required. $\square$

Remark 5.3.9. The transportation formula in Theorem 5.3.8(i) could be seen as stronger than the one given by Theorem 5.3.6, because it says that, given $v \in \breve{\partial} f(\varepsilon_1, x)$, we can translate the point x to an arbitrary point y, and the same vector v belongs to $\breve{\partial} f(\varepsilon, y)$ (as long as the parameter ε is big enough). Item (ii) of this theorem shows how to estimate the value of ε. Theorem 5.3.6, instead, can only be used to express elements in $T^e(\tilde{\varepsilon}, \tilde{x})$ that are convex combinations of other elements $v^i \in T^e(\varepsilon_i, x_i)$, $i = 1, 2$.

5.3.3 The Brøndsted–Rockafellar property

We mentioned at the beginning of this section that, for a closed proper convex function f, the Brøndsted–Rockafellar lemma (see [41]), states that any ε-subgradient of f at a point x_ε can be approximated by some exact subgradient, computed at some x, possibly different from x_ε. This property holds for the enlargement $\breve{\partial} f$ in an arbitrary Banach space. We start this section by proving this fact, as a consequence of Ekeland's variational principle (see Theorem 3.2.2). Then we prove that T^e also satisfies this property in a reflexive Banach space.

Our proof of the Brøndsted–Rockafellar lemma, which was taken from [206, Theorem 3.1], uses a minimax theorem, stated and proved below.

Theorem 5.3.10. *Let X, Y be Banach spaces, $A \subset X$ be a nonempty convex set, and $B \subset Y$ be nonempty, convex, and compact. Let $\psi : A \times B \to \mathbb{R}$ be such that*

(i) $\psi(\cdot, b)$ is convex for all $b \in B$.

(ii) $\psi(a, \cdot)$ is concave and upper-semicontinuous (Usc) in the sense of Definition 3.1.8 for all $a \in A$.

Then

$$\inf_{a \in A} \max_{b \in B} \psi(a, b) = \max_{b \in B} \inf_{a \in A} \psi(a, b).$$

Proof. Note that we always have

$$\inf_{a \in A} \max_{b \in B} \psi(a, b) \geq \max_{b \in B} \inf_{a \in A} \psi(a, b). \tag{5.14}$$

Thus, we must prove the opposite inequality. Take $\eta < \inf_{a \in A} \max_{b \in B} \psi(a, b)$, and fix $a_1, \ldots, a_p \in A$. Applying Lemma 3.4.21(b) with $Z := B$ and $g_i := \psi(a_i, \cdot)$ for all $i = 1, \ldots, p$, we obtain the existence of nonnegative coefficients $\lambda_1, \ldots, \lambda_p$ with $\sum_{i=1}^p \lambda_i = 1$ such that

$$\max_{b \in B} \{\psi(a_1, b) \wedge \cdots \wedge \psi(a_p, b)\} = \max_{b \in B} \{\lambda_1 \psi(a_1, b) + \cdots + \lambda_p \psi(a_p, b)\}.$$

Note that the supremum in Lemma 3.4.21(b) can be replaced by a maximum, because $\psi(a, \cdot)$ is Usc and B is compact. Using now assumption (i) we get

$$\max_{b \in B}\{\psi(a_1, b) \wedge \cdots \wedge \psi(a_p, b)\} \geq \max_{b \in B} \psi(\lambda_1 a_1 + \cdots + \lambda_p a_p, b) > \eta,$$

where the rightmost inequality follows from the definition of η. The above expression clearly implies the existence of some $\bar{b} \in B$ such that $\psi(a_i, \bar{b}) \geq \eta$ for all $i = 1, \ldots, p$. In other words, $\cap_{i=1}^p \{b \in B \mid \psi(a_i, b) \geq \eta\} \neq \emptyset$. The choice $\{a_1, \ldots, a_p\} \subset A$ is arbitrary, thus we proved that the family of sets $\{\{b \in B \mid \psi(a, b) \geq \eta\}\}_{a \in A}$ satisfies the finite intersection property. These sets are closed, because of the upper-semicontinuity of $\psi(a, \cdot)$, and compact because they are subsets of B. Therefore, $\cap_{a \in A}\{b \in B \mid \psi(a, b) \geq \eta\} \neq \emptyset$, which can be also expressed as

$$\max_{b \in B} \inf_{a \in A} \psi(a, b) \geq \eta. \tag{5.15}$$

Because (5.15) holds for any $\eta < \inf_{a \in A} \max_{b \in B} \psi(a, b)$, the result follows combining (5.15) with (5.14). $\square$

Theorem 5.3.11. *If $f : X \to \mathbb{R}$ is a convex lsc function on the Banach space X, then for $x_0 \in \mathrm{Dom}(f)$, $\varepsilon > 0$, $\lambda > 0$, and any $v_0 \in \breve{\partial} f(\varepsilon, x_0)$, there exists $\bar{x} \in \mathrm{Dom}(f)$ and $w \in X^*$ such that*

$$w \in \breve{\partial} f(0, \bar{x}), \quad \|\bar{x} - x_0\| \leq \varepsilon/\lambda \quad \text{and} \quad \|w - v_0\| \leq \lambda.$$

Proof. The assumption on v_0 implies that

$$f(x_0) - \langle v_0, x_0 \rangle \leq f(y) - \langle v_0, y \rangle + \varepsilon$$

for all $y \in X$. Consider the function $g : X \to \mathbb{R}$ defined as $g(y) := f(y) - \langle v_0, y \rangle$. Then $\mathrm{Dom}(g) = \mathrm{Dom}(f)$ and

$$\inf_{y \in X} g(y) \geq f(x_0) - \langle v_0, x_0 \rangle - \varepsilon > -\infty.$$

Also, $g(x_0) < \inf_{y \in X} g(y) + \varepsilon$. Thus, the assumptions of Corollary 3.2.3 are fulfilled for the function g. We claim that if the statement of the theorem holds for all $\lambda \in (0, 1)$, then it holds for all $\lambda > 0$. Indeed, assume that the statement holds for all $\varepsilon' > 0$ and $\lambda' \in (0, 1)$. If $\lambda \geq 1$, then take $\varepsilon' := \varepsilon/2\lambda^2$ and $\lambda' = 1/2\lambda$. Applying the result for these values of ε' and λ' readily gives the conclusion for $\varepsilon > 0$ and $\lambda \geq 1$, establishing the claim. It follows that it is enough to consider the case $\lambda < 1$. By Corollary 3.2.3, there exists $\bar{x} \in \mathrm{Dom}(g) = \mathrm{Dom}(f)$ such that

(a) $\lambda \|\bar{x} - x_0\| \leq \varepsilon$,

(b) $\lambda \|\bar{x} - x_0\| \leq g(x_0) - g(\bar{x})$,

(c) $\lambda \|x - \bar{x}\| > g(\bar{x}) - g(x)$, for all $x \neq \bar{x}$.

Denote $B^* := \{v \in X^* \mid \|v\| \leq \lambda\}$, and define the function $\psi : \mathrm{Dom}(f) \times B^* \to \mathbb{R}$ as

$$\psi(x, v) = g(x) - g(\bar{x}) + \langle v, \bar{x} - x \rangle.$$

Then

$$\max_{v \in B^*} \psi(x, v) = g(x) - g(\bar{x}) + \lambda \|x - \bar{x}\| \geq -\lambda \|x - \bar{x}\| + \lambda \|x - \bar{x}\| = 0,$$

where the last inequality follows from (c). Thus, $\inf_{x \in \mathrm{Dom}(f)} \max_{v \in B^*} \psi(x, v) \geq 0$. From Theorem 5.3.10 with $A := \mathrm{Dom}(f)$ and $B := B^*$, we get

$$\max_{v \in B^*} \inf_{x \in \mathrm{Dom}(f)} \psi(x, v) \geq 0.$$

So there exists $u \in B^*$ such that $\psi(x, u) = g(x) - g(\bar{x}) + \langle \bar{x} - x, u \rangle \geq 0$ for all $x \in \mathrm{Dom}(f)$. In other words, there exists $u \in B^*$ such that

$$f(x) - f(\bar{x}) + \langle u + v_0, \bar{x} - x \rangle \geq \quad 0.$$

This yields $w := u + v_0 \in \breve{\partial} f(0, \bar{x})$ with $\|w - v_0\| = \|u\| \leq \lambda$, and by (a) we also have $\|\bar{x} - x_0\| \leq \varepsilon/\lambda$. $\square$

The previous theorem was used in [189] for establishing maximality of subdifferentials of convex functions. Another consequence of the Brøndsted–Rockafellar lemma is the next result, which is the Bishop–Phelps theorem, established in [30].

Theorem 5.3.12. *Let $f : X \to \overline{\mathbb{R}}$ be proper, convex, and lsc. If $\alpha, \beta > 0$, and $x_0 \in X$ are such that*

$$f(x_0) < \inf_{x \in X} f(x) + \alpha \beta,$$

then there exist $x \in X$ and $w \in \partial f(x)$ such that $\|x - x_0\| < \beta$ and $\|w\| < \alpha$.

Proof. Choose $\varepsilon > 0$ such that $f(x_0) - \inf_{x \in X} f(x) < \varepsilon < \alpha \beta$ and $\lambda \in (\varepsilon/\beta, \alpha)$. Then it is easy to see that $0 \in \breve{\partial} f(\varepsilon, x_0)$. By Theorem 5.3.11, there exist $x \in X$ and $w \in \partial f(x)$ such that

$$\|x - x_0\| < \varepsilon/\lambda \quad \text{and} \quad \|w\| < \lambda.$$

The result follows by noting that $\varepsilon/\lambda < \beta$ and $\lambda < \alpha$. $\square$

The following corollary of the Bishop–Phelps theorem is used in Chapter 4. The proof we present below was taken from [172].

Corollary 5.3.13. *Let C be a convex and closed nonempty subset of a Banach space X. Then the set of points of C at which there exists a supporting hyperplane is dense in the boundary of C.*

Proof. Fix a point $x_0 \in \partial C$ and let $\varepsilon \in (0,1)$. Let $f = \delta_C$ be the indicator function of C. Choose $x_1 \in X \setminus C$ such that $\|x_1 - x_0\| < \varepsilon$. By Theorem 3.4.14(iii), applied for separating $\{x_1\}$ from C, there exists $v_0 \in X^*$ such that $\|v_0\| = 1$ and

$$\sigma_C(v_0) = \sup_{x \in C} \langle x, v_0 \rangle < \langle v_0, x_1 \rangle,$$

which implies that

$$\langle v_0, x \rangle < \langle v_0, x_1 \rangle = \langle v_0, x_1 - x_0 \rangle + \langle v_0, x_0 \rangle \le \varepsilon + \langle v_0, x_0 \rangle$$

for all $x \in C$, so that $\langle v_0, x - x_0 \rangle \le \varepsilon = \varepsilon + f(x) - f(x_0)$. In other words, $v_0 \in \breve{\partial} f(\varepsilon, x_0)$. Applying Theorem 5.3.11 with $\lambda = \sqrt{\varepsilon}$, we conclude that there exists $x \in C = \mathrm{Dom}(f)$ and $v \in \partial f(x) = N_C(x)$ such that $\|x - x_0\| \le \sqrt{\varepsilon}$ and $\|v - v_0\| \le \sqrt{\varepsilon} < 1$. The last inequality implies $\|v\| \neq 0$. Because $v \in N_C(x)$, it is clear that $\sigma_C(v) = \sup_{y \in C} \langle v, y \rangle = \langle v, x \rangle$, implying that the hyperplane $H = \{z \in X \mid \langle v, z \rangle = \langle v, x \rangle\}$ is a supporting hyperplane of C at $x \in C$. Hence, there exists a point in C, arbitrarily close to $x_0 \in \partial C$, at which C has a supporting hyperplane. $\square$

If a point x_0 is not a global minimum of f, can we always find a point $z \in \mathrm{Dom}(f)$ such that f "decreases" from x_0 to z? In other words, is there some $z \in \mathrm{Dom}(f)$ and some $v \in \partial f(z)$ such that

$$f(z) < f(x_0) \quad \text{and} \quad \langle v, z - x_0 \rangle < 0?$$

The positive answer to this question is another corollary of the Bishop–Phelps theorem. This corollary is due to Simons [206], and we used it in Theorem 4.7.1 for proving the maximality of subdifferentials of convex functions. The proof we present below is the one given in [206, Lemma 2.2].

Corollary 5.3.14. *Let $f : X \to \overline{\mathbb{R}}$ be convex, proper, and lsc. If x_0 is such that $\inf_{x \in X} f(x) < f(x_0)$ (it may be $f(x_0) = \infty$), then there exists $(z_0, v_0) \in \mathrm{Gph}(\partial f)$ such that*

$$f(z_0) < f(x_0) \quad \text{and} \quad \langle v_0, z_0 - x_0 \rangle < 0. \tag{5.16}$$

Proof. Choose $\lambda \in \mathbb{R}$ such that $\inf_{x \in X} f(x) < \lambda < f(x_0)$. Define

$$\mu := \sup_{\substack{y \in X \\ y \neq x_0}} \frac{\lambda - f(y)}{\|y - x_0\|}.$$

We claim that $0 < \mu < \infty$. We have $\mu > 0$, because by the definition of λ there exist $y \in \mathrm{Dom}(f)$ such that $f(y) < \lambda$. In order to prove that $\mu < \infty$, define $S := \{y \in X \mid f(y) \le \lambda\}$. If $y \notin S$, then the argument of the supremum above is strictly negative. So it is enough to find an upper bound of this argument for $y \in S$. The existence of this upper bound follows from a separation argument. Indeed, $\mathrm{Dom}(f) \neq \emptyset$, so by Proposition 3.4.17 there exists $u \in X^*$ such that $f^*(u) \in \mathbb{R}$. Because $f^*(u) \ge \langle u, y \rangle - f(y)$ for every $y \in \mathrm{Dom}(f)$, we conclude that there exists a linear functional $\gamma : X \to \mathbb{R}$, $\gamma(y) := \langle u, y \rangle - f^*(u)$ such that $f(y) \geq \gamma(y)$ for all $y \in \mathrm{Dom}(f)$. Take $y \in S$, so that $0 \leq \lambda - f(y) \leq \lambda - \langle u, y \rangle + f^*(u)$ and hence

$$0 \leq \frac{\lambda - f(y)}{\|y - x_0\|} \leq \frac{\lambda - \langle u, y \rangle + f^*(u)}{\|y - x_0\|}$$

$$= \frac{\langle u, x_0 - y \rangle}{\|y - x_0\|} + \frac{\lambda - \langle u, x_0 \rangle + f^*(u)}{\|y - x_0\|}$$

$$\leq \|u\| + \frac{|\lambda - \langle u, x_0 \rangle + f^*(u)|}{d(x_0, S)}. \tag{5.17}$$

Note that the denominator in the rightmost expression of (5.17) is positive, because S is closed and $x_0 \notin S$. The above expression implies that $\mu < \infty$, as we claimed. Fix now $\varepsilon \in (0, 1)$. By definition of μ there exists $z \in X$, $z \neq x_0$ such that

$$\frac{\lambda - f(z)}{\|z - x_0\|} > (1 - \varepsilon)\mu. \tag{5.18}$$

Define the function $\phi : X \to \mathbb{R}$ as $\phi(x) = \mu\|x - x_0\|$. We claim that $\lambda \leq \inf_{x \in X}(f + \phi)$. Indeed, if $y = x_0$ then $\lambda < f(x_0) = (\phi + f)(x_0)$. If $y \neq x_0$ then

$$\frac{\lambda - f(y)}{\|y - x_0\|} \leq \mu,$$

which gives $\lambda \leq \mu\|y - x_0\| + f(y)$. The claim has been proved. Combining this claim with (5.18), we conclude that there exists $z \in X$, $z \neq x_0$ such that

$$(f + \phi)(z) < \inf_{x \in X}(f + \phi)(x) + \varepsilon\phi(z).$$

Using now Theorem 5.3.12, applied to the function $f + \phi$ with $\alpha := \varepsilon\mu$ and $\beta := \|z - x_0\|$, we obtain the existence of $z_0 \in \mathrm{Dom}(f + \phi) = \mathrm{Dom}(f)$ and $w_0 \in \partial(f + \phi)(z_0)$ with $\|w_0\| < \varepsilon\mu$ and $\|z_0 - z\| < \|z - x_0\|$. The latter inequality implies that $\|z_0 - x_0\| > 0$. Note that $\mathrm{Dom}(\phi) = X$, so that the constraint qualification given in Theorem 3.5.7 holds for the sum $f + \phi$, which gives

$$\partial(f + \phi)(z_0) = \partial f(z_0) + \partial\phi(z_0).$$

Therefore, there exist $v_0 \in \partial f(z_0)$ and $u_0 \in \partial\phi(z_0)$ such that $w_0 = v_0 + u_0$. Because $u_0 \in \partial\phi(z_0)$ we have that

$$\langle z_0 - x_0, u_0 \rangle \geq \phi(z_0) - \phi(x_0) = \mu\|z_0 - x_0\|.$$

Thus,

$$\langle x_0 - z_0, v_0 \rangle = \langle x_0 - z_0, w_0 \rangle + \langle z_0 - x_0, u_0 \rangle$$

$$\geq -\|w_0\| \|x_0 - z_0\| + \mu\|z_0 - x_0\| > \mu(1 - \varepsilon)\|z_0 - x_0\| > 0.$$

Therefore, the rightmost inequality in (5.16) holds. For checking the remaining condition in (5.16), combine the above inequality with the fact that $v_0 \in \partial f(z_0)$ getting

$$f(x_0) \geq f(z_0) + \langle v_0, x_0 - z_0 \rangle > f(z_0),$$

and the proof is complete. □

Now we can state and prove for T^e, in reflexive Banach spaces, a property similar to the one established in Theorem 5.3.11 for ε-subdifferentials. The proof of this result relies on a key property of maximal monotone operators. Namely, that the mapping $T + J$ is onto whenever the space is reflexive, where J is the duality mapping defined in Definition 4.4.2. We need the properties of J listed in Proposition 4.4.4.

Theorem 5.3.15. *Let X be a reflexive Banach space. Take $\varepsilon \geq 0$. If $v_\varepsilon \in T^e(\varepsilon, x_\varepsilon)$, then, for all $\eta > 0$ there exists $(x, v) \in G(T)$ such that*

$$\|v - v_\varepsilon\| \leq \frac{\varepsilon}{\eta} \quad \text{and} \quad \|x - x_\varepsilon\| \leq \eta. \tag{5.19}$$

Proof. If $\varepsilon = 0$, then (5.19) holds with $(x, v) = (x_\varepsilon, v_\varepsilon) \in G(T)$. Suppose now that $\varepsilon > 0$. For an arbitrary $\beta > 0$ define the operator $G_\beta : X \rightrightarrows X^*$ as $G_\beta(y) = \beta T(y) + J(y - x_\varepsilon)$, where J is the duality operator of X. Because βT is maximal monotone, G_β is onto by Theorem 4.4.7(i). In particular βv_ε belongs to the image of this operator, and hence there exist $x \in X$ and $v \in T(x)$ such that

$$\beta v_\varepsilon \in \beta v + J(x - x_\varepsilon).$$

Therefore $\beta(v_\varepsilon - v) \in J(x - x_\varepsilon)$. This, together with Proposition 4.4.4(i) and Definition 5.3, yields

$$-\varepsilon \leq \langle v_\varepsilon - v, x_\varepsilon - x \rangle = -\frac{1}{\beta}\|x - x_\varepsilon\|^2 = -\beta\|v - v_\varepsilon\|^2.$$

Choosing $\beta := \eta^2/\varepsilon$, the result follows. □

The following corollary establishes relations among the image, domain, and graph of an operator and its extension T^e.

Corollary 5.3.16. *Let X be a reflexive Banach space. Then,*

(i) $R(T) \subset R(T^e) \subset \overline{R(T)}$.

(ii) $D(T) \subset D(T^e) \subset \overline{D(T)}$.

(iii) If $(x_\varepsilon, v_\varepsilon) \in G(T^e)$ then $d((x_\varepsilon, v_\varepsilon); G(T)) \leq \sqrt{2\varepsilon}$.

Proof. The leftmost inclusions in (i) and (ii) are straightforward from (5.3). The rightmost ones follow from Theorem 5.3.15, letting $\eta \to +\infty$ and $\eta \to 0$ in (i) and (ii), respectively. In order to prove (iii), take $\eta = \sqrt{\varepsilon}$ in (5.19), write

$$d((x_\varepsilon, v_\varepsilon); G(T))^2 \leq \|x - x_\varepsilon\|^2 + \|v - v_\varepsilon\|^2 \leq 2\varepsilon,$$

and take square roots. □

5.4   The family $\mathbb{E}(T)$

Until now, we studied two important examples of enlargements, the enlargement $\breve{\partial}f$ of $T = \partial f$ and the enlargement T^e of an arbitrary maximal monotone operator. We show in this section that each of these enlargements can be regarded as members of a whole family of enlargements of ∂f and T, respectively. For a maximal monotone operator T, $\mathbb{E}(T)$ denotes this family of enlargements. In this section we follow [216] and [52].

Definition 5.4.1. *Let $T : X \rightrightarrows X^*$ be a multifunction. We say that a point-to-set mapping $E : \mathbb{R}_+ \times X \rightrightarrows X^*$ belongs to the family $\mathbb{E}(T)$ when*

(E_1) $T(x) \subset E(\varepsilon, x)$ *for all $\varepsilon \geq 0, x \in X$.*

(E_2) *If $0 \leq \varepsilon_1 \leq \varepsilon_2$, then $E(\varepsilon_1, x) \subset E(\varepsilon_2, x)$ for all $x \in X$.*

(E_3) *The transportation formula holds for $E(\cdot, \cdot)$. More precisely, let $v^1 \in E(\varepsilon_1, x^1)$, $v^2 \in E(\varepsilon_2, x^2)$, and $\alpha \in [0, 1]$. Define*

$$\hat{x} := \alpha x^1 + (1 - \alpha)x^2,$$

$$\hat{v} := \alpha v^1 + (1 - \alpha)v^2,$$

$$\hat{\varepsilon} := \alpha\varepsilon_1 + (1 - \alpha)\varepsilon_2 + \alpha\langle x^1 - \hat{x}, v^1 - \hat{v}\rangle + (1 - \alpha)\langle x^2 - \hat{x}, v^2 - \hat{v}\rangle.$$

Then $\hat{\varepsilon} \geq 0$ and $\hat{v} \in E(\hat{\varepsilon}, \hat{x})$.

Combining this definition with Exercise 8 and Theorem 5.3.6, it follows that when $T = \partial f$, both $\breve{\partial}f$ and ∂f^e belong to $\mathbb{E}(\partial f)$ and that $T^e \in \mathbb{E}(T)$.

The members of the family $\mathbb{E}(T)$ can be ordered in a natural way. Given $E_1, E_2 \in \mathbb{E}(T)$, we say that $E_1 \leq E_2$ if and only if $\text{Gph}(E_1) \subset \text{Gph}(E_2)$.

Theorem 5.4.2. *The enlargement T^e is the largest element of $\mathbb{E}(T)$.*

Proof. Suppose that there exists $E \in \mathbb{E}(T)$ which is not smaller than T^e. In other words, $\text{Gph}(E) \not\subset \text{Gph}(T^e)$. Then there exists $(\eta, z, w) \in \text{Gph}(E)$ which is not in $\text{Gph}(T^e)$; that is, $w \notin T^e(\eta, z)$. From (5.3) it follows that there exist $y \in X$ and $u \in T(y)$ such that

$$\langle w - u, z - y\rangle < -\eta. \tag{5.20}$$

Because E is an enlargement of T and $u \in T(y)$, it holds that $(0, y, u) \in \text{Gph}(E)$. Take some $\theta \in (0, 1)$ and define $\alpha_1 := \theta, \quad \alpha_2 := 1 - \theta$. Using (E_3) with $\alpha = (\alpha_1, \alpha_2)$ and the points

$$(\varepsilon_1, x^1, v^1) := (\eta, z, w) \in \text{Gph}(E),$$

$$(\varepsilon_2, x^2, v^2) := (0, y, u) \in \text{Gph}(E),$$

we conclude that for (ε, x, v) given by

$$x := \alpha_1 z + \alpha_2 y, \qquad v := \alpha_1 w + \alpha_2 u,$$

$$\varepsilon := \alpha_1 \eta + \alpha_1 \langle z - x, w - v\rangle + \alpha_2 \langle y - x, u - v\rangle,$$

it holds that $\varepsilon \geq 0$, and $(\varepsilon, x, v) \in \mathrm{Gph}(E)$. Direct calculation yields

$$\varepsilon = \theta\eta + \theta(1-\theta)\langle w - u, z - y\rangle = \theta\left[\eta + (1-\theta)\langle w - u, z - y\rangle\right].$$

Therefore, because $\theta > 0$, $\eta + (1-\theta)\langle w - u, z - y\rangle \geq 0$. Because θ is an arbitrary number in $(0, 1)$, we conclude that $\eta + \langle w - u, z - y\rangle \geq 0$, contradicting (5.20). □

After proving that $\mathbb{E}(T)$ has a largest element, we can prove the existence of a smallest one.

Theorem 5.4.3. *The family $\mathbb{E}(T)$ has a smallest element, namely $T^{se} : \mathbb{R}_+ \times X \rightrightarrows X^*$, defined as*

$$T^{se}(\varepsilon, x) := \bigcap_{E \in \mathbb{E}(T)} E(\varepsilon, x). \tag{5.21}$$

Proof. First observe that $T^e \in \mathbb{E}(T)$ and hence $\mathbb{E}(T)$ is nonempty. So, T^{se} given by (5.21) is well defined and it is easy to see that it satisfies (E_1)–(E_3). So $T^{se} \in \mathbb{E}(T)$ and $\mathrm{Gph}(T^{se}) \subseteq \mathrm{Gph}(E)$ for any $E \in \mathbb{E}(T)$. □

Remark 5.4.4. Observe that the proofs of Theorem 5.3.4, Lemma 5.3.3, and Theorem 5.3.15 only use the inequality that characterizes elements in T^e. The fact that $\mathrm{Gph}(E) \subset \mathrm{Gph}(T^e)$ for all $E \in \mathbb{E}(T)$, implies that all these results are still valid for all $E \in \mathbb{E}(T)$. In particular, because $\breve{\partial}f(\varepsilon, \cdot) \subset \partial f^e(\varepsilon, \cdot)$, we recover all these facts for ε-subdifferentials. In the case of Theorem 5.3.15, this has not much value inasmuch as the analogous result for ε-subdifferentials holds in arbitrary Banach spaces, not necessarily reflexive. However, it has been pointed out in [219] and in [207, Example 11.5] that Theorem 5.3.15 is not valid in nonreflexive Banach spaces for arbitrary maximal monotone operators (see also [175]).

Lemma 5.4.5. *Take $E \in \mathbb{E}(T)$. Then*

(a) E is convex-valued.

(b) Assume that X is reflexive, so that T^{-1} is maximal monotone. Define

$$\begin{aligned} E^* &: \mathbb{R}_+ \times X^* \rightrightarrows X, \\ E^*(\varepsilon, w) &:= \{x \in X : w \in E(\varepsilon, x)\}. \end{aligned} \tag{5.22}$$

Then $E^* \in \mathbb{E}(T^{-1})$. As a consequence, for all $v \in X^*$, $\varepsilon \in \mathbb{R}_+$, the set

$$\{x \in X \mid v \in E(\varepsilon, x)\}$$

is convex. In this situation, $E^{**} = E$.

(c) Suppose that $\mathrm{Gph}(E)$ is closed. Then, if $A \subset \mathbb{R}_+$ with $\bar{\varepsilon} = \inf A$, it holds that $\cap_{\varepsilon \in A} E(\varepsilon, \cdot) = E(\bar{\varepsilon}, \cdot)$.

(d) If T is maximal monotone, then $\cap_{\varepsilon>0} E(\varepsilon, \cdot) = E(0, \cdot) = T(\cdot)$.

Proof.

(a) This fact follows directly from the transportation formula and is left as an exercise.

(b) The result follows directly from (a) and the definitions of E^* and $\mathbb{E}(\cdot)$.

(c) By (E_2) it holds that $\cap_{\varepsilon \in A} E(\varepsilon, x) \supset E(\bar{\varepsilon}, x)$. For the opposite inclusion take $v \in \cap_{\varepsilon \in A} E(\varepsilon, x)$ and a sequence $\{\varepsilon_n\}$ such that $\varepsilon_n \downarrow \bar{\varepsilon}$. In particular, $v \in \cap_n E(\varepsilon_n, x)$. The assumption on the graph now yields $v \in E(\bar{\varepsilon}, x)$.

(d) By (E_1)–(E_2) and Theorem 5.4.2 we have

$$T(x) \subset E(0, x) \subset \cap_{\varepsilon>0} E(\varepsilon, x) \subset \cap_{\varepsilon>0} T^e(\varepsilon, x) = T(x),$$

using also Lemma 5.2.1(e)–(f).   □

Remark 5.4.6. Let X be a reflexive Banach space and $f : X \to \mathbb{R} \cup \{\infty\}$ convex, proper, and lsc. We proved in Theorem 3.4.18 and Corollary 3.5.5 that in this case $f = f^{**}$ and $(\partial f)^{-1} = \partial f^*$. The latter fact is in a certain sense preserved by taking enlargements. Namely, the inverse of the BR-enlargement of ∂f is the BR-enlargement of $(\partial f)^{-1} = \partial f^*$, as the following proposition establishes.

Proposition 5.4.7. *Let X be a reflexive Banach space and $f : X \to \mathbb{R} \cup \{\infty\}$ convex, proper, and lsc. If $T = \partial f$ and $E = \breve{\partial} f$, then $E^* = \breve{\partial} f^*$.*

Proof. We must prove that $(\breve{\partial} f)^* = \breve{\partial} f^*$. We prove first that $(\breve{\partial} f)^*(\varepsilon, w) \subset \breve{\partial} f^*(\varepsilon, w)$. Take $x \in (\breve{\partial} f)^*(\varepsilon, w)$. By definition, this means that $w \in \breve{\partial} f(\varepsilon, x)$ and hence

$$f(y) \geq f(x) + \langle w, y - x \rangle - \varepsilon, \quad \text{for all} \quad y \in X.$$

Rearrange the last expression and get

$$\langle w, x \rangle - f(x) \geq \langle w, y \rangle - f(y) - \varepsilon \quad \text{for all} \quad y \in X.$$

Take now the supremum over $y \in X$, obtaining

$$\langle w, x \rangle - f(x) \geq f^*(w) - \varepsilon.$$

Using the fact that $f^*(v) - \langle v, x \rangle \geq -f(x)$ for all $v \in X^*$ in the inequality above yields

$$f^*(v) + \langle w - v, x \rangle \geq f^*(w) - \varepsilon, \quad \text{for all} \quad v \in X^*,$$

which implies $x \in \breve{\partial} f^*(\varepsilon, w)$. Conversely, take $x \in \breve{\partial} f^*(\varepsilon, w)$. Then

$$f^*(v) \geq f^*(w) + \langle x, v - w \rangle - \varepsilon, \quad \text{for all} \quad v \in X^*.$$

Rearranging this inequality, we get

$$\langle w, x \rangle - f^*(w) \geq \langle v, x \rangle - f^*(v) - \varepsilon \quad \text{for all} \quad v \in X^*.$$

Take now the supremum over $v \in X^*$ and get

$$\langle w, x \rangle - f^*(w) \geq f^{**}(x) - \varepsilon = f(x) - \varepsilon,$$

using the fact that $f = f^{**}$ in a reflexive Banach space. Because $-\langle w, y \rangle + f(y) \geq -f^*(w)$ for all $y \in X$, the expression above yields $w \in \check{\partial}f(\varepsilon, x)$, or $x \in (\check{\partial}f)^*(\varepsilon, w)$.
□

Condition (E_3) is closely connected to convexity, as we show next. Define $\psi : \mathbb{R}_+ \times X \times X^* \to \mathbb{R}_+ \times X \times X^*$ as

$$\psi(t, x, v) = (t + \langle x, v \rangle, x, v). \tag{5.23}$$

The next lemma establishes that (E_3) for E is equivalent to convexity of $\psi(\mathrm{Gph}(E))$.

Lemma 5.4.8. *Let $E : \mathbb{R}_+ \times X \rightrightarrows X^*$ be an arbitrary point-to-set mapping. E satisfies (E_3) if and only if $\psi(\mathrm{Gph}(E))$ is convex, with ψ as in (5.23).*

Proof. First note that

$$\psi(\mathrm{Gph}(E)) := \{(\varepsilon + \langle x, v \rangle, x, v) \in \mathbb{R}_+ \times X \times X^* : v \in E(\varepsilon, x)\}$$

$$= \{(\varepsilon + \langle x, v \rangle, x, v) \in \mathbb{R}_+ \times X \times X^* : (\varepsilon, x, v) \in \mathrm{Gph}(E)\}. \tag{5.24}$$

For the equivalence, we need some elementary algebra. Take triplets $(\varepsilon_1, x_1, v_1)$, $(\varepsilon_2, x_2, v_2) \in \mathrm{Gph}(E)$ with corresponding points $\psi(\varepsilon_i, x_i, v_i) = (\varepsilon_i + \langle x_i, v_i \rangle, x_i, v_i)$, $i = 1, 2$. Define $t_i := \varepsilon_i + \langle x_i, v_i \rangle$, $i = 1, 2$. Fix $\alpha \in [0, 1]$ and let

$$x := \alpha x_1 + (1 - \alpha)x_2, \qquad v := \alpha v_1 + (1 - \alpha)v_2, \tag{5.25}$$

$$\varepsilon := \big[\alpha t_1 + (1 - \alpha)t_2\big] - \langle x, v \rangle$$

$$= \big[\alpha(\varepsilon_1 + \langle x_1, v_1 \rangle) + (1 - \alpha)(\varepsilon_2 + \langle x_2, v_2 \rangle)\big] - \langle x, v \rangle$$

$$= \alpha(\varepsilon_1 + \langle x_1 - x, v_1 - v \rangle) + (1 - \alpha)(\varepsilon_2 + \langle x_2 - x, v_2 - v \rangle). \tag{5.26}$$

Then

$$\alpha(t_1, x_1, v_1) + (1 - \alpha)(t_2, x_2, v_2) = (\varepsilon + \langle x, v \rangle, x, v). \tag{5.27}$$

Now we proceed with the proof. Assume that E satisfies (E_3). In this case, by (5.25) and (5.26), we get $(\varepsilon, x, v) \in \mathrm{Gph}(E)$. Using (5.27) and the definition of ψ, we conclude that $\alpha(t_1, x_1, v_1) + (1 - \alpha)(t_2, x_2, v_2) \in \psi(\mathrm{Gph}(E))$, yielding the convexity of $\psi(\mathrm{Gph}(E))$. Assume now that $\psi(\mathrm{Gph}(E))$ is convex. By (5.27), we get $(\varepsilon + \langle x, v \rangle, x, v) \in \psi(\mathrm{Gph}(E))$, but the definition of ψ implies $(\varepsilon, x, v) \in \mathrm{Gph}(E)$. Because the points $(\varepsilon_1, x_1, v_1), (\varepsilon_2, x_2, v_2) \in \mathrm{Gph}(E)$ and $\alpha \in [0, 1]$ are arbitrary, we conclude that E satisfies (E_3).   □

5.4.1   Linear skew-symmetric operators are nonenlargeable

Let X be a reflexive Banach space and $L : X \to X^*$ a linear operator. The operator L is said to be

- *Positive semidefinite* when $\langle Lx, x \rangle \geq 0$ for every $x \in X$

- *Skew-symmetric* when $L + L^* \equiv 0$

Our aim now is to extend Example 5.2.5(v) to this more general framework. In order to do this, we need a technical lemma.

Lemma 5.4.9. *Let X be a reflexive Banach space. Take a positive semidefinite linear operator $S : X \to X^*$ with closed range $R(S)$. For a fixed $u \in X^*$, define $\phi : X \to \mathbb{R}$ as $\phi(y) := 1/2 \langle Sy, y \rangle - \langle u, y \rangle$. Then*

(i) If $u \notin R(S)$, then $\operatorname{argmin}_X \phi = \emptyset$ and $\inf_X \phi = -\infty$.

(ii) If $u \in R(S)$, then $\emptyset \neq \operatorname{argmin}_X \phi = S^{-1} u$.

Proof.

   (i) Because $u \notin R(S)$ and $R(S)$ is closed and convex, there exists $\bar{y} \neq 0$, $\bar{y} \in X^{**} = X$ such that $\langle \bar{y}, u \rangle = 1$ and $\langle \bar{y}, Sy \rangle \leq 0$ for all $y \in X$. Because S is positive semidefinite, it holds that $\langle \bar{y}, S\bar{y} \rangle = 0$. Then $\phi(\lambda \bar{y}) := (1/2)\lambda^2 \langle S\bar{y}, \bar{y} \rangle - \lambda \langle u, \bar{y} \rangle = -\lambda$, which tends to $-\infty$ when $\lambda \uparrow \infty$. Thus, $\operatorname{argmin}_X \phi = \emptyset$ and $\inf_X \phi = -\infty$.

   (ii) Observe that $\operatorname{argmin}_X \phi = \{z \mid \nabla \phi(z) = 0\} = \{z \mid S(z) = u\} = S^{-1} u$. This set is nonempty, because $u \in R(S)$.   □

The following proposition extends Example 5.2.5(v).

Proposition 5.4.10. *Let X be a reflexive Banach space and $T : X \to X^*$ a maximal monotone operator of the form $T(x) = Lx + x_0^*$, where L is linear (and hence positive semidefinite, by monotonicity of T) and $x_0^* \in X^*$. Then*

$$T^e(\varepsilon, x) = \{Lx + x_0^* + Sw \mid \langle Sw, w \rangle \leq 2\varepsilon\}, \qquad (5.28)$$

where $S := L + L^*$.

Proof. Call $V(\varepsilon, x)$ the right-hand side of (5.28). Then $V : \mathbb{R}_+ \times X \rightrightarrows X^*$ and our first step is to prove that $T^e(\varepsilon, x) \subset V(\varepsilon, x)$. Then we show that $V \in \mathbb{E}(T)$, and we conclude from Theorem 5.4.2 that $V(\varepsilon, x) \subset T^e(\varepsilon, x)$. In order to prove that $T^e(\varepsilon, x) \subset V(\varepsilon, x)$, we write an element $w \in T^e(\varepsilon, x)$ as $w = Lx + x_0^* + u$, for some $u \in X^*$. We must prove in this step that there exists some $\bar{w}$ such that $S\bar{w} = u$ and $\langle S\bar{w}, \bar{w} \rangle \leq 2\varepsilon$. The definition of T^e implies that for all $z \in X$, it holds that

$$-\varepsilon \leq \langle w - (Lz + x_0^*), x - z \rangle = \langle L(x - z), x - z \rangle + \langle u, x - z \rangle.$$

Rearranging the expression above, we get that

$$-\varepsilon \leq (1/2)\langle S(z - x), z - x \rangle - \langle u, z - x \rangle = \phi(z - x),$$

with ϕ as in Lemma 5.4.9. Because ϕ is bounded below, we get that $u \in R(S)$ and $\operatorname{argmin}_X \phi = x + S^{-1}u$. Thus, for every $\bar{z} \in x + S^{-1}u$ we have

$$-\varepsilon \le \phi(\bar{z} - x) \le \phi(z - x),$$

for all $z \in X$. In other words,

$$-\varepsilon \le \frac{1}{2}\langle S(\bar{z} - x), \bar{z} - x \rangle - \langle u, \bar{z} - x \rangle = -\frac{1}{2}\langle S(\bar{z} - x), \bar{z} - x \rangle,$$

using the fact that $S(\bar{z}-x) = u$. Then $\langle S(\bar{z}-x), \bar{z}-x \rangle \le 2\varepsilon$, and hence there exists $\bar{w} := (\bar{z} - x)$ such that $u = S\bar{w}$ with $\langle S\bar{w}, \bar{w} \rangle \le 2\varepsilon$. The first step is complete. In order to prove that $V \in \mathbb{E}(T)$, it is enough to check (E_3), because (E_1)–(E_2) are easy consequences of the definition of V. Take $\alpha \in [0,1]$ and also $v_1 = Lx_1 + x_0^* + Sw_1 \in V(\varepsilon_1, x_1)$, $v_2 = Lx_2 + x_0^* + Sw_2 \in V(\varepsilon_2, x_2)$, with w_i such that $\langle Sw_i, w_i \rangle \le 2\varepsilon$. We must prove that

$$\tilde{v} := \alpha v_1 + (1-\alpha)v_2 \in V(\tilde{\varepsilon}, \tilde{x}), \tag{5.29}$$

where $\tilde{\varepsilon} := \alpha\varepsilon_1 + (1-\alpha)\varepsilon_2 + \alpha(1-\alpha)\langle v_1 - v_2, x_1 - x_2 \rangle$ and $\tilde{x} := \alpha x_1 + (1-\alpha)x_2$. Direct calculation shows that

$$\tilde{v} = L\tilde{x} + x_0^* + S(\tilde{w}), \tag{5.30}$$

where $\tilde{w} := \alpha w_1 + (1-\alpha)w_2$. Define $q(w) := \frac{1}{2}\langle Sw, w \rangle$. By definition of $V(\varepsilon, x)$, (5.29) is equivalent to establishing that $q(\tilde{w}) \le \tilde{\varepsilon}$. Let us prove this inequality. We claim that

$$-q(w_1 - w_2) \le \langle v_1 - v_2, x_1 - x_2 \rangle, \tag{5.31}$$

and we proceed to prove this claim. Using Lemma 5.4.9 with $u := S(w_2-w_1) \in R(S)$ we get

$$\langle v_1 - v_2, x_1 - x_2 \rangle = \langle L(x_1 - x_2), x_1 - x_2 \rangle - \langle S(w_2 - w_1), x_1 - x_2 \rangle$$

$$= 1/2\langle S(x_1 - x_2), x_1 - x_2 \rangle - \langle S(w_2 - w_1), x_1 - x_2 \rangle = \phi(x_1 - x_2) \ge \phi(z - x_2) \tag{5.32}$$

for all $z \in x_2 + S^{-1}(S(w_2 - w_1))$. Thus, the expression above holds for $z = x_2 + w_2 - w_1$. Hence,

$$\begin{aligned} \phi(z - x_2) &= \phi(w_2 - w_1) = q(w_2 - w_1) - \langle S(w_2 - w_1), w_2 - w_1 \rangle \\ &= -q(w_2 - w_1) = -q(w_1 - w_2). \end{aligned} \tag{5.33}$$

Combining (5.33) with (5.32), we get (5.31), establishing the claim. Because $q(w_i) \le \varepsilon_i$ by assumption, (5.31) implies that

$$\alpha q(w_1) + (1-\alpha)q(w_2) - \alpha(1-\alpha)q(w_1 - w_2)$$

$$\le \alpha\varepsilon_1 + (1-\alpha)\varepsilon_2 + \alpha(1-\alpha)\langle v_1 - v_2, x_1 - x_2 \rangle. \tag{5.34}$$

The definitions of q and S give

$$q(\tilde{w}) = \alpha q(w_1) + (1-\alpha)q(w_2) - \alpha(1-\alpha)q(w_1 - w_2). \tag{5.35}$$

It follows from (5.34), (5.35), and the definition of $\tilde{\varepsilon}$ that

$$q(\tilde{w}) \leq \tilde{\varepsilon}. \tag{5.36}$$

Because (5.29) follows from (5.36) and (5.30), we conclude that (E_3) holds for V, completing the proof. $\square$

An immediate consequence of Proposition 5.4.10 is the characterization of those linear operators T that cannot be actually enlarged by any $E \in \mathbb{E}(T)$ (see Exercise 13).

Corollary 5.4.11. *Let X and T be as in Proposition 5.4.10. Then the following statements are equivalent.*

(i) L is skew-symmetric.

(ii) $T^e(\varepsilon, x) = T(x)$ for every $\varepsilon > 0$.

(iii) $T^e(\bar{\varepsilon}, x) = T(x)$ for some $\bar{\varepsilon} > 0$.

Proof. By Proposition 5.4.10, it is enough to prove that (iii) implies (i). Assume that (iii) holds. The operator $S := L + L^*$ is positive semidefinite. If it were nonnull, then there would exist a positive semidefinite linear operator $R_0 \neq 0$ with $R_0^2 = S$. Take $w \notin Ker(S)$. Thus, $w \notin Ker(R_0)$ and $\langle Sw, w \rangle = \|R_0 w\|^2 > 0$. Let $\bar{w} := (\sqrt{2\bar{\varepsilon}}/\|R_0 w\|)\, w$. Then, $S\bar{w} \neq 0$ and $\langle S\bar{w}, \bar{w} \rangle \leq 2\bar{\varepsilon}$. The latter inequality, together with Proposition 5.4.10 imply that $T(x) + S\bar{w} \in T^e(\bar{\varepsilon}, x)$. This contradicts assumption (iii), which states that the unique element of $T^e(\bar{\varepsilon}, x)$ is $T(x)$. Hence we must have $S = L + L^* = 0$ and (i) holds. $\square$

5.4.2 Continuity properties

We have seen in Remark 5.4.4 that some properties of T^e are inherited by the elements of $\mathbb{E}(T)$. Continuity properties, instead, cannot be obtained from the analysis of T^e alone. We devote this section to establishing these properties.

From Theorem 2.5.4 we have that outer-semicontinuity is equivalent to closedness of the graph. So, we cannot expect any good continuity behavior from elements in $\mathbb{E}(T)$ that do not have a closed graph. We show next that, when X is reflexive, elements of $\mathbb{E}(T)$ that have a strongly closed graph are always (sw)-OSC (see Proposition 5.4.13 below).

Definition 5.4.12. $\mathbb{E}_c(T)$ *denotes the set of elements of $\mathbb{E}(T)$ whose graphs are closed.*

Proposition 5.4.13. *Assume that X is reflexive. Then $E \in E_c(T)$ if and only if E is (sw)-OSC.*

Proof. The "if" statement follows directly from the fact that a (sw)-closed subset of $X \times X^*$ is (strongly) closed. For the "only if" one, assume that $E \in E_c(T)$. By Lemma 5.4.8 and the properties of ψ, $\psi(\mathrm{Gph}(E))$ is closed and convex. By convexity and Corollary 3.4.16 it is closed when both X and X^* are endowed with the weak topology. Because (ww)-closedness implies (sw)-closedness, we conclude that $\mathrm{Gph}(E)$ is (sw)-closed. □

As we have seen in Proposition 4.2.1, a maximal monotone operator T is everywhere (ss)-OSC, but it may fail to be (sw)-OSC at a boundary point of its domain. However, the convexity of $\psi(\mathrm{Gph}(E))$ allows us to establish (sw)-OSC of every (ss)-OSC enlargement E. Moreover, as we have seen in Theorem 4.6.3, maximal monotone operators fail to be ISC, unless they are point-to-point. A remarkable fact is that not only T^e but every element of $\mathbb{E}_c(T)$ is ISC in the interior of its domain. Furthermore, every element of $\mathbb{E}_c(T)$ is Lipschitz continuous in $\mathbb{R}_{++} \times (D(T))^o$ (see Definition 2.5.31).

Lipschitz continuity is closely connected with the transportation formula. In order to prove that the elements of $E(T)$ are Lipschitz continuous, we need two lemmas relying on property (E_3). The first one provides a way of "improving" (i.e., diminishing) the parameter ε of an element $v \in E(\varepsilon, x)$.

Lemma 5.4.14. *Let $E : \mathbb{R}_+ \times X \rightrightarrows X^*$ be an arbitrary point-to-set mapping. Assume that E satisfies (E_1) and (E_3) and take $v \in E(\varepsilon, x)$. Then for all $\varepsilon' < \varepsilon$, there exists $v' \in E(\varepsilon', x)$.*

Proof. Take $u \in T(x)$. Use condition (E_3) with $\alpha := \varepsilon'/\varepsilon$, $x_1 := x$, $x_2 := x$, $v_1 := v$ and $v_2 := u \in T(x) \subset E(0, x)$, for concluding that $v' = \alpha v_1 + (1 - \alpha)u \in E(\varepsilon', x)$. □

The next lemma is needed for proving Lipschitz continuity of enlargements. We point out that Lemma 5.3.3 holds for every $E \in \mathbb{E}(T)$. This fact is used in the following proof.

Lemma 5.4.15. *Take $E \in \mathbb{E}(T)$. Assume that T is bounded on a set $U \subset X$. Let*

$$M := \sup_{\substack{y \in U \\ v \in T(y)}} \|v\|.$$

Take V and $\rho > 0$ as in Lemma 5.3.3, and $\beta > 0$. Then there exist nonnegative constants C_1 and C_2 such that, for all $x_1, x_2 \in V$ and $v_1 \in E(\varepsilon_1, x_1)$, with $\varepsilon_1 \le \beta$, there exist $\varepsilon_2' \ge 0$ and $v_2' \in E(\varepsilon_2', x_2)$ such that

$$\|v_1 - v_2'\| \le C_0 \|x_1 - x_2\|,$$
$$|\varepsilon_1 - \varepsilon_2'| \le C_1 \|x_1 - x_2\|.$$

Proof. Take x_1, x_2, ε_1, and v_1 as above. Take $\lambda := \|x_1 - x_2\|$ and x_3 in the line through x_1 and x_2 satisfying:

$$\|x_3 - x_2\| = \rho, \qquad \|x_3 - x_1\| = \rho + \lambda; \tag{5.37}$$

see Figure 5.2.

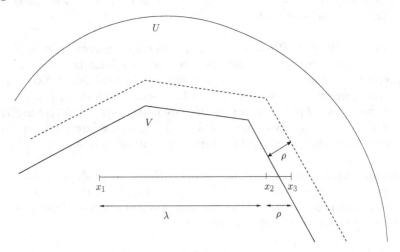

Figure 5.2. *Auxiliary construction*

Define $\theta := \rho/(\rho + \lambda)$, so that $x_2 = \theta x_1 + (1 - \theta)x_3$. Take now $u_3 \in T(x_3)$. By (E_3) we have that

$$v_2' := \theta v_1 + (1 - \theta)u_3 \in E(\tilde{\varepsilon}_2, x_2), \tag{5.38}$$

with

$$0 \le \tilde{\varepsilon}_2 = \theta \varepsilon_1 + \theta(1 - \theta)\langle x_1 - x_3, v_1 - u_3 \rangle. \tag{5.39}$$

From (5.38),

$$\|v_2' - v_1\| = (1 - \theta)\|v_1 - u_3\| \le (1 - \theta)(\|v_1\| + \|u_3\|) \le \lambda/(\rho + \lambda)(\|v_1\| + \|u_3\|)$$

$$\le \lambda/\rho\,(\varepsilon_1/\rho + 2M) = (\beta/\rho + 2M)(1/\rho)\|x_1 - x_2\|,$$

using Lemma 5.3.3 and the definitions of M, θ, and λ. This establishes the first inequality, with $C_0 := (\beta/\rho + 2M)(1/\rho)$. If $\tilde{\varepsilon}_2 \le \varepsilon_1$, then we can take $\varepsilon_2' := \varepsilon_1$, and by (E_2) we have that $v_2' \in E(\tilde{\varepsilon}_2, x_2) \subset E(\varepsilon_2', x_2) = E(\varepsilon_1, x_2)$, so that the second inequality holds trivially. Otherwise, use (5.39) to get

$$\tilde{\varepsilon}_2 \le \varepsilon_1 + \lambda\rho/(\rho + \lambda)^2\|v_1 - u_3\|\,\|x_1 - x_3\| = \varepsilon_1 + \rho/(\rho + \lambda)\|v_1 - u_3\|\,\|x_1 - x_2\|$$

$$\le \varepsilon_1 + (\varepsilon_1/\rho + 2M)\|x_1 - x_2\| \le \varepsilon_1 + (\beta/\rho + 2M)\|x_1 - x_2\|,$$

using the Cauchy–Schwartz inequality and the definition of θ in the first inequality, the fact that $\|x_3 - x_1\| = \rho + \lambda$ and $\|x_2 - x_1\| = \lambda$ in the equality, and Lemma 5.3.3

in the second inequality. Then $|\tilde{\varepsilon}_2 - \varepsilon_1| = \tilde{\varepsilon}_2 - \varepsilon_1 \leq (\beta/\rho + 2M)\|x_1 - x_2\|$, and we get the remaining inequality, with $\varepsilon_2' := \tilde{\varepsilon}_2$ and $C_1 := (\beta/\rho + 2M)$. The proof is complete. $\square$

The next theorem is essential for establishing Lipschitz continuity, in the sense of Definition 2.5.31, of all closed enlargements satisfying (E_1)–(E_3).

Theorem 5.4.16. *Take $E \in \mathbb{E}(T)$. Assume that T is bounded on a set $U \subset X$, with*

$$M := \sup_{\substack{y \in U \\ v \in T(y)}} \|v\|.$$

Let V and $\rho > 0$ be as in Lemma 5.3.3, and take $0 < \alpha \leq \beta$. Then there exist nonnegative constants σ, τ such that, for all x_1, $x_2 \in V$, all ε_1, $\varepsilon_2 \in [\alpha, \beta]$, and all $v_1 \in E(\varepsilon_1, x_1)$, there exists $v_2 \in E(\varepsilon_2, x_2)$ such that

$$\|v_1 - v_2\| \leq \sigma\|x_1 - x_2\| + \tau|\varepsilon_1 - \varepsilon_2|. \tag{5.40}$$

Proof. We claim that (5.40) holds for $\sigma =: C_0 + (C_1/\alpha)[(\beta/\rho) + 2\,M]$ and $\tau := (1/\alpha)[(\beta/\rho) + 2\,M]$, with C_0 and C_1 as in Lemma 5.4.15. In fact, by Lemma 5.4.15, we can take $\varepsilon_2' \geq 0$ and $v_2' \in E(\varepsilon_2', x_2)$ such that

$$\|v_1 - v_2'\| \leq C_0\|x_1 - x_2\|, \tag{5.41}$$

$$\varepsilon_2' \leq \varepsilon_1 + C_1\|x_1 - x_2\|. \tag{5.42}$$

If $\varepsilon_2' \leq \varepsilon_2$ then $v_2' \in E(\varepsilon_2, x_2)$ and (5.40) holds because $\sigma \geq C_0$. Otherwise, define $\eta := \varepsilon_2/\varepsilon_2' < 1$. By Lemma 5.4.14 there exists $v_2 \in E(\varepsilon_2, x_2)$, which can be taken as $v_2 = \eta v_2' + (1 - \eta)u_2$, with $u_2 \in T(x_2)$. We have that

$$\|v_1 - v_2\| \leq \eta \|v_1 - v_2'\| + (1 - \eta) \|v_1 - u_2\|$$

$$\leq \eta C_0 \|x_1 - x_2\| + (1 - \eta) [(\beta/\rho) + 2\,M], \tag{5.43}$$

using (5.41) and affine local boundedness. More precisely, because $x_1, x_2 \in V$, we can use Lemma 5.3.3 and the definition of M to get

$$\|v_1 - u_2\| \leq \|v_1\| + \|u_2\| \leq [(\varepsilon_1/\rho) + M] + M = (\varepsilon_1/\rho) + 2\,M \leq (\beta/\rho) + 2\,M.$$

On the other hand, we get from (5.42)

$$(1 - \eta) = \frac{\varepsilon_2' - \varepsilon_2}{\varepsilon_2'} \leq \frac{\varepsilon_1 + C_1\|x_1 - x_2\| - \varepsilon_2}{\varepsilon_2'}$$

$$\leq \frac{|\varepsilon_1 - \varepsilon_2| + C_1\|x_1 - x_2\|}{\varepsilon_2'} \leq (1/\alpha)|\varepsilon_1 - \varepsilon_2| + (C_1/\alpha)\|x_1 - x_2\|, \tag{5.44}$$

using the fact that $\alpha \leq \varepsilon_2 < \varepsilon_2'$ in the last inequality. Combine (5.44) and (5.43) in order to conclude that (5.40) holds for the above-mentioned values of σ and τ. $\square$

A direct consequence of Theorem 5.4.16 is presented in the following corollary.

Corollary 5.4.17. *Take $E \in \mathbb{E}_c(T)$, and α, β, and V as in Theorem 5.4.16. Then E is Lipschitz continuous on $[\alpha, \beta] \times V$. As a consequence, E is continuous in $\mathbb{R}_{++} \times (D(T))^o$.*

Proof. In view of Definition 2.5.31 and Theorem 5.4.16, we only have to prove the last assertion. Note that F is Lipschitz continuous on $U := (\alpha, \beta) \times V$ for any α, β such that $0 < \alpha \leq \beta$ and V a neighborhood of any $y \in (D(T))^o$. Therefore we can apply Proposition 2.5.32(ii)–(iii) with $F := E$ and $U := (\alpha, \beta) \times V$ to get the continuity of E in $(\alpha, \beta) \times V$. Because this holds for any α, β such that $0 < \alpha \leq \beta$ and the neighborhood of any $y \in (D(T))^o$, we conclude the continuity of E in $\mathbb{R}_{++} \times (D(T))^o$. $\square$

Recall that, in view of Theorem 4.6.3, a maximal monotone operator T cannot be ISC at a boundary point of its domain, and it is only ISC at those interior points of its domain where it is single-valued. A similar behavior is inherited by $E \in \mathbb{E}(T)$. Denote D the domain of T and ∂A the boundary of a set A. We consider in the following theorem inner-semicontinuity with respect to the strong topology in X and the weak* topology in X^*.

Theorem 5.4.18. *Assume that $D^o \neq \emptyset$. Let E be an element of $\mathbb{E}_c(T)$. Then*

(a) E is not ISC at any point of the form $(0, x)$, with $x \in \partial D$.

(b) T is ISC at x if and only if E is ISC at $(0, x)$.

Proof.
   (a) Take $x \in \partial D$ and suppose that E is ISC at $(0, x)$. Fix $\alpha > 0$. Because $D^o \neq \emptyset$, we can find $v' \in N_D(x)$ and $y \in D^o$ such that $\langle v', y - x \rangle < -2\alpha$. Take $t_n \uparrow 1$ and define the sequences $x_n := t_n x + (1 - t_n)y$, $\varepsilon_n := (1 - t_n)\alpha$. For a fixed $v \in T(x)$, it holds that $v + v' \in T(x) = E(0, x)$ (see Theorem 4.2.10(ii-b)). Because E is inner-semicontinuous and the sequence $\{(\varepsilon_n, x_n)\}$ converges to $(0, x)$, we must have $E(0, x) \subset \lim \mathrm{int}_n E(\varepsilon_n, x_n)$. Define the weak* neighborhood of $v + v'$ given by

$$W := \{z \in X^* : |\langle z - (v + v'), y - x \rangle| < \frac{\alpha}{2}\}.$$

By the definition of inner limit there exists $n_0 \in \mathbb{N}$ such that

$$W \cap E(\varepsilon_n, x_n) \neq \emptyset,$$

for all $n \geq n_0$. Choose elements $w_n \in W \cap E(\varepsilon_n, x_n)$ for all $n \geq n_0$. Because $E(\varepsilon_n, x_n) \subset T^e(\varepsilon_n, x_n)$ and $v \in Tx$, we have

$$-\varepsilon_n = -(1 - t_n)\alpha \leq \langle x_n - x, w_n - v \rangle = \langle x_n - x, w_n - (v + v') \rangle + \langle v', x_n - x \rangle$$

$$= (1 - t_n)\langle y - x, w_n - (v + v') \rangle + (1 - t_n)\langle v', y - x \rangle, \qquad (5.45)$$

which simplifies to

$$-\alpha \le \langle y - x, w_n - (v + v') \rangle + \langle v', y - x \rangle < \frac{\alpha}{2} - 2\alpha,$$

where we used the fact that $\{w_n\} \in W$ and the definition of v'. The above inequality yields a contradiction and hence E cannot be inner-semicontinuous at $(0, x)$.

(b) Let $\{(\varepsilon_i, x_i)\}_{i \in I}$ be a net converging to $(0, x)$. If T is ISC at x, then we have $Tx \subset \liminf_i T(x_i)$. Because $T(\cdot) = E(0, \cdot)$ and $T(\cdot) \subset E(\varepsilon, \cdot)$ we can write $E(0, x) = Tx \subset \liminf_i T(x_i) \subset \liminf_i E(\varepsilon_i, x_i)$ which yields the inner-semicontinuity of E at $(0, x)$. Let us prove the converse statement. It is enough, in view of Theorem 4.6.3(ii), to prove that if E is ISC at $(0, x)$, then Tx is a singleton. Assume, for the sake of contradiction, that there exist $v_1, v_2 \in Tx$, with $0 \ne v_1 - v_2 =: w$. Fix $z \in X$ such that $\langle z, w \rangle =: \alpha > 0$. The proof now follows closely the one in Theorem 4.6.3(ii). Define the sequences $x_n := x + (z/n)$ and $\varepsilon_n := \alpha/4\,n$. Because E is ISC at $(0, x)$ we have that $v_2 \in Tx \subset E(0, x) \subset \liminf_n E(\varepsilon_n, x_n)$. Consider the set

$$W_0 := \{w \in X^* : |\langle w - v_2, z \rangle| < \frac{\alpha}{2}\} \in \mathcal{U}_{v_2}.$$

So there exists $n_0 \in \mathbb{N}$ such that $W_0 \cap E(\varepsilon_n, x_n) \ne \emptyset$ for all $n \ge n_0$. For every $n \ge n_0$, take $u_n \in W_0 \cap E(\varepsilon_n, x_n)$. Because $v_1 \in Tx$ we have

$$
\begin{aligned}
-\varepsilon_n = -\frac{\alpha}{4n} &\le \langle u_n - v_1, x_n - x \rangle \\
&= \langle u_n - v_2, x_n - x \rangle + \langle v_2 - v_1, x_n - x \rangle \\
&= \frac{1}{n}[\langle u_n - v_2, z \rangle + \langle v_2 - v_1, z \rangle],
\end{aligned}
$$

which simplifies to

$$-\frac{\alpha}{4} = \langle u_n - v_2, z \rangle + \langle v_2 - v_1, z \rangle < \frac{\alpha}{2} - \alpha,$$

where we used the fact that $u_n \in W_0$. The above expression entails a contradiction, implying that Tx is a singleton. □

5.4.3 Nonenlargeable operators are linear skew-symmetric

We already pointed out that enlargements $E \in \mathbb{E}(T)$ can help us to solve problems involving the original operator T. However, we can only take advantage of this situation as long as E is a "proper" enlargement of T; that is, as long as

$$E(\varepsilon, \cdot) \supsetneq T(\cdot), \quad \text{for all } \varepsilon > 0.$$

This poses a natural question: Which operators T cannot be enlarged by the family $\mathbb{E}(T)$? In other words, we want to determine those operators T for which there exists $\bar{\varepsilon} > 0$ such that

$$E(\bar{\varepsilon}, \cdot) = T(\cdot). \tag{5.46}$$

We show below that if the equality above holds for some $\bar{\varepsilon} > 0$, then it holds for all $\varepsilon > 0$ (see Lemma 5.4.19). By Corollary 5.4.17, $E(\bar{\varepsilon}, \cdot)$ ISC in $(D(T))^\circ$. Therefore, the property above and Theorem 4.6.3 imply that T (and hence $E(\bar{\varepsilon}, \cdot)$) is point-to-point in $(D(T))^o$. In the particular case in which $T = \partial f$, we necessarily have that f is differentiable in the interior of its domain. We established in Corollary 5.4.11 that if T is affine with a skew-symmetric linear part, then (5.46) holds with $E = T^e$. The remaining part of this section is devoted to proving the converse of this fact. First, we show in Theorem 5.4.21 that if (5.46) holds, then T must be affine. In other words, the "nontrivial" elements in $E(\varepsilon, x)$ (i.e., elements in $E(\varepsilon, x) \setminus T(x)$) capture the nonlinear behavior of T. Finally, we prove that if X is reflexive and (5.46) holds with $E = T^e$, then the linear part of T is skew-symmetric.

In order to prove these facts, we need some technical lemmas.

Lemma 5.4.19. *Let $E \in \mathbb{E}(T)$. If there exists $\bar{\varepsilon} > 0$ such that $E(\bar{\varepsilon}, x) = T(x)$ for all $x \in D(T)$, then the same property holds for every $\varepsilon > 0$.*

Proof. Take $x \in D(T)$ and define

$$\varepsilon(x) := \sup\{\varepsilon \geq 0 \mid E(\varepsilon, x) = T(x)\}.$$

By assumption, $\varepsilon(x) \geq \bar{\varepsilon} > 0$. Observe that condition (E_2) implies that $E(\eta, x) = T(x)$ for all $\eta < \varepsilon(x)$. The conclusion of the lemma clearly holds if $\varepsilon(x) = \infty$ for every $x \in D(T)$. Suppose that there exists $x \in D(T)$ for which $\varepsilon(x) < \infty$. Fix $\varepsilon > \varepsilon(x)$. By definition of $\varepsilon(x)$ we can find $w \in E(\varepsilon, x)$ such that $w \notin T(x)$. Consider $v_0 := \mathrm{argmin}\{\|v - w\| \mid v \in T(x)\}$ and take $\alpha \in (0, 1)$ such that $0 < \alpha\varepsilon < \varepsilon(x)$. By (E_3) we have that

$$v_0' := \alpha w + (1 - \alpha)v_0 \in E(\alpha\varepsilon, x) = T(x).$$

Note that $\|v_0' - w\| = (1 - \alpha)\|v_0 - w\| < \|v_0 - w\|$, which contradicts the definition of v_0. Hence, $\varepsilon(x) = \infty$ for every $x \in D(T)$ and the lemma is proved.   □

Lemma 5.4.20. *(i) Let X and Y be vector spaces. If $G : X \to Y$ satisfies*

$$G(\alpha x + (1 - \alpha)y) = \alpha G(x) + (1 - \alpha)G(y) \tag{5.47}$$

for all $x, y \in X$ and all $\alpha \in [0, 1]$, then the same property holds for all $\alpha \in \mathbb{R}$. In this case, G is affine; that is, there exists a linear function $L : X \to Y$ and an element $y_0 \in Y$ such that $G(\cdot) = L(\cdot) + y_0$.

(ii) Let $T : X \to X^*$ be a maximal monotone point-to-point operator such that $D(T)^o \neq \emptyset$. Assume that for all $x, y \in D(T)$, $\alpha \in [0, 1]$, T satisfies that

$$T(\alpha x + (1 - \alpha)y) = \alpha T(x) + (1 - \alpha)T(y).$$

Then T is affine and $D(T) = X$.

Proof.

(i) For $\alpha \in \mathbb{R}$, $x, y \in X$, define $\tilde{x} := \alpha x + (1 - \alpha)y$. If $\alpha > 1$, then $x = (1/\alpha)\tilde{x} + (1 - (1/\alpha))y$. Because $1/\alpha \in [0, 1]$, the assumption on G implies that

$G(x) = (1/\alpha)G(\tilde{x}) + (1 - (1/\alpha))G(y)$. This readily implies $G(\tilde{x}) = \alpha G(x) + (1 - \alpha)G(y)$. Therefore, (5.47) holds for all $\alpha \geq 0$. Assume now that $\alpha < 0$. In this case $y = (\alpha/(\alpha - 1))x + (1/(1 - \alpha))\tilde{x}$. This is an affine combination with nonnegative coefficients, therefore the first part of the proof yields $G(y) = (\alpha/(\alpha - 1))G(x) + (1/(1 - \alpha))G(\tilde{x})$, which readily implies $G(\tilde{x}) = \alpha G(x) + (1 - \alpha)G(y)$. We proved that G is affine, and hence $G(x) - G(0)$ is linear, so that the last statement of the lemma holds for $L(x) := G(x) - G(0)$ and $y_0 := G(0)$. The proof of (i) is complete.

(ii) Note that Theorem 4.6.4 implies that $D(T)$ is open. Without loss of generality, we can assume that $0 \in D(T)$. Otherwise, take a fixed $x_0 \in D(T)$ and consider $T_0 := T(\cdot + x_0)$. The operator T_0 also satisfies the assumption on affine combinations, and its domain is the (open) set $\{y - x_0 \mid y \in D(T)\}$. It is also clear that $D(T_0) = X$ if and only if $D(T) = X$. Thus, we can suppose that $0 \in D(T)$. Because $D(T)$ is open, for every $x \in X$ there exists $\lambda > 0$ such that $\lambda x \in D(T)$. Define $\tilde{T} : X \to X^*$ as

$$\tilde{T}(x) = \frac{1}{\lambda}T(\lambda x) + \left(1 - \frac{1}{\lambda}\right)T(0). \tag{5.48}$$

We claim that

(I) $\tilde{T}(x)$ is well defined and nonempty for every $x \in X$.

(II) $\tilde{T} = T$ in $D(T)$.

(III) $\tilde{T}$ is monotone.

Observe that if (I)–(III) are true, then by maximality of T we get $\tilde{T} = T$ and $D(T) = D(\tilde{T}) = X$. For proving (I), we start by showing that $\tilde{T}$, as defined in (5.48), does not depend on λ. Suppose that $\lambda, \lambda' > 0$ are both such that $\lambda x, \lambda' x \in D(T)$. Without loss of generality, assume that $0 < \lambda' < \lambda$. Then $\lambda' x = (\lambda'/\lambda)\lambda x + (1 - (\lambda'/\lambda))0$. Because $\lambda'/\lambda \in (0, 1)$ we can apply the hypotheses on T to get

$$T(\lambda' x) = \frac{\lambda'}{\lambda}T(\lambda x) + \left(1 - \frac{\lambda'}{\lambda}\right)T(0),$$

which implies that

$$\frac{1}{\lambda}T(\lambda x) + \left(1 - \frac{1}{\lambda}\right)T(0) = \frac{1}{\lambda'}T(\lambda' x) + \left(1 - \frac{1}{\lambda'}\right)T(0),$$

and hence $\tilde{T}(x)$ is well defined. Nonemptiness of $\tilde{T}(x)$ follows easily from the definition, because $\lambda x, 0 \in D(T)$. This gives $D(\tilde{T}) = X$. Assertion (II) follows by taking $\lambda = 1$ in (5.48). By (I) this choice of λ does not change the value of $\tilde{T}$. For proving (III), we claim first that $\tilde{T}$ is affine. Indeed, because $D(\tilde{T}) = X$, this fact follows from item (i) if we show that for all $\lambda \in [0, 1]$, it holds that

$$\tilde{T}(\lambda x + (1 - \lambda)y) = \lambda\tilde{T}(x) + (1 - \lambda)\tilde{T}(y)$$

for all $x, y \in X$. Take $\eta > 0$ small enough to ensure that $\eta x, \eta y, \eta(\lambda x + (1 - \lambda)y) \in D(T)$. Then,

$$\tilde{T}(\lambda x + (1 - \lambda)y) = \tilde{T}\left(\frac{1}{\eta}(\eta\lambda x + \eta(1 - \lambda)y)\right)$$

$$= \frac{1}{\eta}T(\eta\lambda x + \eta(1-\lambda)y) + \left(1 - \frac{1}{\eta}\right)T(0)$$

$$= \frac{\lambda}{\eta}T(\eta x) + \frac{1-\lambda}{\eta}T(\eta y) + \left(1 - \frac{1}{\eta}\right)T(0)$$

$$= \lambda\left[\frac{1}{\eta}T(\eta x) + \left(1 - \frac{1}{\eta}\right)T(0)\right] + (1-\lambda)\left[\frac{1}{\eta}T(\eta y) + \left(1 - \frac{1}{\eta}\right)T(0)\right]$$

$$= \lambda\tilde{T}(x) + (1-\lambda)\tilde{T}(y),$$

using the definition of $\tilde{T}$ in the second and the last equality, and the hypothesis on T in the third one. Then $\tilde{T}$ is affine by item (ii). In this case, monotonicity of $\tilde{T}$ is equivalent to positive semidefiniteness of the linear part $\tilde{L} := \tilde{T}(\cdot) - T(0)$ of $\tilde{T}$. In other words, we must prove that $\langle \tilde{L}x, x \rangle \geq 0$ for all $x \in X$. Indeed, take $x \in X$ and $\lambda > 0$ such that $\lambda x \in D(T)$. Then

$$\langle \tilde{L}x, x \rangle = (1/\lambda^2)\langle \tilde{L}(\lambda x), \lambda x \rangle = (1/\lambda^2)\langle \tilde{T}(\lambda x) - T(0), \lambda x \rangle$$

$$= (1/\lambda^2)\langle T(\lambda x) - T(0), \lambda x \rangle \geq 0,$$

using (II) in the third equality and monotonicity of T in the inequality. Hence, $\tilde{T}$ is monotone and therefore $T = \tilde{T}$. This implies that T is affine and $D(T) = X$. $\square$

Now we are able to prove that if T cannot be enlarged, then it must be affine.

Theorem 5.4.21. *Let X be a reflexive Banach space and $T : X \rightrightarrows X^*$ a maximal monotone operator such that $D(T)^o \neq \emptyset$. If there exist $E \in \mathbb{E}(T)$ and $\bar{\varepsilon} > 0$ such that*

$$E(\bar{\varepsilon}, \cdot) = T(\cdot), \tag{5.49}$$

then

(a) $D(T) = X$, and

(b) *There exists a linear function $L : X \to X^*$ and an element $x_0^* \in X^*$ such that $T(\cdot) = L(\cdot) + x_0^*$.*

In particular, if (5.49) holds for $E = T^e$, then L in (b) is skew-symmetric. Conversely, if (a) and (b) hold with a skew-symmetric L, then $T^e(\varepsilon, \cdot) = T(\cdot)$ for all $\varepsilon \geq 0$.

Proof. The last assertions follow from Corollary 5.4.11. Let us prove (a) and (b). We claim that (5.49) implies that $D(T)$ is open. Indeed, by Theorem 4.6.3(ii) and (5.49), $T(x)$ is a singleton for all x, and hence Theorem 4.6.4 yields the claim.

For proving (a), assume for the sake of contradiction that $D(T) \subsetneq X$. If $\overline{(D(T))} = X$, then by Theorem 4.5.3(ii), it holds that $D(T) = X$. Hence we can assume that $\overline{(D(T))} \subsetneq X$. Take $x, y \in D(T)$ and $\alpha \in [0, 1]$. By (E_3) we have

$$\alpha T(x) + (1-\alpha)T(y) \in E(\hat{\varepsilon}, \hat{x}),$$

where $\hat{\varepsilon} := \alpha(1 - \alpha)\langle T(x) - T(y), x - y \rangle$ and $\hat{x} := \alpha x + (1 - \alpha)y$. By Lemma 5.4.19, we get that $E(\hat{\varepsilon}, \hat{x}) = T(\hat{x})$, and inasmuch as T is point-to-point, it holds that

$$\alpha T(x) + (1 - \alpha)T(y) = T(\hat{x}).$$

We conclude from Lemma 5.4.20(ii) that T is affine and $D(T) = X$. This proves (a) and (b). Assume now that $E = T^e$. We have already proved that T is affine, and so we can apply Corollary 5.4.11, which asserts that (5.49) is equivalent to L being skew-symmetric. □

Remark 5.4.22. Condition $D(T)^o \neq \emptyset$ is necessary for Theorem 5.4.21 to hold. For showing this, let $X = \mathbb{R}^n$, $0 \neq a \in \mathbb{R}^n$ and $T = N_V$, where $V := \{y \in \mathbb{R}^n \mid \langle a, y \rangle = 0\}$. It is clear that T is not affine, because its domain is not the whole space and it is nowhere point-to-point. Namely,

$$T(x) = \begin{cases} V^\perp = \{\lambda a \mid \lambda \in \mathbb{R}\} & \text{if } x \in V, \\ \emptyset & \text{otherwise.} \end{cases}$$

We claim that $T^e(\varepsilon, \cdot) = T(\cdot)$ for all $\varepsilon > 0$. For $x \notin V$ this equality follows from the fact that $V = D(T) \subset D(T^e(\varepsilon, \cdot)) \subset \overline{(D(T))} = V$. In this case both mappings have empty value at x. Suppose now that there exist $x \in V$ and $\varepsilon > 0$ such that $T^e(\varepsilon, \cdot) \not\subset T(\cdot)$. Then there exists $b \in T^e(\varepsilon, x)$ such that $b \notin V^\perp$. In this case, $b = ta + w$ for some $t \in \mathbb{R}$ and $0 \neq w \in V$. Because $b \in T^e(\varepsilon, x)$, for $y = x + 2\varepsilon w/|w|^2 \in V$ we have

$$-\varepsilon \leq \langle b - \lambda a, x - y \rangle = 2\langle (t - \lambda)a + w, -\varepsilon w/|w|^2 \rangle = -2\varepsilon,$$

which is a contradiction.

Before starting with the applications of the enlargements dealt with so far in this chapter, it is worthwhile to attempt an educated guess about the future developments on this topic. It seems safe to predict that meaningful progress will be achieved by some appropriate relaxation of the notion of monotonicity. Let us mention that enlargements related to the ones discussed in this book have already been proposed for nonmonotone operators; see, for example, [196] and [229]. However, a general theory of enlargements in the nonmonotone case is still lacking. On the other hand, there are strong indications that the tools for such a theory are currently becoming available, through the interplay of set-valued analysis and a new field, which is starting to be known as variational analysis.

Let us recall that the basic tools for extending the classical apparatus of calculus to nonsmooth convex functions were displayed in the form of an articulated theory by Rockafellar in 1970, in his famous book, *Convex Analysis* [187]. The next fundamental step, namely the extension of such tools (e.g., the subdifferential, the normal cone, etc.) to nonconvex sets and functions, was crystallized in 1990 in Clarke's book [66]. The following milestone consisted in the extension of the theory beyond the realm of real-valued functions and their differentiation, moving instead into variational inequalities associated with operators, and was achieved in two essential books. The first one, *Variational Analysis*, [194], written by R.T. Rockafellar and R.J.B. Wets in 1998, gave the subject a new name (as had been the case with Rockafellar's 1970 book). It dealt only with the finite-dimensional case. The second one, by B.S. Mordukhovich [156], expands the theory to infinite-dimensional spaces and provides a very general framework for the concept of differentiation. We refrain nevertheless from presenting the consequences of these new developments upon our subject, namely enlargements of point-to-set mappings, because they are still at a very preliminary stage. We emphasize that both [194] and [156] are the natural references for a deep study of the relationships among set-valued analysis and modern aspects of variational analysis and generalized differentiation, especially from the viewpoint of applications to optimization and control.

5.5   Algorithmic applications of T^e

Throughout this section, $T : \mathbb{R}^n \rightrightarrows \mathbb{R}^n$ is a maximal monotone operator whose domain is the whole space and $C \subset \mathbb{R}^n$ is a closed and convex set. Our aim now is to show how the enlargement T^e can be used for devising schemes for solving the variational inequality problem associated with T and C. This problem, described in Example 1.2.5 and denoted VIP(T, C), can be formulated as follows. Find $x^* \in C$ and $v^* \in T(x^*)$ such that

$$\langle v^*, y - x^* \rangle \geq 0, \quad \text{for every } y \in C. \tag{5.50}$$

Denote S^* the set of solutions of VIP(T, C). Because $D(T) = \mathbb{R}^n$, (5.50) is equivalent to finding x^* such that $0 \in T(x^*) + N_C(x^*)$, where N_C is the normality operator associated with C (see Definition 3.6.1). Problem VIP(T, C) is a natural generalization of the constrained convex optimization problem. Indeed, recall that x^* satisfies the (necessary and sufficient) optimality conditions for the problem of minimizing a convex function f in C if and only if $0 \in \partial f(x^*) + N_C(x^*)$. In VIP$(T, C)$ the term ∂f is replaced by an arbitrary maximal monotone operator T. Given the wide variety of methods devised to solve the constrained optimization problem, a natural question to pose is if and how a given method for solving this problem can be adapted for also solving VIP(T, C). We discuss next the case of the *extragradient method* for monotone variational inequalities, and describe how the enlargement T^e can be used for defining an extension that solves VIP(T, C). We start by recalling the projected gradient method for constrained optimization. Given a differentiable function $f : \mathbb{R}^n \to \mathbb{R}$, the projected gradient method is an algorithm devised for minimizing f on C, which takes the form

$$x^{k+1} = P_C(x^k - \gamma_k \nabla f(x^k)), \tag{5.51}$$

where P_C is the orthogonal projection onto C and $\{\gamma_k\}$ is a sequence of positive real numbers called *stepsizes*. The stepsizes γ_k are chosen so as to ensure sufficient decrease of the functional value. If ∇f is Lipschitz continuous with constant L, a suitable value for γ_k can be chosen in terms of L. When such an L is not available, an *Armijo search* (see [5]) does the job: an initial value β is given to γ_k, and the functional value at the resulting point $P_C(x^k - \beta \nabla f(x^k))$ is evaluated and compared to $f(x^k)$. If there is enough decrease, according to some criterion, then γ_k is taken as β and x^{k+1} is computed according to (5.51); otherwise β is halved, the functional value at the new trial point $x^k - (\beta/2)\nabla f(x^k)$ is compared to $f(x^k)$, and the process continues until the first time that the trial point satisfies the decrease criterion. Assuming this happens at the jth attempt (i.e., at the jth iteration of the inner loop), then γ_k is taken as $2^{-j}\beta$ and again x^{k+1} is given by (5.51).

If f is convex and has minimizers, then the sequence defined by (5.51), with an Armijo search for determining the stepsizes, is globally convergent to a minimizer if f, without further assumptions (see, e.g., [227, 101]). The rationale behind this result is that for γ_k small enough, the point x^{k+1} given by (5.51) is closer than x^k to any minimizer of f; that is, the sequence $\{x^k\}$ is Fejér convergent to the solution set of the problem (cf. Definition 5.5.1).

In the point-to-point case, an immediate extension of this scheme to VIP(T, C) is

$$x^{k+1} = P_C(x^k - \gamma_k T(x^k)). \tag{5.52}$$

The convergence properties of the projected gradient method, however, do not extend to the sequence given by (5.52) in such an immediate way. In order to establish convergence, something else is required besides existence of solutions of VIP(T, C) and monotonicity of T (equivalent to convexity of f in the case $T = \nabla f$). In [28], for instance, global convergence of the sequence given by (5.52) is proved for *strongly monotone* T (i.e., $\langle x - y, T(x) - T(y) \rangle \geq \theta \|x - y\|^2$ for some $\theta > 0$ and all $x, y \in \mathbb{R}^n$). In [1], the same result is achieved demanding *uniform monotonicity* of T (i.e. $\langle x - y, T(x) - T(y) \rangle \geq \psi(\|x - y\|)$ for all $x, y \in \mathbb{R}^n$, where $\psi : \mathbb{R} \to \mathbb{R}$ is a continuous increasing function such that $\psi(0) = 0$). Note that strong monotonicity is a particular case of uniform monotonicity, corresponding to taking $\psi(t) = \theta t^2$. In [141] some form of cocoercivity of T is demanded. On the other hand, when T is plainly monotone, there is no hope: no choice of γ_k will make the sequence convergent, because the Fejér convergence to the solution set S^* is lost.

The prototypical examples are the problems of finding zeroes of the rotation operators in the plane, for instance, VIP(T, C) with $C = \mathbb{R}^n$ and $T(x) = Ax$, where

$$A = \begin{pmatrix} 0 & -1 \\ 1 & 0 \end{pmatrix}.$$

The unique solution of VIP(T, C) is the unique zero of T, namely $x = 0$. It is easy to check that $\|x - \gamma T(x)\| > \|x\|$ for all $x \in \mathbb{R}^n$ and all $\gamma > 0$; that is, the sequence moves farther away from the unique solution at each iteration, independently of the choice of the stepsize (see the position of z in Figure 5.3). Note that in this example P_C is the identity, because C is the whole space. We also mention, parenthetically, that rotation operators are monotone (because the corresponding matrices A are skew-symmetric), but nothing more: they fail to satisfy any stronger monotonicity or coercivity property.

The rotation example also suggests a simple fix for this drawback, presented for the first time by Korpelevich in [123], with the name of *extragradient method*. The idea is to move away from x^k not in the *gradient-like* direction $-T(x^k)$, but

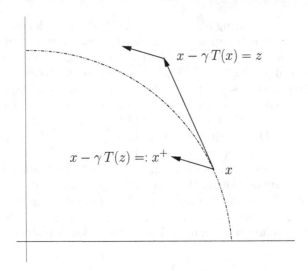

Figure 5.3. *Extragradient direction for rotation operators*

in an *extragradient* direction $-T(z^k)$, with z^k close enough to x^k, in which case
the Fejér convergence to the solution set is preserved. In other words, the method
performs first a *projected gradient-like* step for finding the auxiliary point z^k,

$$z^k := P_C(x^k - \gamma_k T(x^k)), \qquad (5.53)$$

and this auxiliary point defines the search direction for obtaining the new iterate:

$$x^{k+1} := P_C(x^k - \gamma_k T(z^k)). \qquad (5.54)$$

See Figure 5.3. If T is point-to-point and Lipschitz continuous with constant L,
then for $\gamma_k \in (0, 1/2L)$, the sequence $\{x^k\}$ converges to a solution of VIP(T, C)
(see [123]). This method has been extensively studied for point-to-point monotone
operators, and several versions of it can be found, for example, in [98, 117, 140] and
[107]. In most of these references, no Lipschitz constant is assumed to be available
for T, and instead an Armijo-type search for the stepsize γ_k is performed: an
initial value β is tried, and this value is successively halved until the resulting point
$P_C(x^k - 2^{-j}\beta T(x^k))$ satisfies a prescribed inequality, which is then used to ensure
the Fejér monotonicity of the sequence $\{x^k\}$ to the set of solutions of VIP(T, C).

However, the scheme (5.53)–(5.54) cannot be automatically extended to point-
to-set mappings. Indeed, if we just replace the directions $T(x^k)$ and $T(z^k)$ by
$u^k \in T(x^k)$ and $v^k \in T(z^k)$, respectively, then the algorithm might be unsuccessful.
An example is given in Section 2 of [104], taking $C = \mathbb{R}$ and $T = \partial f$, where

$$f(x) = \begin{cases} 0 & \text{if } x \leq 0 \\ x & \text{if } x \in [0, 1] \\ \theta x - \theta + 1 & \text{if } x \geq 1 \end{cases}$$

with $\theta > 1$. In this case VIP(T,C) reduces to minimizing f on $\mathbb{R}$ and the solution set is the nonnegative halfline. T is constant with value 0 in the negative halfline, 1 in $(0,1)$, and θ in $(1,\infty)$, with "jumps" of size 1 at $x = 0$, and of size $\theta - 1$ at $x = 1$. It is shown in [104] that, starting with $x^0 > 1$, the sequence $\{x^k\}$ will stay in the halfline $(1,\infty)$, unable to jump over the "inner discontinuity" of T at $x = 1$, and converging eventually to $x^* = 1$ (which is not a solution), unless the initial trial stepsize β in the Armijo search is taken large enough compared to θ, but the right value for starting the search cannot be determined beforehand, and even more so for an arbitrary T.

The reason for this failure is the lack of inner-semicontinuity of arbitrary maximal monotone operators. This motivates the idea in [104], where the direction u^k is chosen in $T^e(\varepsilon_k, x^k)$, instead of in $T(x^k)$, and convergence is established using the fact that the enlargement T^e *is* inner-semicontinuous.

However, the scheme proposed in [104], called EGA (for *Extragradient Algorithm*), is not directly implementable: for convergence, it is required that the direction u^k be chosen with some additional properties. The element of minimal norm in $T^e(\varepsilon_k, x^k)$ does the job, but of course this introduces a highly nontrivial additional minimization problem. EGA determines u^k through an inexact norm minimization in $T^e(\varepsilon_k, x^k)$, but still such an inexact solution is hard to find unless a characterization of the whole set $T^e(\varepsilon_k, x^k)$ is available, which in general is not the case. After a suitable auxiliary point z^k is found in the direction u^k, an Armijo search is performed in the segment between x^k and z^k, finding a point y^k, and a suitable $v^k \in T(y^k)$ which is, finally, the appropriate direction: x^{k+1} is obtained by moving from x^k in the direction v^k, with an endogenously determined stepsize μ_k.

We mention, incidentally, that this problem does not occur for the pure projected gradient-like method given by (5.52), which can indeed be extended to the point-to-set realm, taking

$$x^{k+1} = P_C(x^k - \gamma_k u^k),$$

where u^k is any point in $T(x^k)$. However, as in the case of the iteration (5.52), T must be uniformly monotone (i.e., $\langle u - v, x - y \rangle \geq \psi(\|x - y\|)$ for all $u \in T(x)$ and all $v \in T(y)$, with ψ as above) and γ_k must be chosen through a procedure linked to ψ (see [1]).

In order to overcome the limitation of EGA, in connection with the practical computation of u^k, a much more sophisticated technique must be used. Presented for the first time in [48], it is inspired by the *bundle methods* for nonsmooth convex optimization, which employ the ε-subdifferential of the objective function. In these methods the set $\partial_\varepsilon f(x^k)$ is approximated by a polyhedron, namely the convex hull of arbitrary subgradients at carefully selected points near x^k (these subgradients form the "bundle").

In [48], a similar idea is proposed for approximating $T^e(\varepsilon, x^k)$ instead of $\partial_\varepsilon f(x^k)$, when the problem at hand is that of finding zeroes of a monotone point-to-set operator T instead of minimizers of a convex function f. We point out that the convergence analysis of the method presented in [48], called IBS (for *implementable bundle strategy*), strongly relies on the transportation formula for T^e, which allows us to construct suitable polyhedral approximations of the sets $T^e(\varepsilon, x^k)$. Algorithms EGA and IBS are the subject of the next two sections. We recall that inner-semicontinuity of T^e is essentially embedded in the transportation formula. Again, it is this inner-semicontinuity, which T^e always enjoys but T itself may lack, that makes EGA and IBS work.

5.5.1 An extragradient algorithm for point-to-set operators

We proceed to state formally the method studied in [104]. The algorithm requires the exogenous constants $\varepsilon, \delta^-, \delta^+, \alpha^-, \alpha^+$ and β satisfying $\varepsilon > 0$, $0 < \delta^- < \delta^+ < 1, 0 < \alpha^- < \alpha^+$, and $\beta \in (0, 1)$. It also demands two exogenous sequences $\{\alpha_k\}, \{\beta_k\}$, satisfying $\alpha_k \in [\alpha^-, \alpha^+]$ and $\beta_k \in [\beta, 1]$ for all k.

We briefly describe the roles of these constants: ε is the initial value of the enlargement parameter; the direction u^k belongs to $T^e(\varepsilon_k, x^k)$, with $\varepsilon_0 = \varepsilon$; α^- and α^+ are lower and upper bounds for α_k, which is the stepsize for determining u^k; δ^+ is the precision parameter in the inexact norm minimization in the selection of $u^k \in T^e(\varepsilon_k, x^k)$; δ^- is the precision parameter in the Armijo search; and β is a lower bound for the initial trial stepsize β_k of the Armijo search.

Basic Algorithm

1. **Initialization:** Take
$$x^0 \in C, \quad \text{and} \quad \varepsilon_0 = \varepsilon. \tag{5.55}$$

2. **Iterative Step.** $(k \geq 1)$ Given x^k,

   (a) **Selection of ε_k:**
   $$\varepsilon_k := \min\{\varepsilon_{k-1}, \|x^{k-1} - z^{k-1}\|^2\}. \tag{5.56}$$

   (b) **Selection of u^k:** Find
   $$u^k \in T^e(\varepsilon_k, x^k), \tag{5.57}$$
   such that
   $$\langle w, x^k - z^k \rangle \geq \frac{\delta^+}{\alpha_k} \|x^k - z^k\|^2, \quad \forall\, w \in T^e(\varepsilon_k, x^k), \tag{5.58}$$
   where
   $$z^k = P_C(x^k - \alpha_k u^k). \tag{5.59}$$

   (c) **Stopping Criterion:** If $z^k = x^k$ stop, otherwise

   (d) **Armijo Search:** For $i = 0, 1, \ldots$ define
   $$y^{k,i} := \left(1 - \frac{\beta_k}{2^i}\right) x^k + \frac{\beta_k}{2^i} z^k \tag{5.60}$$

   $$i(k) := \min\left\{ i \geq 0 \;\middle|\; \begin{array}{c} \exists\, v^{k,i} \in T(y^{k,i}) \text{ such that} \\ \langle v^{k,i}, x^k - z^k \rangle \geq \frac{\delta^-}{\alpha_k} \|x^k - z^k\|^2 \end{array} \right\}. \tag{5.61}$$

Set

$$\lambda_k := \frac{\beta_k}{2^{i(k)}}, \tag{5.62}$$

$$v^k := v^{k,i(k)}. \tag{5.63}$$

(e) **Extragradient Step:**

$$\mu_k := \frac{\lambda_k \langle v^k, x^k - z^k \rangle}{\|v^k\|^2}, \tag{5.64}$$

$$x^{k+1} := P_C(x^k - \mu_k v^k). \tag{5.65}$$

Before the formal analysis of the convergence properties of Algorithm EGA, we comment, in a rather informal way, on the behavior of the sequences defined by EGA; that is, by (5.55)–(5.65).

It is easy to check that if we use in (5.59) the minimum norm vector in $T^e(\varepsilon_k, x^k)$ instead of u^k, the resulting z^k is such that the inequality (5.58) holds. In fact, (5.58) can be seen as a criterion for finding an approximate minimum norm vector in $T^e(\varepsilon_k, x^k)$, with precision given by δ^+. In (5.61), we perform a backward search in the segment between x^k and z^k, starting at $x^k + \beta_k(z^k - x^k)$, looking for a point close to x^k (it is the point $y^{k,i(k)}$, according to (5.60), (5.61)), such that some $v^k \in T(y^{k,i(k)})$ satisfies an Armijo-type inequality with parameter δ^-.

At this point, a hyperplane H_k normal to v^k, separating x^k from the solution set S^* has been determined. The coefficient μ_k, given by (5.64) is such that $x^k - \mu_k v^k$ is the orthogonal projection of x^k onto H_k, and x^{k+1} is the orthogonal projection onto C of this projection. Because $S^* \subset C \cap H_k$, both projections decrease the distance to any point in S^*, ensuring the Fejér convergence of $\{x^k\}$ to S^*, and ultimately the convergence of the whole sequence $\{x^k\}$ to an element of S^*; that is, to a solution of VIP(T,C).

We proceed to the convergence analysis of EGA, for which we recall the well-known concept of Fejér convergence.

Definition 5.5.1. *Given a nonempty set $M \subset \mathbb{R}^n$, a sequence $\{p^k\} \subset \mathbb{R}^n$ is said to be* Fejér convergent *to M if for all k large enough it holds that $\|p^{k+1} - q\| \le \|p^k - q\|$ for all $q \in M$.*

The following result concerning Fejér convergent sequences is well known and useful for establishing convergence. Its proof can be found in [98, Proposition 8].

Lemma 5.5.2. *Let $\emptyset \ne M \subset \mathbb{R}^n$, and assume that $\{p^k\} \subset \mathbb{R}^n$ is Fejér convergent to M. Then*

(i) *$\{p^k\}$ is bounded.*

(ii) *If $\{p^k\}$ has a cluster point in M, then the whole sequence converges to that point in M.*

The following facts, established in [104], do not rely upon the properties of T^e, and hence they are stated without proof.

Proposition 5.5.3. *Assume that $S^* \neq \emptyset$. If $\{x^k\}$ and $\{z^k\}$ are defined as in (5.55)–(5.65), then*

(i) *If $x^k \neq z^k$ then there exists $i(k)$ verifying (5.61).*

(ii) *The sequences $\{x^k\}$ and $\{u^k\}$ are well defined and $\{x^k\} \subset C$.*

(iii) *If $S^* \neq \emptyset$, then the sequences $\{x^k\}$, $\{y^k\}$, $\{z^k\}$, and $\{\mu_k\}$ are bounded.*

(iv)

$$\lim_{k \to 0} \mu_k v^k = 0. \tag{5.66}$$

(v) *The sequence $\{x^k\}$ is Fejér convergent to the set S^*.*

Proof. See Lemmas 5.1–5.3 in [104]. □

Because $D(T) = \mathbb{R}^n$, we can combine the result above with Theorems 4.2.10 and 5.3.4, and conclude that the sequences $\{u^k\}$ and $\{v^k\}$ are also bounded.

We need now the concept of approximate solutions of $\text{VIP}(T, C)$, for which we recall some technical tools from [44]. Define $h : D(T) \cap C \to \mathbb{R} \cup \{\infty\}$ as

$$h(x) := \sup_{\substack{v \in T(y), \\ y \in C \cap D(T)}} \langle v, x - y \rangle. \tag{5.67}$$

The relation between the function h and the solutions of $\text{VIP}(T,C)$ is given in the following lemma.

Lemma 5.5.4. *Assume that $ri(D(T)) \cap ri(C) \neq \emptyset$, and take h as in (5.67). Then h has zeroes if and only if $S^* \neq \emptyset$ and in such a case S^* is precisely the set of zeroes of h.*

Proof. See Lemma 4 in [44]. □

We define next the concept of *approximate solutions* of $\text{VIP}(T, C)$.

Definition 5.5.5. *A point $x \in D(T) \cap C$ is said to be an ε-solution of $\text{VIP}(T, C)$ if and only if $h(x) \leq \varepsilon$ (another motivation for this definition can be found in Exercise 14).*

Theorem 5.5.6. *Assume that $S^* \neq \emptyset$.*

(a) *If the sequence $\{x^k\}$ is infinite, then $\bar{\varepsilon} := \lim_{k \to \infty} \varepsilon_k = 0$. Moreover, the sequence converges to a solution of $\text{VIP}(T, C)$.*

(b) *If the sequence is finite, the last point x^k of the sequence is an ε_k-solution of VIP(T, C).*

Proof.

(a) We start by proving convergence of $\{x^k\}$ to a solution. Due to Lemma 5.5.2 and Proposition 5.5.3(v), it is enough to prove that there exists a cluster point in S^*. Because $\{x^k\}$, $\{z^k\}$, $\{u^k\}$, $\{v^k\}$, $\{\alpha_k\}$, $\{\lambda_k\}$, and $\{\mu_k\}$ are bounded (see Proposition 5.5.3(iii)), they have limit points. We consider a subset of $\mathbb{N}$ such that the corresponding subsequences of these seven sequences converge, say to $\bar{x}$, $\bar{z}$, $\bar{u}$, $\bar{v}$, $\bar{\alpha}$, $\bar{\lambda}$, and $\bar{\mu}$, respectively. For the sake of a simpler notation, we keep the superindex (or subindex) k for referring to these subsequences. By (5.65), and using continuity of P_C, we get

$$\bar{x} = P_C(\bar{x} - \bar{\mu}\bar{u}).$$

Recall that

$$\langle a - P_C(a), y - P_C(a)\rangle \leq 0 \tag{5.68}$$

for all $a \in \mathbb{R}^n$, $y \in C$. Taking $a := \bar{x} - \bar{\mu}\bar{u}$ we obtain

$$0 \geq \langle(\bar{x} - \bar{\mu}\bar{u}) - \bar{x}, y - \bar{x}\rangle = \bar{\mu}\langle\bar{u}, \bar{x} - y\rangle \tag{5.69}$$

for all $y \in C$. Hence $\bar{\mu}\langle\bar{u}, \bar{x} - y\rangle \leq 0$. By (5.61), (5.63), and (5.64), we have that $\bar{\mu} \geq 0$. We consider two cases: $\bar{\mu} > 0$ and $\bar{\mu} = 0$. In the first case, by Proposition 5.5.3(iv) we get $\bar{v} = 0$. Using (5.61), we conclude that $\bar{x} = \bar{z}$. Because T is OSC and $v^k \in T((1 - \lambda_k)x^k + \lambda_k z^k)$, we get $0 = \bar{v} \in T(\bar{x}) \subset (T + N_C)(\bar{x})$, and hence $\bar{x} \in S^*$ (we also use here that $D(T) = \mathbb{R}^n$ and $\bar{x} \in C$, which follows from (5.65)). Using now that $\varepsilon_k \leq \|x^{k-1} - z^{k-1}\|^2$ and taking limits we conclude that $\bar{\varepsilon} = 0$. Assume now that $\bar{\mu} = 0$. Using (5.64) and the fact that $\{v^k\}$ is bounded, $\bar{\mu} = 0$ can happen only if $\bar{\lambda} = 0$ or $\langle\bar{v}, \bar{x} - \bar{z}\rangle = 0$. We start by assuming that $\bar{\lambda} = 0$. We prove first that $\bar{\varepsilon} = 0$. Consider the sequence $\{y^{k,i(k)-1}\}$. Because $\bar{\lambda} = 0$, $\lim_k i(k) = \infty$ and hence $\lim_k y^{k,i(k)-1} = \bar{x}$. Because $D(T) = \mathbb{R}^n$ and $\{y^{k,i(k)-1}\}$ is bounded, there exists a bounded sequence $\{\xi^k\}$ such that $\xi^k \in T(y^{k,i(k)-1})$. Hence the sequence $\{\xi^k\}$ has a subsequence that converges to some $\bar{\xi}$. Again, we keep the superindex k for referring to the subsequence. By outer-semicontinuity of T we conclude that $\bar{\xi} \in T(\bar{x})$. We know by (5.61) and definition of ξ_k that $\langle\xi^k, x^k - z^k\rangle < (\delta^-/\alpha_k)\|x^k - z^k\|^2$. This implies that

$$\langle\bar{\xi}, \bar{x} - \bar{z}\rangle \leq \frac{\delta^-}{\bar{\alpha}}\|\bar{x} - \bar{z}\|^2. \tag{5.70}$$

If

$$\bar{\varepsilon} > 0, \tag{5.71}$$

then $T^e(\bar{\varepsilon}, \cdot)$ is ISC. Because $\bar{\xi} \in T^e(\bar{\varepsilon}, \bar{x})$, there exists a sequence $w^k \in T^e(\bar{\varepsilon}, x^k)$ with $\lim_k w^k = \bar{\xi}$. By definition, $\varepsilon_k \geq \bar{\varepsilon}$, and hence $w^k \in T^e(\varepsilon_k, x^k)$. By (5.58),

$$\langle w^k, x^k - z^k\rangle \geq \frac{\delta^+}{\alpha_k}\|x^k - z^k\|^2,$$

and taking limits we get

$$\langle \bar{\xi}, \bar{x} - \bar{z} \rangle \geq \frac{\delta^+}{\bar{\alpha}} \| \bar{x} - \bar{z} \|^2. \tag{5.72}$$

Combining (5.70), (5.72), and the assumptions on the parameters of the algorithm, we conclude that $\bar{x} = \bar{z}$. So $\bar{\varepsilon} = \lim_k \varepsilon_k \leq \lim_k \| x^{k-1} - z^{k-1} \|^2 = 0$. This contradicts (5.71) and hence we must have $\bar{\varepsilon} = 0$. We claim that in this case $\bar{x} = \bar{z}$. Indeed, $\bar{\varepsilon} = \lim_k \varepsilon_k = 0$, thus we must have a subsequence of $\{\varepsilon_k\}$ that is strictly decreasing. This implies that for an infinite number of indices the minimum in the definition of ε_k is equal to $\| x^{k-1} - z^{k-1} \|^2$. Combining this with the fact that $\lim_{k \to \infty} \varepsilon_k = \bar{\varepsilon} = 0$, we conclude that $\bar{x} = \bar{z}$. Because T^e is osc, $\bar{u} \in T^e(\bar{\varepsilon}, \bar{x}) = T^e(0, \bar{x}) = T(\bar{x})$. By definition of z^k, we get $\bar{x} = \bar{z} = P_C(\bar{x} - \bar{\alpha}\bar{u})$. Altogether, we obtain

$$h(\bar{x}) = \sup_{\substack{v \in T(y) \\ y \in C \cap D(T)}} \langle v, \bar{x} - y \rangle \leq \sup_{y \in C \cap D(T)} \langle \bar{u}, \bar{x} - y \rangle$$

$$= \frac{1}{\bar{\alpha}} \sup_{y \in C \cap D(T)} \langle (\bar{x} - \bar{\alpha}\bar{u}) - \bar{x}, y - \bar{x} \rangle \leq 0,$$

using (5.68). Then $\bar{x}$ is a solution of VIP(T, C). Consider now the case $\langle \bar{v}, \bar{x} - \bar{z} \rangle = 0$. Using (5.61) and taking limits, we get $0 = \langle \bar{v}, \bar{x} - \bar{z} \rangle \geq (\delta^- / \bar{\alpha}) \| \bar{x} - \bar{z} \|^2$ and hence again $\bar{x} = \bar{z} = P_C(\bar{x} - \bar{\alpha}\bar{u})$. In the same way as before, we obtain $\bar{x} \in S^*$ and $\bar{\varepsilon} = 0$.

We proceed now to prove assertion (b) of the theorem. If the algorithm stops at iteration k, then by the stopping criterion we have $z^k = x^k = P_C(x^k - \alpha_k u^k)$. Using again the fact that $u^k \in T^e(\varepsilon_k, x^k)$, combined with (5.68), we have

$$h(x^k) = \sup_{\substack{v \in T(y), \\ y \in C \cap D(T)}} \langle v, x^k - y \rangle \leq \sup_{y \in C \cap D(T)} \langle u^k, x^k - y \rangle + \varepsilon_k$$

$$= \frac{1}{\alpha_k} \sup_{y \in C \cap D(T)} \langle (x^k - \alpha_k u^k) - x^k, y - x^k \rangle + \varepsilon_k \leq \varepsilon_k.$$

Thus, x^k is an ε_k-solution of VIP(T, C). $\square$

5.5.2   A bundle-like algorithm for point-to-set operators

We present in this section the method introduced in [48]. As mentioned above, a monotone variational inequality problem can be reduced to a problem of finding zeroes of a maximal monotone operator T; that is, the problem of finding $x^* \in \mathbb{R}^n$ such that $0 \in T(x^*)$. Indeed, the solution set of VIP(T,C), given by (5.50), coincides with the set of zeroes of the maximal monotone operator $\hat{T} = T + N_C$, where N_C is the normalizing operator of C, as defined in (3.39). On the other hand, when working, as we do in this section, within the framework of zeroes of operators rather than variational inequalities (i.e., ignoring the specific form of a feasible set C), we refrain from taking advantage of the structure of C. More specifically, Algorithm EGA, dealt with in the previous section, uses explicitly the orthogonal projection onto C. When such a projection has an easy formula (e.g., when C is a ball or a box), that approach seems preferable. The algorithms discussed in this section are also amenable to a variational inequality format, but we prefer to avoid explicit reference to the feasible set C for the sake of simplicity of the exposition: the bundle method we are starting to discuss is involved enough in the current context.

We assume that $D(T) = \mathbb{R}^n$ and that the solution set $T^{-1}(0) \neq \emptyset$. As it is standard in this context, we also suppose that we have an *oracle* (also called *black box*), that at each point $x \in \mathbb{R}^n$, provides an element $u \in T(x)$. Our departing point for understanding the scheme proposed in [48] is the following simple remark: given an arbitrary y and $v \in T(y)$, the monotonicity of T implies that $T^{-1}(0)$ is contained in the half-space

$$H(y, v) := \{z \in \mathbb{R} : \langle z - y, v \rangle \leq 0\}. \tag{5.73}$$

Before introducing the implementable bundle strategy, which overcomes the limitations of EGA in terms of computational implementation, we present a conceptual method for finding zeroes of T, introduced in [47], which is in a certain sense similar but in fact simpler than EGA, and consequently even less implementable, but which is useful for understanding the much more involved implementable bundle procedure introduced later.

First, we describe somewhat informally this theoretical method, called CAS, (as in *Conceptual Algorithmic Scheme*), for finding zeroes of a maximal monotone point-to-set operator T. For a closed and convex $C \subset \mathbb{R}^n$, let $P_C : \mathbb{R}^n \to \mathbb{R}^n$ denote the orthogonal projection onto C. Take a current iterate $x^k \notin T^{-1}(0)$. First, find y^k and $v^k \in T(y^k)$ such that $x^k \notin H(y^k, v^k)$, with H as in (5.73). Then, project x^k onto $H(y^k, v^k) \supseteq T^{-1}(0)$ to obtain a new iterate:

$$x^{k+1} := P_{H(y^k, v^k)}(x^k) = x^k - \frac{\langle x^k - y^k, v^k \rangle}{\|v^k\|^2} v^k. \tag{5.74}$$

By an elementary property of orthogonal projections, x^{k+1} is closer to any point in $T^{-1}(0)$ than x^k. However, in order to have a significant progress from x^k to x^{k+1}, adequate choices of (y^k, v^k) are required. Because $v^k \in T(y^k)$ is given by an oracle, we can only control the selection of y^k. In [47], y^k is chosen as

$$y^k = x^k - \gamma_k u^k, \tag{5.75}$$

where $\{\gamma_k\} \subset \mathbb{R}_{++}$, and u^k is the minimum norm vector in $T^e(\varepsilon_k, x^k)$.

The scheme (5.74)–(5.75) is useful for theoretical purposes. Actually, it allows us to clearly identify the crucial elements to obtain convergence: (5.74) and (5.75) have to be combined for driving ε_k to 0, in order to generate a convergent sequence. At the same time, ε^k should not go to 0 too fast, because in such a case the resulting multifunction $x \mapsto T^e(\varepsilon^k, x)$ would not be smooth enough. When coming to implementation concerns, it appears that u^k in (5.75) cannot be computed without having full knowledge of $T^e(\varepsilon_k, x^k)$, a fairly bold (if not impossible) assumption, as discussed in the previous section. Instead, we only assume that an oracle, giving one element in $T(z)$ for each z, is available. Then u^k can be approached by finding the minimum norm element of a polyhedral approximation of $T^e(\varepsilon_k, x^k)$. A suitable polyhedral approximation is obtained by using the transportation formula for T^e, together with *bundle techniques* (see, e.g., [94, Volume II]): having at the current iteration a raw "bundle" with all the oracle information collected so far, $\{(z^i, w^i)\}$, $(0 \leq i \leq p)$, with $w^i \in T(z^i)$, the convex hull of some selected w^is is a good approximation of $T^e(\varepsilon_k, x^k)$. Again, special attention has to be given when selecting the sub-bundle, in order to control ε_k and preserve convergence. Note that the oracle provides only elements in the graph of T, not of T^e, and the set to be approximated belongs to the range of T^e.

We start by a simple consequence of the transportation formula, stated in a way which is useful in our analysis. Namely, we use the transportation formula in its (equivalent) form stated in Exercise 7, and with $\varepsilon_i = 0$, that is, when $w^i \in T(z^i)$ (given by the oracle for each z^i).

Corollary 5.5.7. *Consider the notation of Exercise 7 for $E := T^e$. Suppose that $\varepsilon_i = 0$ for all $i \leq m$. Take $\tilde{x} \in \mathbb{R}^n$ and $\rho > 0$ such that $\|z^i - \tilde{x}\| \leq \rho$ for all $i \leq m$. Then, the convex sum $(\hat{x}, \hat{u}) := \left(\sum_{i=1}^m \alpha_i z^i, \sum_{i=1}^m \alpha_i w^i \right)$ satisfies:*

$$\|\hat{x} - \tilde{x}\| \leq \rho,$$
$$\hat{u} \in T^e(\hat{\varepsilon}, \hat{x}) \quad with \quad \hat{\varepsilon} := \sum_{i=1}^m \alpha_i \langle z^i - \hat{x}, w^i - \hat{u} \rangle \leq 2\rho M,$$

where $M := \max\{\|w^i\| \mid i = 1, \ldots, m\}$.

Proof. The convexity of the norm implies that $\|\hat{x} - \tilde{x}\| \leq \rho$, and also that $\|\hat{u}\| \leq M$. Exercise 7 establishes that $\hat{u}$ belongs to $T^e(\hat{\varepsilon}, \hat{x})$, for $\hat{\varepsilon} = \sum_{i=1}^m \alpha_i \langle z^i - \hat{x}, w^i - \hat{u} \rangle$. The last expression can be rewritten as follows.

$$\sum_{i=1}^m \alpha_i \langle z^i - \hat{x}, w^i - \hat{u} \rangle = \sum_{i=1}^m \alpha_i \langle z^i - \tilde{x}, w^i - \hat{u} \rangle + \sum_{i=1}^m \alpha_i \langle \tilde{x} - \hat{x}, w^i - \hat{u} \rangle$$

$$= \sum_{i=1}^m \alpha_i \langle z^i - \tilde{x}, w^i - \hat{u} \rangle,$$

using the definition of $\hat{u}$ in the second equality. Inasmuch as $\|\hat{u}\| \leq M$, the result follows from the Cauchy–Schwartz inequality. $\qquad \square$

In order to make the implementable version of the above-described theoretical scheme more clear, we present first the formal statement of CAS, where we consider explicitly sets of the form $T^e(\varepsilon, x)$. Further on, we use a bundle technique to build adequate polyhedral approximations of such sets. CAS can be described as follows.

Initialization:
    Choose parameters $\tau, \beta, \varepsilon > 0$ and $\sigma \in (0, 1)$. Take $x^0 \in \mathbb{R}^n$.

Iterative step: Given x^k,
    Step 0: If $0 \in T(x^k)$, then STOP.   **(stopping test)**
    Step 1:   **(computing search direction)**
        Compute $u^k := \operatorname{argmin}\{\|v\|^2 \mid v \in T^e(\varepsilon 2^{-j}, x^k)\}$, where $j \in \mathbb{Z}_+$ is such that $\|u^k\| > \tau 2^{-j}$.

Step 2:   **(line search)**

Define $y^k := x^k - \beta 2^{-\ell} u^k$ and take $v^k \in T(y^k)$, for $\ell \in \mathbb{Z}_+$ such that $\langle v^k, u^k \rangle > \sigma \|u^k\|^2$.

Step 3:   **(projection step)**

Define

$$\mu_k := \frac{\langle v^k, x^k - y^k \rangle}{\|v^k\|^2},$$

$$x^{k+1} := x^k - \mu_k v^k.$$

Observe that informally speaking, CAS is a simplified version of EGA: instead of the inexact norm minimization of (5.58) in EGA, step 1 of CAS demands the exact norm minimizer point in $T^e(\varepsilon_k, x^k)$; step 2 in CAS is basically the Armijo search in step (d) of EGA (more specifically (5.60) and (5.61)); step 3 of CAS is equivalent to step (e) in EGA (i.e., the extragradient step), taking into account that in the case of CAS we have $C = \mathbb{R}^n$, so that P_C is the identity mapping.

In view of the similarity between EGA and CAS, presentation of the latter might be considered superfluous; we keep it because it is the conceptual skeleton behind the much more involved bundle method IBS, considered below.

In [47] it is proved that CAS is well defined, with no infinite loops in steps 1 or 2. CAS is also convergent in the following sense. It either stops with a last $x^k \in T^{-1}(0)$, or it generates an infinite sequence $\{x^k\}$ converging to a zero of T. We show later on that the same convergence result holds for the implementable version of CAS (cf. Theorem 5.5.12).

Now it is time to overcome the announced computational limitation of both EGA and CAS: the impossibility of evaluating u^k when we do not have an explicit characterization of $T^e(\varepsilon_k, x^k)$, or, even worse, we only have an oracle providing just one element in $T(z)$ for each $z \in \mathbb{R}^n$. We intend to use the so-called *dual-bundle methods* (see, for instance, [133, 214] and also Chapters XI and XIII in [94]). However, these methods, devised for convex optimization, make extensive use of the objective functional values $f(x^k)$. Inasmuch as we do not have functional values in the wider context of finding zeroes of monotone operators, we have to adapt the "bundle" ideas to this more general context. Before proceeding to this adaptation we give a quick description of the main ideas behind the "bundle concept".

It is well-known that the direction $-\nabla f(x)$ is a descent one for a differentiable function f at the point x. When dealing with a convex but nonsmooth f, we have the set $\partial f(x)$ instead of the vector $\nabla f(x)$, and an important property is lost in the extension: it is not true that $-v$ is a descent direction at x for all $v \in \partial f(x)$.

It is not difficult to check that the minimum norm subgradient of f at x gives indeed a descent direction, but in general it is not practical to solve the implicit minimization problem, because $\partial f(x)$ cannot be described in a manageable way (e.g., through a finite set of equations or inequalities). Assuming that we only have an oracle for generating subgradients, as described above, one could think of replacing the set $\partial f(x)$ by the convex hull of a finite set of subgradients at points close to x, but here we confront the lack of inner-semicontinuity of ∂f: in general a subgradient at x cannot be approximated by subgradients at nearby points. On the other hand, the ε-subdifferential $\partial_\varepsilon f$ is inner-semicontinuous: subgradients at points close to x are ε-subgradients at x for an adequate ε.

Starting from this fact, bundle methods typically begin with a point x^0 and a subgradient v^0 at the point, provided by the oracle. A finite line search is performed along the direction $-v^0$: if enough decrease in functional values is attained, we get a new iterate x^1. Otherwise, we keep x^0 as the current iterate, but in both cases v^0 is saved to form the bundle. After m steps (in some of which a new iterate x^k is produced, whereas in some others this does not occur), we have the bundle of subgradients $v^1, \ldots, v^m$, all of them provided by the oracle along the m steps, but the trial direction is not one of them: rather it is the minimum norm vector in the convex hull of the v^is. Such a convex hull approximates $\partial_\varepsilon f$ at the current iterate for a suitable ε, thus this trial direction has a good chance of being indeed an appropriate descent direction, and eventually it is, ensuring convergence to a minimizer of f under reasonable assumptions. Also, finding such a minimum norm vector requires just the unconstrained minimization of a convex quadratic function of m variables; that is, a rather modest computational task, if m is not too big. At each step, a line search is performed along the trial direction starting at the current iterate: if successful, a new iterate is produced; otherwise the current iterate is left unchanged, but in both cases a subgradient at the last point found in the search, provided by the oracle, is added to the bundle.

Some safeguards are needed: it is necessary, for example, to refine the bundle, eliminating "old" subgradients at points far from the current iterate, but the basic scheme is the one just described. We proceed to extend this idea to the realm of monotone operators.

We denote as

$$\Delta_I := \left\{ \lambda \in \mathbb{R}_+^I : \sum_{i \in I} \lambda_i = 1 \right\}$$

the unit simplex associated with a set of indices I.

We present next the formal statement of the bundle strategy. As usual in bundle methods, we suppose that an oracle is available, which computes some $v \in T(x)$ for each $y \in \mathbb{R}^n$. Immediately after this formal statement, and before the convergence analysis, we explain the rationale behind it.

Implementable Bundle Strategy (IBS)

Initialization: Choose parameters $\tau, \beta > 0$ and $\sigma \in (0,1)$. Set $k := 0$, $p := 0$ and take $x^0 \in \mathbb{R}^n$.

Implementable iterative step: Given x^k,

    Step 0:

       (0.a) Ask the oracle to provide $u^k \in T(x^k)$; if $u^k = 0$, then STOP. (**stopping test**)

       (0.b) Else, set $p := p + 1$, $(z^p, w^p) := (x^k, u^k)$. Set $m := 0$.

    Step 1:   (**computing search direction**)

       (1.a) Set $j := 0$.

       (1.b) Define $I(k, m, j) := \{ 1 \le i \le p \mid \|z^i - x^k\| \le \beta 2^{-j} \}$.

       (1.c) Compute $\alpha^{k,m,j} := \operatorname{argmin}\{ \| \sum_{i \in I(k,m,j)} \alpha_i w^i \|^2 \mid \alpha \in \Delta_{I(k,m,j)} \}$.

(1.d) Define $s^{k,m,j} := \sum_{i \in I(k,m,j)} \alpha_i^{k,m,j} w^i$.

(1.e) If $\|s^{k,m,j}\| \leq \tau 2^{-j}$ then set $j := j + 1$ and LOOP to (1.b).

(1.f) Else, define $j(k,m) := j$, $s^{k,m} := s^{k,m,j(k,m)}$.

Step 2: (line search)

(2.a) Set $\ell := 0$.

(2.b) Define $y^{k,m,\ell} := x^k - (\beta 2^{-\ell}/\|s^{k,m}\|)s^{k,n}$ and ask the oracle to provide some $v^{k,m,\ell} \in T(y^{k,m,\ell})$.

(2.c) If

$$\langle v^{k,m,\ell}, s^{k,m} \rangle \leq \sigma \|s^{k,m}\|^2 \tag{5.76}$$

and $\ell < j(k,m) + 1$, then
Set $\ell := \ell + 1$ and **Loop** to (2.b).

(2.d) Else, define $\ell(k,m) := \ell$, $y^{k,n} := y^{k,m,\ell(k,m)}$, $v^{k,n} := v^{k,m,\ell(k,m)}$.

Step 3: (evaluating the pair (y,v))

(3.a) If

$$\langle v^{k,m}, s^{k,m} \rangle \leq \sigma \|s^{k,m}\|^2, \tag{5.77}$$

then (null step)
Set $p := p + 1$, $(z^p, w^p) := (y^{k,m}, v^{k,m})$.
Set $m := m + 1$ and **Loop** to (1.b).

(3.b) Else (serious step)
Define $m(k) := m$, $j(k) := j(k, m(k))$, $\ell(k) := \ell(k, m(k))$,
$s^k := s^{k,m(k)}$, $y^k := y^{k,m(k)}$, $v^k := v^{k,m(k)}$.
Define $\mu_k := \langle v^k, x^k - y^k \rangle v^k / \|v^k\|^2$.
Define $x^{k+1} := x^k - \mu_k v^k$.

We try to describe next in words what is going on along the iterative process defined by IBS. We perform two tasks simultaneously: the generation of the sequence of points $\{x^k\}$ and the construction of the bundle $\{(z^1, w^1) \ldots (z^p, w^p)\}$. At each step, we look at the convex hull Q_p of $w^1, \ldots, w^p$ as a polyhedral approximation of $T^e(\varepsilon_k, x^k)$. This double task gives rise to two kinds of steps: the serious ones (step (3.b) of IBS), indexed in k, where a new point x^{k+1} is produced, such as the whole iterative step in CAS or EGA, and the null steps (step (3.a) of IBS), indexed in m, where the generated point $y^{k,m}$ with its companion vector $v^{k,m} \in T(y^{k,m})$ produced by the oracle, fail to pass the test given by the inequality in (5.77). In this case, we do not abandon our current point x^k, but the pair $(y^{k,m}, v^{k,m})$, renamed (z^p, w^p) in step (3.a) of IBS, is added to the bundle, meaning that the polyhedral approximation of $T^e(\varepsilon_k, x^k)$ is modified (and it is hoped, improved).

This polyhedral approximation (namely the set Q_p defined above) is used to generate the search direction as in EGA or CAS but with an enormous computational advantage: whereas both the inexact norm minimization in $T^e(\varepsilon_k, x^k)$ (step (d) of EGA) and the exact one (step 1 of CAS), are almost unsurmountable computational tasks in most instances of VIP(T,C), finding the minimum norm vector in the polyhedron Q_p demands the unconstrained minimization of a convex quadratic function in variables $\alpha_1, \ldots, \alpha_p$ (see step (1.c) of IBS). However, as p increases and x^k moves away from x^0, Q_p would become too big, and unrelated to the current point x^k. Thus the bundle is "pruned". Instead of approximating $T^e(\varepsilon_k, x^k)$

with the full collection $w^0, \ldots, w^k$, we discard those w^is associated with vectors z^i that are far from the current point x^k. This is achieved through the definition of the index set $I(k, m, j)$ in step (1.b) of IBS. We construct Q_p only with vectors $w^i \in T(z^i)$ such that z^i belongs to a ball of radius $\beta 2^{-j}$ around x^k, where the radius is successively halved through an inner loop in the index j, which continues until the minimum norm solution ($s^{k,m,j}$ in step (1.d) of IBS) has a norm greater than $\tau 2^{-j}$. As the radius of the ball gets smaller, the size of the set $I(k, m, j)$ (i.e., the cardinality of the set of vectors that span the polyhedron Q_p) gets reduced. This procedure keeps the "active" bundle small enough.

We emphasize that when a serious step is performed, the generated pair is also added to the bundle. When an appropriate pair $(y^{k,m}, v^{k,m})$ is found, after minimizing the (possibly several) quadratic functions in step (1.c), the algorithm proceeds more or less as in EGA or CAS: an Armijo-like line search (step 2 of IBS) is performed along the direction $s^{k,m}$ from x^k, starting with a stepsize $\beta / \|s^{k,m}\|$, which is halved along the loop in the index ℓ, until the point $v^{k,m,\ell}$ given by the oracle satisfies the inequality in (5.76), or the counting index for this inner loop attains the value $j(k, m)$, which guarantees a priori the finiteness of the line search loop.

Finally, in the serious step (3.b), the vector y^k is projected onto the separating hyperplane, exactly as in step (e) of EGA or step 3 of CAS. We now complement this explanation of the IBS iteration with some additional remarks.

Remark 5.5.8.

- In step 0 a stopping test of the kind $\|u^k\| \leq \eta$ for some $\eta \geq 0$ could have been used in (0.a), instead of the computationally disturbing test $u^k = 0$. In such a case, the convergence analysis would essentially remain the same, with very minor variations.

- As explained above, at each iteration of the inner loop in step 1, we have indeed two bundles: the *raw bundle*, consisting of $\{(z^0, w^0), \ldots, (z^p, w^p)\}$, and the *reduced* one, formed by $\{(z^i, w^i) : i \in I(k, m, j)\}$, which is the one effectively used for generating the polyhedron Q_p. Observe that, by definition of the index set $I(k, m, j)$, the pair (x^k, u^k), with $u^k \in T(x^k)$, is always in the reduced bundle. Hence the reduced bundle is always nonempty, ensuring that the vector $\alpha^{k,m,j}$ obtained in step (1.b) is well defined. Lemma 5.5.10 below ensures that if x^k is not a zero of T, then the loop in step 1 (i.e., the loop in the index j), is always finite.

- Regarding the line search in step 2, as mentioned above it always ends (it cannot perform more than $j(k, m)$ steps). The two possible endings are a null step, which increases the bundle but keeps the current iterate x^k unchanged, or a serious step, which both increases the bundle and updates the sequence $\{x^k\}$, by producing a new x^{k+1}. In the case of the null step, the inequality in (5.76) guarantees that not only the raw bundle grows in size, but also the reduced one does, in view of the definition of $I(k, m, j)$ in (1.b). Proposition 5.5.9(ii) shows that, when x^k is not a solution, the number of null steps in iteration k is finite, ensuring that a new serious step will take place, generating a new iterate x^{k+1}.

- As commented on above, in serious steps $v^{k,m}$ defines a hyperplane separating x^k from the solution set $T^{-1}(0)$. In fact such a hyperplane is far enough from x^k, in the sense that $\|x^k - x^{k+1}\|$ is not too small as compared to the distance from x^k to S^*. In addition, note that at step (3.b) the following relations hold.

$$\ell_k \leq j_k + 1, \qquad \langle v^k, s^k \rangle > \sigma \|s^k\|^2, \qquad \|s^k\| > \tau 2^{-j(k)},$$

$$y^k = x^k - \left(\beta 2^{-\ell(k)} / \|s^k\| \right) s^k. \tag{5.78}$$

5.5.3 Convergence analysis

We show in this section that either IBS generates a finite sequence, with the last iterate being a solution, or it generates an infinite sequence, converging to a solution.

The bundle strategy IBS provides us with a constructive device for CAS. For this reason the convergence analysis is close to the one in [47]. The main difference appears when analyzing null steps, which lead to an enrichment of the bundle.

We state below without proof several results that do not depend on the properties of the enlargement T^e. All of them are proved in [48].

Proposition 5.5.9. *Let $\{x^k\}$ be the sequence generated by IBS. Then*

(i) *If x^k is a solution, then either the oracle answers $u^k = 0$ and IBS stops in (0.a), or IBS loops forever after this last serious step, without updating k.*

(ii) *Else, x^k is not a solution, and IBS reaches step (3.b) after finitely many inner iterations. Furthermore,*

$$\|s^{k,m^*(k),j(k)-1}\| \leq \tau 2^{-j(k)+1}, \tag{5.79}$$

where $m^*(k)$ is the smallest value of m such that $j(k,m) = j(k)$, whenever $j(k) > 0$.

(iii) *For all $x^* \in T^{-1}(0)$, $\|x^{k+1} - x^*\|^2 \leq \|x^k - x^*\|^2 - \|x^{k+1} - x^k\|^2$.*

(iv) *If the sequence $\{x^k\}$ is infinite, then $\lim_{k \to \infty} j(k) = +\infty$.*

As a consequence of Proposition 5.5.9, the sequence of serious points $\{x^k\}$ generated by IBS is either finite, ending at a solution; or infinite, with no iterate being a solution, and in the second case there is always a finite number of null steps after each serious step. We show next how the transportation formula for T^e is used to prove that the inner loop in step 1 (in the index j) is always finite when x^k is not a solution. In other words, the process of constructing a suitable polyhedral approximation of $T^e(\varepsilon_k, x^k)$, using the point $w^i \in T(z^i)$ provided by the oracle, always ends satisfactorily.

Lemma 5.5.10. *Let $\{x^k\}$ be the sequence generated by IBS. If x^k is not a zero of T, then in step 1 there exists a finite $j = j(k, m)$ such that*

$$\|s^{k,m}\| > \tau 2^{-j(k,m)} \tag{5.80}$$

in which case (1.f) is reached. Furthermore, the loop in the linesearch of step 2 is also finite: step (2.d) is reached with $\ell(k, m) \le j(k, m) + 1$.

Proof. By assumption, $0 \notin T(x^k)$. Suppose, for the sake of contradiction, that the result does not hold. Then an infinite sequence $\{s^{k,m,j}\}_{j \in \mathbb{N}}$ is generated, satisfying $\|s^{k,m,j}\| \le \tau 2^{-j}$ for all j. Therefore, there exist subsequences $\{m_q\}$, $\{j_q\}$ such that

$$\|s^{k,m_q,j_q}\| \le \tau 2^{-j_q}, \tag{5.81}$$

with $\lim_{q \to \infty} j_q = \infty$. Define $\hat{I}_q := I(k, m_q, j_q)$. By step (1.b), $\|z^i - x^k\| \le \beta 2^{-j_q}$ for all $i \in I_q$. Consider the convex sum induced by the vector $\hat{\alpha}^q := \alpha^{k,m_q,j_q}$, which solves the norm minimization problem in step (1.c), namely

$$(\hat{x}^q, \hat{s}^q) := \left(\sum_{i \in I_q} \hat{\alpha}_i^q z^i, \sum_{i \in I_q} \hat{\alpha}_i^q s^{k,m_q,j_q} \right).$$

Corollary 5.5.7 applies, with $\rho = \beta 2^{-j_q}$ and $x = x^k$, and we have

$$\hat{s}^q \in T^e(\hat{\varepsilon}_q, \hat{x}^q), \tag{5.82}$$

with $\hat{\varepsilon}_q \le 2\beta 2^{-j_q} M$, where $M := \sup \left\{ \|u\| \mid u \in T\left(\overline{B(x^k, \beta)} \right) \right\}$. In addition,

$$\|\hat{x}^q - x^k\| \le \beta 2^{-j_q}. \tag{5.83}$$

Altogether, letting $q \to \infty$ in (5.81), (5.82) and (5.83), outer-semicontinuity of T^e yields:

$$\lim_{q \to \infty} (\hat{\varepsilon}_q, \hat{x}^q, \hat{s}^q) = (0, x^k, 0)$$

with $\hat{s}^q \in T^e(\hat{\varepsilon}_q, \hat{x}^q)$, implying that $0 \in T^e(0, x^k) = T(x^k)$, which is a contradiction. Hence, there exists a finite j at which the inner loop of step 1 ends; that is, an index $j(k, m)$ such that the inequality in (5.80) holds. The statement about the loop in step 2 is immediate, in view of the upper bound on ℓ given in step (2.c).   □

The boundedness of the variables generated by IBS now follows from the Fejér convergence of $\{x^k\}$ to $T^{-1}(0)$; that is, from Lemma 5.5.2 and Proposition 5.5.9(iii). We state the results, omitting some technical details of the proof.

Lemma 5.5.11. *Assume that $T^{-1}(0) \neq \emptyset$. Then all the variables generated by IBS, namely x^k, s^k, y^k, v^k, $\{(y^{k,m}, v^{k,m})\}$, and $\{(z^p, w^p)\}$ are bounded.*

Proof. See Proposition 4.3 and 4.4(ii), and Lemma 4.6 in [48]. $\square$

Along the lines of [47, Section 3.3], convergence of IBS is proved using Lemma 5.5.2(ii), by exhibiting a subsequence of triplets $(\hat{\varepsilon}_q, \hat{x}^q, \hat{s}^q) \in T^e(\varepsilon_q, \hat{x}^q)$ tending to $(0, \bar{x}, 0)$, as q is driven by j_k to infinity.

Theorem 5.5.12. *The sequence $\{x^k\}$ generated by IBS is either finite with the last element in $T^{-1}(0)$, or it converges to a zero of T.*

Proof. We already dealt with the finite case in Proposition 5.5.9(i). If there are infinitely many x^ks, keeping Lemma 5.5.2(ii) in mind, we only need to establish that some accumulation point of the bounded sequence $\{x^k\}$ is a zero of T. Let $\{x^{k_q}\}$ be a convergent subsequence of $\{x^k\}$, with limit point $\bar{x}$. Because of Proposition 5.5.9(iv), we can assume that $j(k_q) > 0$, for q large enough. Then Proposition 5.5.9(ii) applies: for $m^*(k)$ defined therein, we have

$$\|s^{k_q, m^*(k_q), j(k_q)-1}\| \leq \tau 2^{-j(k_q)+1}. \tag{5.84}$$

Consider the associated index set $I_q := I(k_q, m^*(k_q), j(k_q) - 1))$. As in the proof of Lemma 5.5.10, define

$$\begin{aligned}
\tilde{\alpha}^q &:= \alpha^{k_q, m^*(k_q), j(k_q)-1}, \\
\hat{x}^q &:= \sum_{i \in I_q} \tilde{\alpha}_i^q z^i, \\
\hat{s}^q &:= s^{k_q, m^*(k_q), j(k_q)-1} = \sum_{i \in I_q} \tilde{\alpha}_i^q w^i.
\end{aligned}$$

We have that

$$\|\hat{x}^q - x^{k_q}\| \leq \beta 2^{-j(k_q)+1}. \tag{5.85}$$

By Lemma 5.5.11, there exists an upper bound M of $\|w^1\|, \ldots, \|w^p\|$. Then Corollary 5.5.7 yields

$$\hat{s}^q \in T^e(\hat{\varepsilon}_q, \hat{x}^q), \tag{5.86}$$

with $\hat{\varepsilon}_q \leq 2\beta 2^{-j(k_q)+1} M$. Using Proposition 5.5.9(iv), we have $\lim_{q \to \infty} j_{k_q} = \infty$. Hence, by (5.84), (5.85), and (5.86) we conclude that

$$\lim_{q \to \infty} (\hat{\varepsilon}_q, \hat{x}^q, \hat{s}^q) = (0, \bar{x}, 0),$$

with $\hat{s}^q \in T^e(\hat{\varepsilon}_q, \hat{x}^q)$. By outer-semicontinuity of T^e, we conclude that $0 \in T(\bar{x})$. $\square$

From the proofs of Lemma 5.5.10 and Theorem 5.5.12, we see that both inner and outer-semicontinuity are key ingredients for establishing convergence of the sequence $\{x^k\}$ generated by IBS. Inner-semicontinuity is implicitly involved in the transportation formula, which is essential in the analysis of IBS.

5.6   Theoretical applications of T^e

5.6.1   An alternative concept of sum of maximal monotone operators

It has been pointed out before (see Section 4.8), that the sum of two maximal monotone operators is always monotone. However, this sum is not maximal in general. Theorem 4.8.3 in Chapter 4 ensures that this property holds when X is a reflexive Banach space and the domain of one of the operators intersects the interior of the domain of the other. A typical example (and a most important one, in fact) for which the sum of two maximal monotone operators fails to be maximal, occurs when the sum of the subdifferentials of two lsc convex and proper functions is strictly included in the subdifferential of the sum of the original functions. This motivated the quest for a different definition of sum such that

(a) The resulting operator is maximal, even when the usual sum is not.

(b) When the usual sum is maximal, both sums coincide.

(c) The operators are defined in an arbitrary Banach space.

Items (a) and (b) have been successfully addressed in [10]. This work presents, in the setting of Hilbert spaces, the *variational sum* of maximal monotone operators. This concept was extended later on in [175] to reflexive Banach spaces. However, item (c) cannot be fulfilled by the variational sum, because its definition relies heavily on Yosida approximations of the given operators, and these approximations are only well defined in the context of reflexive Banach spaces.

The aim of this section is to show how T^e (and also $E(T)$) can be used for defining an alternative sum, which satisfies items (a), (b), and (c). This notion of sum, which relies on the enlargements T^e of the original operators, was defined for the first time in [175], in the context of reflexive Banach spaces. It was then extended to arbitrary Banach spaces in [176], and is called an *extended sum*. Throughout this section, we follow essentially the analysis in [176].

We point out that the extended sum coincides with the variational sum in reflexive Banach spaces, when the closure of the sum of the operators is maximal. They also coincide when the operators are subdifferentials, and in this case both sums are equal to the subdifferential of the sum.

The extended sum is defined as follows.

Definition 5.6.1. *Let T_1, $T_2 : X \rightrightarrows X^*$ be two maximal monotone operators. The extended sum of T_1 and T_2 is the point-to-set mapping $T_1 +_{ext} T_2 : X \rightrightarrows X^*$ given by*

$$(T_1 +_{ext} T_2)(x) := \bigcap_{\varepsilon > 0} \overline{T_1^e(\varepsilon, x) + T_2^e(\varepsilon, x)}^{w^*}, \qquad (5.87)$$

where $\overline{C}^{\,w^*}$ stands for the w^*-closure of $C \subset X^*$.

Before proving the properties of the sum above, we must recall some well-known facts on the w^*-topology in X^*. A subbase of neighborhoods of $v_0 \in X^*$ in this topology is given by the sets $\{V_{\varepsilon,x}(v_0)\}_{x\in X, \varepsilon>0}$ defined as

$$V_{\varepsilon,x}(v_0) := \{v \in X^* : |\langle v - v_0, x\rangle| < \varepsilon\}. \tag{5.88}$$

For a multifunction $F : X \rightrightarrows X^*$, denote $\overline{F}$ the multifunction given by $\overline{F}(x) := \overline{F(x)}^{w^*}$. Throughout this section, this closure symbol refers to the weak* closure.

Lemma 5.6.2. *Let $T : X \rightrightarrows X^*$ be monotone. Then $\overline{T}^e(\varepsilon, \cdot) = T^e(\varepsilon, \cdot)$.*

Proof. Because $\mathrm{Gph}(T) \subset \mathrm{Gph}(\overline{T})$, it is clear from the definition of T^e that $\overline{T}^e(\varepsilon, \cdot) \subset T^e(\varepsilon, \cdot)$. Fix $x \in X$ and take $v^* \in T^e(\varepsilon, x)$. Let $(z, w) \in \mathrm{Gph}(\overline{T})$. We must show that

$$\langle w - v^*, z - x\rangle \geq -\varepsilon.$$

Because $w \in \overline{T}(z)$, we have that $T(z) \cap V_{\rho, x-z}(w) \neq \emptyset$ for all $\rho > 0$. Thus, for all $\rho > 0$ there exists $u_\rho \in T(z)$ with $|\langle u_\rho - w, x - z\rangle| < \rho$. Hence,

$$\langle w - v^*, z - x\rangle = \langle w - u_\rho, z - x\rangle + \langle u_\rho - v^*, z - x\rangle$$

$$\geq -\varepsilon + \langle w - u_\rho, z - x\rangle \geq -\varepsilon - \rho,$$

using the definition of T^e in the first inequality and the definition of u_ρ in the second one. Because $\rho > 0$ is arbitrary, we obtained the desired conclusion. $\square$

We recall now a formula proved in [95]. It considers an expression similar to (5.87), but using the $\breve{\partial}f$-enlargement. This formula can be seen as a special sum between subdifferentials, which generates as a result a maximal operator: the subdifferential of the sum.

Theorem 5.6.3. *Let X be a Banach space and take $f, g : X \to \mathbb{R} \cup \{\infty\}$ proper, convex, and lsc. Then for all $x \in \mathrm{dom}\, f \cap \mathrm{dom}\, g$ it holds that*

$$\partial(f + g)(x) = \bigcap_{\varepsilon>0} \overline{\breve{\partial}f(\varepsilon, x) + \breve{\partial}g(\varepsilon, x)}. \tag{5.89}$$

As a consequence of Proposition 5.2.4, the right-hand side of (5.89) is contained in the extended sum of the subdifferentials of f and g. We show below that both coincide when the weak* closure of the sum of the subdifferentials is maximal.

An important property of the extended sum is that it admits a formula similar to (5.89), and it gives rise to a maximal monotone operator, which is the subdifferential of the sum. In order to establish that fact, we need a characterization of the subdifferential of the sum of two proper, convex, and lsc functions. This characterization has been established in [170].

Theorem 5.6.4. *Let X be a Banach space and take $f, g : X \to \mathbb{R} \cup \{\infty\}$ proper, convex, and lsc. Then $y^* \in \partial(f + g)(y)$ if and only if there exist two nets $\{(x_i, x_i^*)\}_{i\in I} \subset \mathrm{Gph}(\partial f)$ and $\{(z_i, z_i^*)\}_{i\in I} \subset \mathrm{Gph}(\partial g)$ such that*

(i) $x_i \to^s y$ and $z_i \to^s y$.

(ii) $x_i^* + z_i^* \to^{w^*} y^*$.

(iii) $\langle x_i - y, x_i^* \rangle \to 0$ and $\langle z_i - y, z_i^* \rangle \to 0$.

5.6.2   Properties of the extended sum

We start by proving that, when $\overline{T_1 + T_2}$ is maximal monotone, the extended sum can be also defined using arbitrary elements in $\mathbb{E}(T_1)$ and $\mathbb{E}(T_2)$.

Proposition 5.6.5. *Let* $T_1, T_2 : X \rightrightarrows X^*$ *be two maximal monotone operators such that* $\overline{T_1 + T_2}$ *is maximal. Then*

$$(T_1 +_{ext} T_2)(x) = \bigcap_{\varepsilon > 0} \overline{E_1(\varepsilon, x) + E_2(\varepsilon, x)} = \overline{(T_1 + T_2)}(x), \qquad (5.90)$$

where $E_1 \in \mathbb{E}(T_1)$ *and* $E_2 \in \mathbb{E}(T_2)$ *are arbitrary.*

Proof. We have that

$$\bigcap_{\varepsilon > 0} \overline{E_1(\varepsilon, x) + E_2(\varepsilon, x)} \supset (T_1 + T_2)(x).$$

Taking now the weak* closure in the expression above, we get

$$\bigcap_{\varepsilon > 0} \overline{E_1(\varepsilon, x) + E_2(\varepsilon, x)} \supset \overline{(T_1 + T_2)}(x).$$

On the other hand, we obtain from Lemmas 5.2.1(b) and 5.6.2,

$$(T_1 + T_2)(x) \subset E_1(\varepsilon, x) + E_2(\varepsilon, x) \subset T_1^e(\varepsilon, x) + T_2^e(\varepsilon, x)$$

$$\subset (T_1 + T_2)^e(2\varepsilon, x) = (\overline{T_1 + T_2})^e(2\varepsilon, x).$$

Taking again the weak* closure and computing the intersection for all $\varepsilon > 0$ in the expression above we conclude that

$$\overline{(T_1 + T_2)}x \subset \bigcap_{\varepsilon > 0} \overline{E_1(\varepsilon, x) + E_2(\varepsilon, x)} \subset \bigcap_{\varepsilon > 0} \overline{T_1^e(\varepsilon, x) + T_2^e(\varepsilon, x)}$$

$$\subset \bigcap_{\varepsilon > 0} \overline{(T_1 + T_2)^e(2\varepsilon, x)} = \overline{(T_1 + T_2)}x,$$

using maximality of $\overline{T_1 + T_2}$ and Lemma 5.2.1(e)–(f) in the last equality. The result follows then from (5.87).   □

When T_1 and T_2 are subdifferentials, Proposition 5.6.5 provides a case in which the extended sum coincides with the subdifferential of the sum.

Corollary 5.6.6. *Let $f, g : X \to \mathbb{R} \cup \{\infty\}$ be two convex, proper, and lsc functions such that $\overline{\partial f + \partial g}$ is maximal monotone. Then*

$$(\partial f +_{ext} \partial g)(x) = \bigcap_{\varepsilon > 0} \overline{\breve{\partial} f(\varepsilon, x) + \breve{\partial} g(\varepsilon, x)} = \partial(f + g)(x). \qquad (5.91)$$

Proof. The first equality is a direct consequence of Proposition 5.6.5, where we choose $E_1 = \breve{\partial} f$ and $E_2 = \breve{\partial} g$ in (5.90). The second equality follows from (5.89). □

Another direct consequence of Proposition 5.6.5 is that the extended sum coincides with the usual one when the latter is maximal.

Corollary 5.6.7. *Let $T_1, T_2 : X \rightrightarrows X^*$ be two maximal monotone operators such that $T_1 + T_2$ is maximal. Then*

$$(T_1 +_{ext} T_2)(x) = \bigcap_{\varepsilon > 0} [E_1(\varepsilon, x) + E_2(\varepsilon, x)] = (T_1 + T_2)(x), \qquad (5.92)$$

where $E_1 \in \mathbb{E}(T_1)$ and $E_2 \in \mathbb{E}(T_2)$ are arbitrary.

Proof. Observe that the maximality of $T_1 + T_2$ implies that $T_1 + T_2 = \overline{T_1 + T_2}$, and hence the previous proposition yields $(T_1 +_{ext} T_2)(x) = (T_1 + T_2)(x)$. In order to prove the remaining equality, observe that

$$(T_1 + T_2)(x) = (T_1 +_{ext} T_2)(x) \supset \bigcap_{\varepsilon > 0} [E_1(\varepsilon, x) + E_2(\varepsilon, x)] \supset (T_1 + T_2)(x).$$

□

A straightforward consequence of Corollary 5.6.7 is a formula similar to (5.89) for arbitrary enlargements $E_f \in \mathbb{E}(\partial f)$ and $E_g \in \mathbb{E}(\partial g)$, where the weak* closures can be removed. However, we must assume that the sum of the subdifferentials is maximal, which means that this sum is equal to the subdifferential of the sum of the original functions.

Corollary 5.6.8. *Let $f, g : X \to \mathbb{R} \cup \{\infty\}$ be two convex, proper, and lsc functions such that $\partial f + \partial g$ is maximal. Then*

$$(\partial f +_{ext} \partial g)(x) = \bigcap_{\varepsilon > 0} [E_f(\varepsilon, x) + E_g(\varepsilon, x)] = \partial(f + g)(x), \qquad (5.93)$$

where $E_f \in \mathbb{E}(\partial f)$ and $E_g \in \mathbb{E}(\partial g)$ are arbitrary.

Our aim now is to prove that the extended sum of two subdifferentials is always maximal, and hence equal to the subdifferential of the sum of the original functions. Before we establish this result, we need a technical lemma.

Lemma 5.6.9. *Let $f, g : X \to \mathbb{R} \cup \{\infty\}$ be two convex, proper, and lsc functions. Then for all $x \in \operatorname{dom} f \cap \operatorname{dom} g$ it holds that*

$$\bigcap_{\varepsilon > 0} \overline{E_f(\varepsilon, x) + E_g(\varepsilon, x)} \subset \partial(f + g)(x),$$

for all $E_f \in \mathbb{E}(\partial f)$, $E_g \in \mathbb{E}(\partial g)$.

Proof. Define the point-to-set mapping $F : X \rightrightarrows X^*$ as

$$F(x) := \bigcap_{\varepsilon > 0} \overline{E_f(\varepsilon, x) + E_g(\varepsilon, x)}.$$

We must prove that for all $x \in \operatorname{dom} f \cap \operatorname{dom} g$ it holds that $F(x) \subset \partial(f + g)(x)$. Take $x^* \in F(x)$. Using the maximality of $\partial(f + g)$, it is enough to prove that (x, x^*) is monotonically related to the graph of $\partial(f + g)(x)$; that is,

$$\langle x - y, x^* - y^* \rangle \geq 0, \tag{5.94}$$

for all $(y, y^*) \in \operatorname{Gph}(\partial(f + g))$. Fix $(y, y^*) \in \operatorname{Gph}(\partial(f + g))$ and $\delta > 0$. Take $\rho > 0$ such that $\rho < \delta/8$. Because $x^* \in F(x) \subset \overline{E_f(\rho, x) + E_g(\rho, x)}$, we have $V_{\rho, x-y}(x^*) \cap [E_f(\rho, x) + E_g(\rho, x)] \neq \emptyset$. Thus, there exist $v_\rho \in E_f(\rho, x)$ and $w_\rho \in E_g(\rho, x)$ such that

$$|\langle x - y, x^* - (v_\rho + w_\rho) \rangle| < \rho. \tag{5.95}$$

Using now Theorem 5.6.4, there exist nets $\{(x_i, x_i^*)\}_{i \in I} \subset \operatorname{Gph}(\partial f)$ and $\{(z_i, z_i^*)\}_{i \in I} \subset \operatorname{Gph}(\partial g)$ such that conditions (i)–(iii) of this theorem hold. So we can take i large enough such that

$$|\langle y - x_i, v_\rho \rangle| < \rho, \ |\langle y - z_i, w_\rho \rangle| < \rho, \tag{5.96}$$

$$|\langle x - y, x_i^* + z_i^* - y^* \rangle| < \rho, \tag{5.97}$$

$$|\langle x_i - y, x_i^* \rangle| < \rho, \ |\langle z_i - y, z_i^* \rangle| < \rho. \tag{5.98}$$

By (5.95)–(5.98), we have

$$
\begin{aligned}
\langle x - y, x^* - y^* \rangle &= \langle x - y, x^* - (v_\rho + w_\rho) \rangle + \langle x - y, (v_\rho + w_\rho) \\
&\quad - (x_i^* + z_i^*) \rangle + \langle x - y, (x_i^* + z_i^*) - y^* \rangle \\
&\geq -2\rho + \langle x - y, (v_\rho + w_\rho) - (x_i^* + z_i^*) \rangle \\
&= -2\rho + \langle x - x_i, v_\rho \rangle + \langle x_i - y, v_\rho \rangle + \langle x - z_i, w_\rho \rangle + \langle z_i - y, w_\rho \rangle \\
&\quad + \langle y - x_i, x_i^* \rangle + \langle x_i - x, x_i^* \rangle + \langle y - z_i, z_i^* \rangle + \langle z_i - x, z_i^* \rangle \\
&\geq -6\rho + \langle x - x_i, v_\rho - x_i^* \rangle + \langle x - z_i, w_\rho - z_i^* \rangle \geq -8\rho > -\delta,
\end{aligned}
$$

using the fact that $v_\rho \in E_f(\rho, x) \subset \partial f^e(r, x)$ and $w_\rho \in E_g(\rho, x) \subset \partial g^e(\rho, x)$ in the second-to-last inequality. Because $\delta > 0$ is arbitrary, (5.94) holds and the left-hand side is contained in the right-hand side.   □

Now we are able to prove the maximality of the extended sum of subdifferentials.

Corollary 5.6.10. *Let $f, g : X \to \mathbb{R} \cup \{\infty\}$ be two convex, proper, and lsc functions. Then for all $x \in \operatorname{dom} f \cap \operatorname{dom} g$ it holds that*

$$(\partial f +_{ext} \partial g)(x) = \partial(f + g)(x). \tag{5.99}$$

Proof. Choose $E_f := \partial f^e$ and $E_g := \partial g^e$ in Lemma 5.6.9 to get $(\partial f +_{ext} \partial g)(x) \subset \partial(f + g)(x)$. On the other hand, by Proposition 5.2.4, we have that $\partial(f + g)(x) \subset (\partial f +_{ext} \partial g)(x)$. Hence both operators coincide, which yields the maximality of $\partial f +_{ext} \partial g$. ☐

5.6.3 Preservation of well-posedness using enlargements

We start this section by recalling some important notions of well-posedness. Suppose that we want to solve a "difficult" problem P, with solution set S. Inasmuch as we cannot solve P directly, we embed it in a parameterized family of problems $\{P_\mu\}_{\mu \in \mathbb{R}_+}$, such that

(i) $P_0 = P$.

(ii) For every $\mu > 0$, problem P_μ is "simpler" than P.

A sequence $\{x(\mu)\}_{\mu > 0}$ such that $x(\mu)$ solves P_μ for all $\mu > 0$ is called a *solving sequence*. Denote $S(\mu)$ the solution set of P_μ, for $\mu > 0$ and define $S(0) := S$. Whether and how a solving sequence $\{x(\mu)\}$ approaches the original solution set S, depends on the family $\{P_\mu\}_{\mu \in \mathbb{R}_+}$ and on the original problem P.

The classical notion of well-posedness is due to Tykhonov [221]. A problem P is *Tykhonov well-posed* when it has a unique solution and every solving sequence of the problem converges to it. The concept of *Hadamard well-posedness* incorporates the condition of continuity of the solutions with respect to the parameter μ. A notion that combines both Tykhonov and Hadamard well-posedness has been studied by Zolezzi in [237] and [238], and is called *well-posedness with respect to perturbations*. In this section, we study how the enlargements help in defining families of well-posed problems in the latter sense. We follow the theory and technique developed in [130]. As in Section 5.5.2, our analysis focuses on the problem of finding zeroes of a maximal monotone operator:

$$(P) \quad \text{Find } x^* \in X, \text{ such that } 0 \in T(x^*),$$

where X is a reflexive Banach space, and $T : X \rightrightarrows X^*$ is a maximal monotone operator. We consider now "perturbations" of problem (P), replacing the original operator T by suitable approximations $\{T_\mu\}_{\mu > 0}$, where $T_\mu : X \rightrightarrows X^*$ are point-to-set mappings. The corresponding family of perturbed problems are defined as

$$(P_\mu) \quad \text{Find } x^* \in X, \text{ such that } 0 \in T_\mu(x^*).$$

Ideally, the perturbed problems $\{P_\mu\}_{\mu > 0}$ must be "close" to the original problem, but easier to solve. Nevertheless, solving (P_μ) exactly is not practical, because it may be too expensive, and we know in advance that the result is not the solution of the original problem. So the concept of solving sequence given above is replaced by the following weaker notion. Given a sequence $\{\mu_n\}_{n\in\mathbb{N}}$ with $\mu_n \downarrow 0$, a sequence $\{x(\mu_n)\}_{n\in\mathbb{N}}$ is said to be an *asymptotically solving sequence* associated with $\{P_\mu\}_{\mu>0}$ when

$$\lim_{n\to\infty} d(0, T_{\mu_n}(x(\mu_n))) = 0. \tag{5.100}$$

5.6.4   Well-posedness with respect to the family of perturbations

We are now able to offer a formal statement of the first concept of well-posedness. As mentioned above, all the definitions and most proofs in this section have been taken from [130].

Definition 5.6.11. *Problem* (P) *is* well posed with respect to the family of perturbations $\{P_\mu\}_{\mu>0}$ *if and only if*

(a) *Problem* (P) *has solutions; that is,* $T^{-1}(0) \neq \emptyset$.

(b) *If* $\mu_n \downarrow 0$, *then every asymptotically solving sequence* $\{x(\mu_n)\}_{n\in\mathbb{N}}$ *associated with* $\{P_\mu\}_{\mu>0}$ *satisfies*

$$\lim_{n\to\infty} d(x(\mu_n), T^{-1}(0)) = 0. \tag{5.101}$$

When (a) and (b) hold, we also say that *the family of perturbations* $\{P_\mu\}_{\mu>0}$ *is well posed for* (P).

A usual choice for the family of perturbed operators $\{T_\mu\}_{\mu>0}$ requires that T_μ be maximal monotone for all $\mu > 0$. An important question is whether the well-posedness with respect to these perturbations is preserved, when replacing each T_μ by one of its enlargements $E_\mu \in \mathbb{E}(T_\mu)$. More precisely, assume each T_μ is maximal monotone and consider the family of perturbed problems $\{\tilde{P}_\mu\}_{\mu>0}$ defined as

$$(\tilde{P}_\mu) \quad \text{Find } x^* \in X \text{ such that } 0 \in E_\mu(\mu, x^*),$$

where $E_\mu \in \mathbb{E}(T_\mu)$ for all $\mu > 0$. Because $E_\mu(\mu, x^*) \supset T_\mu(x^*)$, the latter problem is in principle easier to solve than (P_μ). We show below that well-posedness with respect to the perturbations $\{\tilde{P}_\mu\}_{\mu>0}$ is equivalent to well-posedness with respect to the perturbations $\{P_\mu\}_{\mu>0}$. In a similar way as in (5.100), we define asymptotically solving sequences for this family of problems. Given a sequence $\{\mu_n\}_{n\in\mathbb{N}}$ with $\mu_n \downarrow 0$, a sequence $\{\tilde{x}(\mu_n)\}_{n\in\mathbb{N}}$ is said to be an *asymptotically solving sequence* associated with $\{\tilde{P}_\mu\}_{\mu>0}$ when

$$\lim_{n\to\infty} d(0, E_{\mu_n}(\mu_n, x(\mu_n))) = 0. \tag{5.102}$$

The preservation of well-posedness is a consequence of the Brøndsted–Rockafellar property for the family of enlargements.

Theorem 5.6.12. *Assume that X is a reflexive Banach space. Then the family of problems $\{P_\mu\}_{\mu>0}$ is well posed for (P) if and only if $\{\tilde{P}_\mu\}_{\mu>0}$ is well posed for (P).*

Proof. Condition (a) is the same for both families, so we consider only condition (b). Suppose first that $\{\tilde{P}_\mu\}_{\mu>0}$ is well posed for (P), and take sequences $\mu_n \downarrow 0$ and $\{x(\mu_n)\}_{n\in\mathbb{N}}$ such that $\lim_n d(0, T_{\mu_n}(x(\mu_n))) = 0$. We need to prove that $\lim_{n\to\infty} d(x(\mu_n), T^{-1}(0)) = 0$. Because $E_{\mu_n}(\mu_n, x(\mu_n)) \supset T_{\mu_n}(x(\mu_n))$, we have that

$$\lim_n d(0, E_{\mu_n}(\mu_n, x(\mu_n))) \leq \lim_n d(0, T_{\mu_n}(x(\mu_n))) = 0,$$

and hence $\{x(\mu_n)\}_{n\in\mathbb{N}}$ is also asymptotically solving for $\{\tilde{P}_\mu\}_{\mu>0}$. Now we can use the assumption for concluding that $\lim_{n\to\infty} d(x(\mu_n), T^{-1}(0)) = 0$ (note that this implication holds without requiring reflexivity of the space). Assume now that $\{P_\mu\}_{\mu>0}$ is well posed for (P), and take sequences $\mu_n \downarrow 0$ and $\{\tilde{x}(\mu_n)\}_{n\in\mathbb{N}}$ such that

$$\lim_n d(0, E_{\mu_n}(\mu_n, \tilde{x}(\mu_n))) = 0. \tag{5.103}$$

We must show that

$$\lim_{n\to\infty} d(\tilde{x}(\mu_n), T^{-1}(0)) = 0. \tag{5.104}$$

By (5.103), there exists a sequence $\varepsilon_n \downarrow 0$ such that $d(0, E_{\mu_n}(\mu_n, \tilde{x}(\mu_n))) < \varepsilon_n$ for all n. Take $v^n \in E_{\mu_n}(\mu_n, \tilde{x}(\mu_n))$ with $\|v^n\| < \varepsilon_n$ for all n. By Theorem 5.3.15, for every fixed n there exists $(z^n, u^n) \in \mathrm{Gph}(T_{\mu_n})$ such that $\|v^n - u^n\| < \sqrt{\mu_n}$ and $\|\tilde{x}(\mu_n) - z^n\| < \sqrt{\mu_n}$. Hence

$$\lim_n d(0, T_{\mu_n}(z^n)) \leq \lim_n \|u^n\| \leq \lim_n [\|v^n - u^n\| + \|v^n\|]$$

$$\leq \lim_n [\sqrt{\mu_n} + \varepsilon_n] = 0.$$

This implies that $\{z^n\}_{n\in\mathbb{N}}$ is asymptotically solving for $\{P_\mu\}_{\mu>0}$. Now we can use the assumption for concluding that $\lim_{n\to\infty} d(z^n, T^{-1}(0)) = 0$. Using the latter fact and the definition of $\{z^n\}$ we get

$$\lim_n d(\tilde{x}(\mu_n), T^{-1}(0)) \leq \lim_n \left[\|\tilde{x}(\mu_n) - z^n\| + d(z^n, T^{-1}(0))\right]$$

$$\leq \lim_n \left[\sqrt{\mu_n} + d(z^n, T^{-1}(0))\right] = 0.$$

Hence (5.104) is established and the proof is complete. $\square$

Exercises

5.1. Prove Lemma 5.2.1.

5.2. Let $T : X \rightrightarrows X^*$ a point-to-set mapping. Prove that for all $\alpha, \varepsilon \geq 0$

(i) $\alpha\left(T^e(\varepsilon, x)\right) \subset (\alpha T)^e(\alpha\varepsilon, x), \forall x \in X$.

(ii) If $\alpha \in [0,1]$ and $\varepsilon_1, \varepsilon_2 \geq 0$, then

$$\alpha T_1^e(\varepsilon_1, x) + (1 - \alpha) T_2^e(\varepsilon_2, x) \subset (\alpha T_1 + (1 - \alpha)T_2)^e(\alpha\varepsilon_1 + (1 - \alpha)\varepsilon_2, x).$$

5.3. Let $f : X \rightarrow \mathbb{R} \cup \{\infty\}$ be a proper convex function. Assume that $x \in (\mathrm{Dom}(f))^o$ is not a minimizer of f. Let $v = \mathrm{argmin}_{u \in \partial f(x)} \|u\|$. Prove that $f(x - tv) < f(x)$ for all t in some interval $(0, \delta)$ (in words, the direction opposite to the minimum norm subgradient at a point x provides a descent direction for a convex function at x). Hint: use the facts that $\partial f(x)$ is closed and convex, and that the directional derivative of f at x in a given direction z is equal to $\sigma_{\partial f(x)}(z)$, with σ as in Definition 3.4.19.

5.4. Prove all the assertions in Example 5.2.5. Generalize item (iv) of that example to $X = \mathbb{R}^n$. Namely, take $f(x) = -\alpha \sum_{j=1}^n \log x_j$ with $\alpha > 0$. Prove that $0 \in \partial f^e(\varepsilon, x)$ for all $x > 0$ and $\varepsilon \geq \alpha n$, and $0 \notin \partial_\varepsilon f(x)$ for any $x > 0$ and any $\varepsilon > 0$. Therefore, if $\varepsilon \geq \alpha n$, $\partial f^e(\varepsilon, x) \not\subset \partial_{\varepsilon'} f(x)$ for every $x > 0$ and every $\varepsilon' > 0$.

5.5. Given $T : X \rightrightarrows X^*$, let T_c be the multifunction given by $T_c(x) = T(x) + c$. Prove that $T_c^e(\varepsilon, \cdot) = T^e(\varepsilon, \cdot) + c$ for all $\varepsilon > 0$. The converse of this fact is also true if T is maximal monotone. Prove in this case that if the above equality holds for all $\varepsilon > 0$, then $T_c(x) = T(x) + c$. Hint: use Lemma 5.2.1(e)–(f).

5.6. Prove Theorem 5.2.3.

5.7. Prove that E as in Definition 5.4.1 satisfies condition (E_3) if and only if it satisfies the same condition for m elements in the graph. In other words, fix $\{\alpha_i\}_{i=1,\ldots,m}$ such that $\alpha_i \geq 0$ and $\sum_{i=1}^m \alpha_i = 1$. For an arbitrary set of m triplets on the graph of E,

$$(\varepsilon_i, z^i, w^i) \quad (1 \leq i \leq m),$$

define

$$\hat{x} := \sum_{i=1}^m \alpha_i z^i$$

$$\hat{s} := \sum_{i=1}^m \alpha_i w^i$$

$$\hat{\varepsilon} := \sum_{i=1}^m \alpha_i \varepsilon_i + \sum_{i=1}^m \alpha_i \langle z^i - \hat{x}, w^i - \hat{s} \rangle.$$

Then $\hat{\varepsilon} \geq 0$ and $\hat{s} \in E(\hat{\varepsilon}, \hat{x})$.

5.8. Prove that the ε-subdifferential satisfies (E_3).

5.9. Using the notation of Theorem 5.3.8, prove that this theorem can be expressed as

$$v \in \breve{\partial} f(0, x) \Longleftrightarrow v \in \breve{\partial} f\left([f(y) - f(x) - \langle v, y - x \rangle], y\right) \quad \text{for all } y \in X.$$

5.10. Prove items (a) and (b) of Lemma 5.4.5.

5.11. Assume that X is reflexive. With the notation of Lemma 5.4.5(b), prove that

(i) $E = E^{**}$.

(ii) If $E_1 \subset E_2$, then $E_1^* \subset E_2^*$.

(iii) $(T^e)^* = (T^{-1})^e$.

(iv) $(T^{se})^* = (T^{-1})^{se}$.

5.12. Take the quadratic function q used in the proof of Proposition 5.4.10. Prove that

$$q(\tilde{w}) = \alpha q(w_1) + (1 - \alpha)q(w_2) - \alpha(1 - \alpha)q(w_1 - w_2),$$

where $\tilde{w} := \alpha w_1 + (1 - \alpha)w_2$.

5.13. Let $T : \mathbb{R}^2 \to \mathbb{R}^2$ be the $\pi/2$ rotation. Prove directly (i.e., without invoking the results of Section 5.4.3) that

(a) T is maximal monotone.

(b) $T^e(\varepsilon, x) = T(x)$ for all $\varepsilon \geq 0$ and all $x \in \mathbb{R}^2$.

5.14. Take T and C as in Section 5.5 and assume also that $\mathrm{ri}(D(T)) \cap \mathrm{ri}(C) \neq \emptyset$. Consider h as in (5.67) and fix $\varepsilon \geq 0$. Prove that the following statements are equivalent.

(i) $x \in D(T) \cap C$ verifies $h(x) \leq \varepsilon$.

(ii) $0 \in (T + N_C)^e(\varepsilon, x)$.

5.7 Historical notes

The notion of the ε-subdifferential, which originated the concept of enlargement of a maximal monotone operator, was introduced in [41]. In [225], a so-called ε-*monotone* operator was defined, which is to some extent related to the T^e enlargement, but without presenting any idea of enlargement of a given operator. In [144], the set $\partial f^e(\varepsilon, x)$ appears explicitly for the first time, as well as the inclusion $\partial_\varepsilon f(x) \subset \partial f^e(\varepsilon, x)$. Lipschitz continuity of the ε-subdifferential for a fixed positive ε was proved first by Nurminskii in [164]. Later on Hiriart-Urruty established Lipschitz continuity of $\partial_\varepsilon f(x)$ as a function of both x and ε (with $\varepsilon \in (0, \infty)$), in [93]. The transportation formula for the ε-subdifferential appeared in Lemaréchal's thesis [131].

The enlargement T^e for an arbitrary maximal monotone T in a finite-dimensional space was introduced in [46], where it was used to formulate an inexact version of the proximal point method. The notion was extended to Hilbert spaces in [47] and [48], and subsequently to Banach spaces in [50]. Theorems 5.3.4, 5.3.6, 5.3.15, and 5.4.16 were proved in [47] for the case of Hilbert spaces and in [50] for the case of Banach spaces. A result closely related to Theorem 5.3.15 had been proved earlier in [219].

The family of enlargements denoted $\mathbb{E}(T)$ was introduced by Svaiter in [216], where Theorem 5.4.2 was proved, and further studied in [52]. The results in Sections

5.4.1 and 5.4.3, related to the case in which T cannot be enlarged within the family $\mathbb{E}(T)$, have been taken from [45].

Enlargements related to the ones discussed in this book have also been proposed for nonmonotone operators; see [196] and [229].

The extragradient algorithm for finding zeroes of point-to-point monotone operators was introduced in [123], and improved upon in many articles (see, e.g., [98, 117, 140] and [107]). The extragradient-like method for the point-to-set case, called EGA, was presented in [104]. Algorithm CAS appeared in [47] and Algorithm IBS in [48].

The notion of the extended sum of monotone operators, studied in Section 5.6.1, started with the work by Attouch, Baillon, and Théra [10], and was developed in [175] and [176].

The classical notion of well-posedness is due to Tykhonov (see [221]). The concept of well-posedness with respect to perturbations has been studied by Zolezzi in [237] and [238]. The use of the ε-subdifferential for defining families of well-posed problems with respect to perturbations originates in [130]. The extension to general enlargements, as in Section 5.6.3, is new.

Chapter 6

Recent Topics in Proximal Theory

6.1 The proximal point method

The proximal point method was presented in [192] as a procedure for finding zeroes of monotone operators in Hilbert spaces in the following way. Given a Hilbert space H and a maximal monotone point-to-set operator $T : H \rightrightarrows H$, take a bounded sequence of regularization parameters $\{\lambda_k\} \subset \mathbb{R}_{++}$ and any initial $x^0 \in H$, and then, given x^k, define x^{k+1} as the only $x \in H$ such that

$$\lambda_k(x^k - x) \in T(x). \tag{6.1}$$

The essential fact that x^{k+1} exists and is unique is a consequence of the fundamental result due to Minty, which states that $T + \lambda I$ is onto if and only if T is maximal monotone (cf. Corollary 4.4.7 for the case of Hilbert spaces). This procedure can be seen as a dynamic regularization of the (possibly ill-conditioned) operator T. For instance, for $H = \mathbb{R}^n$ and T point-to-point and affine, say $T(x) = Ax - b$, with $A \in \mathbb{R}^{n \times n}$ positive semidefinite and $b \in \mathbb{R}^n$, (6.1) demands solution of the linear system $(A + \lambda_k I)x = b + \lambda_k x^k$, whereas the original problem of finding a zero of T demands solution of the system $Ax = b$. If A is singular, this latter system is ill-conditioned, and the condition number of $A + \lambda_k I$ becomes arbitrarily close to 1 for large enough λ_k. The interesting feature of iteration (6.1) is that it is not necessary to drive the regularization parameters λ_k to 0 in order to find the zeroes of T; in fact, it was proved in [192] that when T has zeroes the sequence $\{x^k\}$ defined by (6.1) is weakly convergent to a zero of T, independent of the choice of the bounded sequence $\{\lambda_k\}$.

A very significant particular case of the problem of finding zeroes of maximal monotone operators, is the *convex optimization problem*, namely minimizing a convex $\phi : \mathbb{R}^n \to \mathbb{R}$ over a closed convex set $C \subset \mathbb{R}^n$, or equivalently, minimizing $\phi + \delta_C(x) : \mathbb{R}^n \to \mathbb{R} \cup \{+\infty\}$, where δ_C is the indicator function associated with the set C given in Definition 3.1.3; that is,

$$\delta_C(x) = \begin{cases} 0 & \text{if } x \in C, \\ +\infty & \text{otherwise.} \end{cases}$$

As we saw in Proposition 3.6.2(i), the subdifferential of δ_C is the normality operator given in Definition 3.6.1. Hence we can write $\partial\phi(x) + \partial\delta_C(x) = \partial\phi(x) + N_C(x)$ for every $x \in C \cap \text{Dom}(\phi)$. Now, assume that $\partial(\phi + \delta_C)(x) = \partial\phi(x) + \partial\delta_C(x)$ for every $x \in C \cap \text{Dom}(\phi)$ (for instance, if the constraint qualification (CQ) holds for ϕ and δ_C). Then the problem of minimizing ϕ on the set C can be recast in terms of maximal monotone operators as that of finding a zero of the operator $T := \partial\phi + N_C$, where N_C is the normality operator defined in (3.39). If we apply the proximal point method (6.1) with this T, the resulting iteration can be rewritten in terms of the original data ϕ, C as

$$x^{k+1} = \text{argmin}_{x \in C}\left\{\phi(x) + \frac{\lambda_k}{2}\|x - x^k\|^2\right\}. \tag{6.2}$$

We remark that the proximal point method is a conceptual scheme, rather than a specific computational procedure, where a possibly ill-conditioned problem is replaced by a sequence of subproblems of a similar type, but better conditioned. As such, the computational effectiveness of the method (e.g., vis-á-vis other algorithms) depends on the specific procedure chosen for the computational solution of the subproblems. We refrain from discussing possible choices for this procedure inasmuch as our goal is limited to the extension of the method to Banach spaces under some variants that allow for inexact solutions of the subproblems; that is, of (6.1), with rather generous bounds on the error.

If we have a Banach space X, instead of H, in which case we must take $T : X \rightrightarrows X^*$, where X^* is the dual of X, then an appropriate extension of (6.1) is achieved by taking x^{k+1} as the unique $x \in X$ such that

$$\lambda_k[f'(x^k) - f'(x)] \in T(x), \tag{6.3}$$

where $f : X \to \mathbb{R} \cup \{\infty\}$ is a convex and Gâteaux differentiable function satisfying certain assumptions (see H1–H4 in Section 6.2), and f' is its Gâteaux derivative. When $f(x) = \frac{1}{2}\|x\|_X^2$ and X is a Hilbert space, (6.3) reduces to (6.1). We mention parenthetically that an alternative extension of (6.1), namely $\lambda_k[f'(x^k - x)] \in T(x)$, leads to a procedure which fails to exhibit any detectable convergence properties.

The use of a general regularizing function f instead of the square of the norm is rather natural in a Banach space, where the square of the norm loses the privileged role it enjoys in Hilbert spaces. Other choices of f may lead to easier equations to be solved when implementing (6.3). For instance, it is not hard to verify that in the spaces ℓ_p or $\mathcal{L}^p$ ($1 < p < \infty$), (6.3) becomes simpler with $f(x) = \|x\|_p^p$ than with $f(x) = \|x\|_p^2$.

Under assumptions H1–H4, it was proved in [49] that if T has zeroes then the sequence $\{x^k\}$ is bounded and all its weak cluster points are zeroes of T, and that there exists a unique weak cluster point when f' is weak-to-weak* continuous. Functions satisfying H1–H4 exist in most Banach spaces. This is the case, for instance, of $f(x) = \|x\|_X^r$ with $r > 1$, where X is any uniformly smooth and uniformly convex Banach space, as we show in Section 6.2.

Up to now our description of the method demands exact solution of the subproblems; that is, the inclusion given in (6.3) is assumed to be exactly solved in order to find x^{k+1}. Clearly, an analysis of the method under such a requirement fails to cover any practical implementation, because for almost any operator T, (6.3) must be solved by a numerical iterative procedure generating a sequence converging to the solution of the inclusion, and it does not make sense to go too far with the iterations of such an inner loop, inasmuch as the exact solution of the subproblem has no intrinsic meaning in connection with a solution of the original problem; it only provides the next approximation to such a solution in the outer loop, given by (6.3). Thus, it is not surprising that inexact variants in Hilbert spaces were considered as early as in [192], where the iteration

$$e^k + \lambda_k(x^k - x^{k+1}) \in T(x^{k+1}), \qquad (6.4)$$

is analyzed, as an inexact variant of (6.3). Here e^k is the error in the kth iteration, and convergence of the sequence $\{x^k\}$ to a zero of T is ensured under a "summability" assumption on the errors; that is, assuming that $\sum_{k=0}^{\infty} \|x^{k+1} - \hat{x}^k\| < \infty$, where x^{k+1} is given by (6.4) and $\hat{x}^k$ is the unique solution of (6.1). This summability condition cannot be verified in actual implementations, and also demands that the precision in the computations increases with the iteration index k.

New error schemes, accepting constant relative errors, and introducing computationally checkable inequalities such that any vector that satisfies them can be taken as the next iterate, have been recently presented in [209, 210] for the case of quadratic f in Hilbert spaces and in [211, 51] for the case of nonquadratic f in finite-dimensional spaces (in [51], with a regularization different from the scheme given by (6.1)). In all these methods, first an approximate solution of (6.1) is found. This approximate solution is not taken as the next iterate x^{k+1}, but rather as an auxiliary point, which is used either for defining a direction that in a certain sense points toward the solution set, or for finding a hyperplane separating x^k from the solution set. Then, an additional *nonproximal* step is taken from x^k, using the direction or separating hyperplane found in the *proximal* step. Usually, the computational burden of this extra step is negligible, as compared with the proximal step.

We discuss next the error scheme in [210], which uses the enlargements of monotone operators introduced in Chapter 5, and which we extend to Banach spaces in Section 6.3.

Observe first that the inclusion in (6.1) can be rewritten, with two unknowns, namely $x, v \in H$, as

$$0 = v + \lambda_k(x - x^k), \qquad (6.5)$$

$$v \in T(x). \qquad (6.6)$$

In [210], the proximal step consists of finding an approximate solution of (6.3), defined as a pair $(y^k, v^k) \in H \times H$ such that

$$e^k = v^k + \lambda_k(y^k - x^k), \qquad (6.7)$$

$$v^k \in T^e(\varepsilon_k, y^k), \qquad (6.8)$$

where $e^k \in H$ is the error in solving (6.5), T^e is the enlargement of T presented in Section 5.2, and $\varepsilon_k \in \mathbb{R}_+$ allows for an additional error in solving the inclusion

(6.6). Then the next iterate is obtained in the nonproximal step, given by

$$x^{k+1} = x^k - \lambda_k^{-1} v^k. \tag{6.9}$$

The bound for the errors e^k and ε_k is given by

$$\|e^k\|^2 + 2\lambda_k \varepsilon_k \leq \sigma \lambda_k^2 \|y^k - x^k\|^2, \tag{6.10}$$

where $\sigma \in [0, 1)$ is a relative error tolerance.

We emphasize two important features of the approximation scheme presented above: no summability of the errors is required, and admission of a pair (y^k, v^k) as an appropriate approximate solution of (6.1) reduces to computing e^k through (6.7), finding ε_k so that (6.8) holds, and finally checking that e^k and ε_k satisfy (6.10) for the given tolerance σ.

It has been proved in [210] that a sequence $\{x^k\}$ defined by (6.7)–(6.10) is weakly convergent to a zero of T whenever T has zeroes. Convergence is strong, and the convergence rate is indeed linear, when T^{-1} is Lipschitz continuous around zero (this assumption entails that T has indeed a unique zero).

When $e^k = 0$ and $\varepsilon_k = 0$, which occurs only when y^k is the exact solution of (6.1), (6.10) holds with strict inequality for any $y^k \neq x^k$, so that this inequality can be seen, under some assumptions on T, as an Armijo-type rule which will be satisfied by any vector close enough to the exact solution of the subproblem (see Proposition 6.3.3 below).

The constant σ can be interpreted as a relative error tolerance, measuring the ratio of a measure of proximity between the candidate point y^k and the exact solution, and a measure of proximity between the same candidate point and the previous iterate.

Note that if $\sigma = 0$, then $e^k = 0$ and $\varepsilon_k = 0$, so that (6.7) and (6.9) imply that $x^{k+1} = y^k$ and (y^k, v^k) satisfy (6.5)–(6.6), and hence x^{k+1} is indeed the exact solution of (6.1). If we take $\varepsilon_k = 0$; that is, we do not allow error in the inclusion (6.6), then the algorithm reduces essentially to the one studied in [209]. On the other hand, if we take $e^k = 0$, and replace (6.10) by $\sum_{k=0}^{\infty} \varepsilon_k < \infty$, we obtain the approximate algorithm studied in [46]. This option (errors allowed in the inclusion (6.6), but not in Equation (6.5)), is possibly too restrictive: as seen in Corollary 5.4.11, for an affine operator $T(x) = Ax + b$ in $\mathbb{R}^n$, with skew-symmetric A, it holds that $T^\varepsilon = T$ for all ε, in which case (6.6) and (6.8) are the same, and the inexact algorithm becomes exact.

It is worthwhile to mention that the method in [210] indeed allows for larger errors than the approximation scheme in [192], and that the nonproximal step (6.9) is essential. There exist examples for which the method without (6.9) generates a divergent sequence, even taking $\varepsilon_k = 0$ for all k, meaning that the errors are not summable. An example of this situation with a linear skew-symmetric operator T can be found in [210]; another one, in which T is the subdifferential of a convex function defined in ℓ_2, appeared in [86].

We extend the algorithm given by (6.7)–(6.10) to reflexive Banach spaces, providing a complete convergence analysis that encompasses both the method in [210] and the one in [49]. First we must discuss appropriate regularizing functions in Banach spaces, which substitute for the square of the norm.

6.2 Existence of regularizing functions

In this section, X is a reflexive Banach space with the norm denoted $\|\cdot\|$ or $\|\cdot\|_X$, X^* its topological dual (with the operator norm denoted $\|\cdot\|_*$ or $\|\cdot\|_{X^*}$), and $\langle\cdot,\cdot\rangle$ denotes the duality pairing in $X^* \times X$ (i.e., $\langle\phi,x\rangle = \phi(x)$ for all $\phi \in X^*$ and all $x \in X$). Given Banach spaces X_1, X_2, we consider the Banach space $X_1 \times X_2$ with the product norm $\|(x,y)\|_{X_1 \times X_2} = \|x\|_{X_1} + \|y\|_{X_2}$.

We denote $\mathcal{F}$ the family of functions $f : X \to \mathbb{R} \cup \{+\infty\}$ that are strictly convex, lower-semicontinuous, and Gâteaux differentiable in the interior of their domain. For $f \in \mathcal{F}$, f' denotes its Gâteaux derivative.

We need in the sequel the *modulus of total convexity* $\nu_f : (\text{Dom}(f))^o \times \mathbb{R}_+ \to \mathbb{R}$ defined as

$$\nu_f(x,t) = \inf_{y \in U(x,t)} D_f(y,x), \tag{6.11}$$

where $U(x,t) = \{y \in X : \|y - x\| = t\}$, and $D_f : \text{Dom}(f) \times (\text{Dom}(f))^o \to \mathbb{R}$, given by

$$D_f(x,y) = f(x) - f(y) - \langle f'(y), x - y \rangle, \tag{6.12}$$

is the *Bregman distance* induced by f.

We use the following result on the modulus of total convexity, taken from [54].

Proposition 6.2.1. *If x belongs to $(\text{Dom}(f))^o$ then*

(i) *The domain of $\nu_f(x,\cdot)$ is an interval $[0,\tau_f(x))$ or $[0,\tau_f(x)]$ with $\tau_f(x) \in (0,+\infty]$.*

(ii) *$\nu_f(x,\eta t) \geq \eta\nu_f(x,t)$ for all $\eta \geq 1$ and all $t \geq 0$.*

(iii) *$\nu_f(x,\cdot)$ is nondecreasing for all $x \in (\text{Dom}(f))^o$.*

Proof.

(i) Take $x \in (\text{Dom}(f))^o$ and $t \geq 0$ such that $\nu_f(x,t) < +\infty$. By definition of ν_f, there exists a point $y_t \in \text{Dom}(f)$ such that $\|y_t - x\| = t$. Because $\text{Dom}(f)$ is convex, the segment $[x, y_t]$ is contained in $\text{Dom}(f)$. Thus, for all $s \in [0,t]$ there exists a point $y_s \in \text{Dom}(f)$ such that $\|y_s - x\| = s$. Consequently, $(0,t) \subset \text{Dom}(\nu_f(x,\cdot))$ whenever $\nu_f(x,t) < +\infty$. The conclusion follows.

(ii) If $\eta = 1$, or $\eta = 0$, or $\nu_f(x,\eta t) = +\infty$, then the result is obvious. Otherwise, let ε be a positive real number. By (6.11), there exists $u \in \text{Dom}(f)$ such that $\|u - x\| = \eta t$ and

$$\nu_f(x,\eta t) + \varepsilon > D_f(u,x) = f(u) - f(x) - \langle f'(x), u - x \rangle. \tag{6.13}$$

For $\alpha \in (0,1)$, let $u_\alpha := \alpha u + (1-\alpha)x$. Let $\beta = \eta^{-1}$ and observe that $\|u_\beta - x\| = \beta\|u - x\| = t$. Note that, for all $\alpha \in (0,1)$,

$$\frac{\alpha}{\beta}u_\beta + \left(1 - \frac{\alpha}{\beta}\right)x = \frac{\alpha}{\beta}[\beta u + (1-\beta)x] + \left(1 - \frac{\alpha}{\beta}\right)x = u_\alpha. \tag{6.14}$$

By convexity of f, the function $\phi_f(x, y, t) := t^{-1}(f(x+ty) - f(x))$ is nondecreasing in t for $t \in (0, +\infty)$. It follows that $\phi_f(x, u - x, \cdot)$ is also nondecreasing in t for $t \in (0, 1)$. Therefore, in view of (6.11) and (6.13), we have

$$\nu_f(x, \eta t) + \varepsilon > f(u) - f(x) - \frac{f(x + \alpha(u - x)) - f(x)}{\alpha},$$

for all $\alpha \in (0, 1)$. As a consequence,

$$
\begin{aligned}
\nu_f(x, \eta t) + \varepsilon \;\; &> \;\; \frac{\alpha f(u) + (1 - \alpha)f(x) - f(x + \alpha(u - x))}{\alpha} \\
&= \;\; \frac{\alpha f(u) + (1 - \alpha)f(x) - \frac{\alpha}{\beta}f(u_\beta) - (1 - \frac{\alpha}{\beta})f(x)}{\alpha} \\
&\quad + \frac{\frac{\alpha}{\beta}f(u_\beta) + (1 - \frac{\alpha}{\beta})f(x) - f(u_\alpha)}{\alpha} \\
&= \;\; \frac{\beta f(u) + (1 - \beta)f(x) - f(u_\beta)}{\beta} \\
&\quad + \frac{\frac{\alpha}{\beta}f(u_\beta) + (1 - \frac{\alpha}{\beta})f(x) - f(\frac{\alpha}{\beta}u_\beta + (1 - \frac{\alpha}{\beta})x)}{\alpha},
\end{aligned}
\tag{6.15}
$$

using (6.14) in the last equality. The first term in the rightmost expression of (6.15) is nonnegative by convexity of f. Thus,

$$
\begin{aligned}
\nu_f(x, \eta t) + \varepsilon &> \frac{\frac{\alpha}{\beta}f(u_\beta) + (1 - \frac{\alpha}{\beta})f(x) - f(\frac{\alpha}{\beta}u_\beta + (1 - \frac{\alpha}{\beta})x)}{\alpha} \\
&= \frac{1}{\beta}\left[f(u_\beta) - f(x) - \frac{f(x + \frac{\alpha}{\beta}(u_\beta - x)) - f(x)}{\frac{\alpha}{\beta}} \right].
\end{aligned}
\tag{6.16}
$$

Letting $\alpha \to 0+$ in (6.16), and taking into account (6.12) and the definition of f', we conclude that

$$\nu_f(x, \eta t) + \varepsilon > \eta D_f(u_\beta, x) \geq \eta \nu_f(x, t).$$

Because ε is an arbitrary positive real number, the result follows.

(iii) Take $t_1 > t_2 > 0$, so that $\theta := t_1 / t_2 > 1$. Then

$$\nu_f(x, t_1) = \nu_f(x, \theta t_2) \geq \theta \nu_f(x, t_2) \geq \nu_f(x, t_2),$$

using (ii) in the first inequality. $\quad\square$

The convergence results for our algorithm use, as an auxiliary device, regularizing functions $f \in \mathcal{F}$ that satisfy some or all of the following assumptions.

H1: The level sets of $D_f(x, \cdot)$ are bounded for all $x \in \mathrm{Dom}(f)$.

H2: $\inf_{x \in C} \nu_f(x, t) > 0$, for every bounded set $C \subset (\mathrm{Dom}(f))^o$ and all $t \in \mathbb{R}_{++}$.

H3: f' is uniformly continuous on bounded subsets of $(\mathrm{Dom}(f))^o$.

H4: For all $y \in X^*$, there exists $x \in (\mathrm{Dom}(f))^o$ such that $f'(x) = y$.

H5: $f' : (\mathrm{Dom}(f))^o \to X^*$ is weak-to-weak* continuous.

We use later on the following equivalent formulation of H2.

Proposition 6.2.2. *For $f \in \mathcal{F}$, the following conditions are equivalent.*

(i) f satisfies H2.

(ii) If $\{x^k\}$, $\{y^k\} \subset X$ are sequences such that $\lim_{k\to\infty} D_f(y^k, x^k) = 0$ and $\{x^k\}$ is bounded, then $\lim_{k\to\infty} \|x^k - y^k\| = 0$.

Proof.
(i) $\Rightarrow$ (ii). Suppose, by contradiction, that there exist two sequences $\{x^k\}$, $\{y^k\} \subset X$ such that $\{x^k\}$ is bounded and $\lim_{k\to\infty} D_f(y^k, x^k) = 0$, but $\{\|y^k - x^k\|\}$ does not converge to zero. Then, there exist $\alpha > 0$ and subsequences $\{x^{i_k}\}$ $\{y^{i_k}\}$ such that $\alpha \le \|y^{i_k} - x^{i_k}\|$ for all k. Because $E := \{x^k\}$ is bounded, we have, for all k,

$$D_f(y^{i_k}, x^{i_k}) \ge \nu_f\left(x^{i_k}, \|y^{i_k} - x^{i_k}\|\right) \ge \nu_f(x^{i_k}, \alpha) \ge \inf_{x \in E} \nu_f(x, \alpha),$$

using Proposition 6.2.1(iii) in the first inequality. Because $\lim_{k\to\infty} D_f(x^k, y^k) = 0$ by hypothesis, we conclude that $\inf_{x \in E} \nu_f(x, \alpha) = 0$, contradicting H2.

(ii) $\Rightarrow$ (i) Suppose, by contradiction, that there exists a nonempty bounded subset $E \subset X$ such that $\inf_{x \in E} \nu_f(x, t) = 0$ for some $t > 0$. Then, there exists a sequence $\{x^k\} \subset E$ such that, for all k,

$$\frac{1}{k} > \nu_f(x^k, t) = \inf \left\{ D_f(y, x^k) : \|y - x^k\| = t \right\}.$$

Therefore, there exists a sequence $\{y^k\} \subset E$ such that $\|y^k - x^k\| = t$ and $D_f(y^k, x^k) < 1/k$. The sequence $\{x^k\}$ is bounded because it is contained in E. Also, we have $\lim_{k\to\infty} D_f(y^k, x^k) = 0$. Hence, $0 < t = \lim_{k\to\infty} \|y^k - x^k\| = 0$, which is a contradiction. $\quad\Box$

It is important to exhibit functions that satisfy these properties in as large a class of Banach spaces as possible, and we focus our attention on functions of the form $f_r(x) = (1/r)\|x\|_X^r$ with $r > 1$, for which we introduce next some properties of duality mappings in Banach spaces.

We recall that X is *uniformly smooth* if and only if $\lim_{t\to 0} t^{-1} \rho_X(t) = 0$, where the *modulus of smoothness* ρ_X is defined as

$$\rho_X(t) = \sup\{\|(x+y)/2\| + \|(x-y)/2\| - 1 : \|x\| = 1, \|y\| = t\}.$$

A *weight function* φ is a continuous and strictly increasing function $\varphi : \mathbb{R}_+ \to \mathbb{R}_+$ such that $\varphi(0) = 0$ and $\lim_{t\to\infty} \varphi(t) = \infty$. The *duality mapping of weight φ* is the mapping $J_\varphi : X \rightrightarrows X^*$, defined by $J_\varphi x = \{x^* \in X^* | \langle x^*, x \rangle = \|x^*\|_* \|x\|, \|x^*\|_* = \varphi(\|x\|)\}$ for all $x \in X$. When $\varphi(t) = t$, J_φ is the *normalized* duality mapping denoted J in Definition 4.4.2. If J_φ is the duality mapping of weight φ in a reflexive Banach space X, then the following properties hold.

J1: $J_\varphi x = \partial \psi(\|x\|)$ for all $x \in X$, where $\psi(t) = \int_0^t \varphi(s)ds$.

J2: If there exists φ^{-1}, the inverse function of the weight φ, then φ^{-1} is a weight function too and $J^*_{\varphi^{-1}}$, the duality mapping of weight φ^{-1} on X^*, is such that $x^* \in J_\varphi x$ if and only if $x \in J^*_{\varphi^{-1}} x^*$.

J3: If J_1 and J_2 are duality maps of weights φ_1 and φ_2, respectively, then, for all $x \in X$, $\varphi_2(\|x\|)J_1 x = \varphi_1(\|x\|)J_2 x$.

J4: If X is uniformly smooth, then J is uniformly continuous on bounded subsets of X.

Properties J1–J3 have been proved in [65] and property J4 in [235]. We prove next that J4 holds for duality mappings with an arbitrary weight φ.

Proposition 6.2.3. *Let X be a uniformly smooth Banach space and J_φ a duality mapping of weight φ on X. Then J_φ is uniformly continuous on bounded subsets of X (considering the strong topology both in X and X^*).*

Proof. If the result does not hold, then we can find a bounded subset A of X such that J_φ is not uniformly continuous on A; that is, there exist $\epsilon > 0$, and sequences $\{x^k\}$, $\{y^k\} \subset A$ such that $\lim_{k\to\infty} \|x^k - y^k\| = 0$ and $\|J_\varphi x^k - J_\varphi y^k\|_{X^*} \geq \epsilon$ for all k.

Note, first, that if $\lim_{k\to\infty} \|x^k\| = 0$, then $\lim_{k\to\infty} \|y^k\| = 0$ too, and from the definition of J_φ it follows that $\lim_{k\to\infty} \|J_\varphi x^k - J_\varphi y^k\|_* = 0$, which is a contradiction. Thus, we can assume, without loss of generality, that $0 < m \leq \|x^k\| \leq M$ and that $\|y^k\| \leq M$, for some $m, M \in \mathbb{R}$. Let $\bar{\varphi}(t) = \varphi(t)/t$, and $\bar{M} = \max_{t \in [m,M]} \bar{\varphi}(t)$. Using J3 for $\varphi_1 := \varphi$ and $\varphi_2(t) = t$ and the definition of J, we have

$$
\begin{aligned}
\epsilon \ &\leq \ \|J_\varphi x^k - J_\varphi y^k\|_* = \left\| \frac{\varphi(\|x^k\|)}{\|x^k\|} J x^k - \frac{\varphi(\|y^k\|)}{\|y^k\|} J y^k \right\|_* \\
&\leq \ \bar{\varphi}(\|x^k\|)\|J x^k - J y^k\|_* + \left| \bar{\varphi}(\|x^k\|) - \bar{\varphi}(\|y^k\|) \right| \|J y^k\|_* \\
&\leq \ \bar{M}\|J x^k - J y^k\|_* + \left| \bar{\varphi}(\|x^k\|) - \bar{\varphi}(\|y^k\|) \right| \|y^k\|.
\end{aligned}
$$

Because $\lim_{k\to\infty} \left| \bar{\varphi}(\|x^k\|) - \bar{\varphi}(\|y^k\|) \right| = 0$ by uniform continuity of $\bar{\varphi}$ in $[m, M]$, and $\{y^k\}$ is bounded, it follows that $\|J x^k - J y^k\|_* \geq \epsilon/(2\bar{M})$ for large enough k, contradicting J4. $\quad\square$

We analyze the validity of H1–H5 for $f_r(x) = (1/r)\|x\|^r_X$. We need some preliminary material. We denote $\theta_X : [0, 2] \to [0, 1]$ the *modulus of uniform convexity* of the space X, given by

$$
\theta_X(t) = \begin{cases} \inf \left\{ 1 - \frac{1}{2}\|x + y\| : \|x\| = \|y\| = 1, \|x - y\| \geq t \right\} & \text{if } t > 0, \\ 0 & \text{if } t = 0. \end{cases}
$$

Recall that X is called *uniformly convex* if $\theta_X(t) > 0$ for all $t > 0$, and is called *smooth* if the function $\|\cdot\|$ is differentiable at each point $x \neq 0$.

In the following result we do not require that X be smooth, in which case powers of the norm might fail to be differentiable. Thus, we need a definition of the Bregman distance D_f without assuming differentiability of f.

Given a convex $f : X \to \mathbb{R} \cup \{+\infty\}$, we define the Bregman distance D_f : $\mathrm{Dom}(f) \times (\mathrm{Dom}(f))^o \to \mathbb{R}$, as

$$D_f(x, y) = f(x) - f(y) - \inf_{v \in \partial f(y)} \langle v, x - y \rangle, \qquad (6.17)$$

where $\partial f(y)$ denotes the subdifferential of f at y. Note that this concept reduces to the previously defined Bregman distance when f is differentiable.

We start with the following proposition, taken from [55].

Proposition 6.2.4. *If X is uniformly convex, then $f_r(x) = r^{-1}\|x\|^r$ satisfies H1 and H2 for all $r > 1$.*

Proof. We prove first that f_r satisfies H2. Note that ∂f_r is precisely the duality mapping of weight t^{r-1} on the space X (see e.g., page 25 in [65]).

We need now the following inequality, established in Theorem 1 of [230]: if X is uniformly convex, then there exists $\Gamma > 0$ such that

$$\langle u - v, x - y \rangle \geq \Gamma \left[\max\{\|x\|, \|y\|\}\right]^r \theta_X \left(\frac{\|x - y\|}{2\max\{\|x\|, \|y\|\}}\right), \qquad (6.18)$$

for all $x, y \in X$ with $\|x\| + \|y\| \neq 0$, all $u \in \partial f_r(x)$, and all $v \in \partial f_r(y)$.

Fix $z \in X$ and $t \in (0, \infty)$, take $x \in X$ such that $\|x - z\| = t$, and let $y = x - z$. Then $\|y\| = t$ and

$$D_{f_r}(x, z) = D_{f_r}(y + z, z) = \frac{\|z + y\|^r - \|z\|^r}{r} - \inf\{\langle u, y \rangle : u \in \partial f_r(z)\}, \qquad (6.19)$$

by (6.17). Define $\varphi : [0, \infty) \to [0, \infty)$ as $\varphi(\tau) = \|z + \tau y\|^r$. Then, we have

$$\|z + y\|^r - \|z\|^r = \varphi(1) - \varphi(0). \qquad (6.20)$$

We now make the following claim.

Claim. If, for each $\tau \in [0, 1]$, we choose $u(\tau) \in \partial f_r(z + \tau y)$, then the following integral exists and satisfies

$$\int_0^1 \langle u(\tau), y \rangle d\tau = \varphi(1) - \varphi(0). \qquad (6.21)$$

We proceed to establish the claim. Observe first that, if $g : X \to \mathbb{R}$ is a convex and continuous function, then the function $\psi : \mathbb{R} \to \mathbb{R}$ defined by $\psi(\tau) = g(u + \tau v)$ is also convex and continuous for all $u, v \in X$. In view of Theorem 3.4.22, ψ is locally Lipschitz and, according to Rademacher's theorem (e.g., page 11 in [171]), almost everywhere differentiable. Consequently, if u is a selector of the point-to-set mapping $W(\tau) = \partial g(u + \tau v)$, it follows from the definition of ψ that $\psi'(\tau) = \langle u(\tau), v \rangle$ for almost all $\tau \in \mathbb{R}$. Hence, for all $\alpha, \beta \in \mathbb{R}$, with $\alpha \leq \beta$, we have

$$\psi(\beta) - \psi(\alpha) = \int_\alpha^\beta \psi'(\tau) d\tau = \int_\alpha^\beta \langle u(\tau), v \rangle d\tau, \qquad (6.22)$$

for any choice of $u(\tau) \in W(\tau)$. Now, we apply (6.22) to the case of $\psi = \varphi$, $g = f_r$, $u = z$, $v = y$, $\alpha = 0$, and $\beta = 1$, and we conclude that (6.21) holds.

Now we use (6.21) in order to complete the proof of the proposition. Fix a selector u of the point-to-set mapping $R(\tau) = \partial f_r(z + \tau y)$. By (6.21) and (6.20), we have

$$\|z + y\|^r - \|z\|^r = \int_0^1 \langle u(\tau), y \rangle d\tau.$$

Therefore, for any $v \in \partial f_r(z)$, we have

$$\|z + y\|^r - \|z\|^r - r\langle v, y \rangle = r \left[\int_0^1 \langle u(\tau), y \rangle d\tau - \langle v, y \rangle \right]$$

$$= r \int_0^1 \langle u(\tau) - v, y \rangle d\tau = r \int_0^1 \frac{1}{\tau} \langle u(\tau) - v, \tau y \rangle \, d\tau. \tag{6.23}$$

Because, for any $\tau \in [0, 1]$, $u(\tau)$ belongs to $\partial f_r(z + \tau y)$, and $\tau y = (z + \tau y) - z$, we conclude from (6.23) and (6.18) that

$$\|z + y\|^r - \|z\|^r - r\langle v, y \rangle \geq$$

$$r\Gamma \int_0^1 \frac{[\max\{\|z + \tau y\|, \|z\|\}]^r}{\tau} \theta_X \left(\frac{\|\tau y\|}{2 \max\{\|z + \tau y\|, \|z\|\}} \right) d\tau, \tag{6.24}$$

for all $v \in \partial f_r(z)$, where the last integral exists because θ_X is nondecreasing. Dividing both sides of (6.24) by r, taking the infimum over $v \in \partial f_r(z)$ in the right-hand side of (6.24), and using (6.17), we get

$$D_{f_r}(x, z) \geq \Gamma \int_0^1 \frac{[\max\{\|z + \tau y\|, \|z\|\}]^r}{\tau} \theta_X \left(\frac{\|\tau y\|}{2 \max\{\|z + \tau y\|, \|z\|\}} \right) d\tau. \tag{6.25}$$

Clearly, $\max\{\|z + \tau y\|, \|z\|\} \leq \|z\| + \tau\|y\|$. Using again the fact that θ_X is nondecreasing, we deduce that

$$\theta_X \left(\frac{\|\tau y\|}{2 \max\{\|z + \tau y\|, \|z\|\}} \right) \geq \theta_X \left(\frac{\|\tau y\|}{2 (\|z\| + \tau\|y\|)} \right)$$

$$= \theta_X \left(\frac{\tau t}{2 (\|z\| + \tau t)} \right), \tag{6.26}$$

because $\|y\| = t$. Now, we claim that

$$\max\{\|z + \tau y\|, \|z\|\} \geq \frac{\tau}{2}\|y\|.$$

This is clearly the case if $\|z\| \geq (\tau/2)\|y\|$. Otherwise, we have $\|z\| < (\tau/2)\|y\|$, and therefore,

$$\|z + \tau y\| \geq |\tau| \, \|y\| - \|z\| \geq \tau\|y\| - \frac{\tau}{2}\|y\| = \frac{\tau}{2}\|y\|, \tag{6.27}$$

and the result also holds. Replacing (6.27) and (6.26) in (6.25), we get, for all $x \in X$ such that $\|x - z\| = t$,

$$D_{f_r}(x, z) \geq \Gamma \left(\frac{t}{2}\right)^r \int_0^1 \tau^{r-1} \theta_X \left(\frac{\tau t}{2(\|z\| + \tau t)}\right) d\tau. \tag{6.28}$$

Taking the infimum over x in the left-hand side of (6.28) and using the definition (6.11) of the modulus of total convexity ν_{f_r}, we obtain

$$\nu_{f_r}(z, t) \geq \Gamma \left(\frac{t}{2}\right)^r \int_0^1 \tau^{r-1} \theta_X \left(\frac{\tau t}{2(\|z\| + \tau t)}\right) d\tau > 0, \tag{6.29}$$

where the rightmost inequality holds because X is uniformly convex.

Now, if $C \subset (\mathrm{Dom}(f))^\circ$ is bounded (say $\|z\| \leq \eta$ for all $z \in C$), we get from (6.29), using again that θ_X is nondecreasing,

$$\inf_{z \in C} \nu_{f_r}(z, t) \geq \Gamma \left(\frac{t}{2}\right)^r \int_0^1 \tau^{r-1} \theta_X \left(\frac{\tau t}{2(\eta + \tau t)}\right) d\tau > 0,$$

which establishes that f_r satisfies H2.

We prove now that f_r satisfies H1. We must verify that for all $\alpha \geq 0$ and for all $x \in X$, the set $S_\alpha(x) = \{y \in X : D_{f_r}(x, y) \leq \alpha\}$ is bounded. Suppose that for some $\alpha \geq 0$ and for some $x \in X$, there exists a unbounded sequence $\{y^k\} \subset S_\alpha(x)$. Note that for all k there exists some $v^k \in \partial f_r(y^k)$ such that

$$\begin{aligned}
\alpha \geq D_{f_r}(x, y^k) &= f_r(x) - f_r(y^k) - \langle v^k, x - y^k \rangle \\
&= f_r(x) - f_r(y^k) + \langle v^k, y^k \rangle - \langle v^k, x \rangle \\
&= f_r(x) - f_r(y^k) + \|y^k\|^r - \langle v^k, x \rangle \\
&\geq r^{-1}\left(\|x\|^r - \|y^k\|^r\right) + \|y^k\|^r - \|x\|\|y^k\|^{r-1} \\
&= r^{-1}\|x\|^r + \|y^k\|^{r-1}\left[(1 - r^{-1})\|y^k\| - \|x\|\right],
\end{aligned} \tag{6.30}$$

where the second equality holds because $\partial f_r(x)$ is the duality mapping of weight t^{r-1}. Because $\{y^k\}$ is unbounded and $r > 1$, letting $k \to \infty$ in the rightmost expression of (6.30) leads to a contradiction. □

We summarize our results on f_r in connection with H1–H5 in the following corollary.

Corollary 6.2.5.

(i) If X is a uniformly smooth and uniformly convex Banach space, then $f_r(x) = (1/r)\|x\|_X^r$ satisfies H1, H2, H3, and H4 for all $r > 1$.

(ii) If X is a Hilbert space, then $f_2(x) = (1/2)\|x\|^2$ satisfies H5. If $X = \ell_p$ $(1 < p < \infty)$ then $f_p(x) = (1/p)\|x\|_p^p$ satisfies H5.

Proof. By definition of smoothness of X, for all $r > 1$, f_r is Gâteaux differentiable when X is smooth, and by Proposition 6.2.4 it satisfies H1 and H2 when X is uniformly convex.

Consider the weight function $\varphi(t) = t^{r-1}$. Because $\int_0^t \varphi(s)ds = (1/r)t^r$, we get from J1 that $J_\varphi = f_r'$ when X is smooth. Proposition 6.2.3 and J1 with this weight φ easily imply that f_r satisfies H3 for all $r > 1$. Note that φ is invertible for $r > 1$, so that J_φ is onto by J2. It follows that f_r satisfies H4, and thus (i) holds. For (ii), note that in the case of a Hilbert space, f_2' is the identity, which is certainly weak-to-weak continuous. The result for f_p in ℓ_p has been proved in [43, Proposition 8.2]. $\square$

Unfortunately, it has been proved in [54] that for $X = \ell_p$ or $X = \mathcal{L}^p[\alpha, \beta]$ the function $f_r(x) = (1/r)\|x\|_p^r$ does not satisfy H5, except in the two cases considered in Corollary 6.2.5(ii). We remark that, as we show, properties H1–H4 are required for establishing existence and uniqueness of the iterates of the algorithm under consideration, and also boundedness of the generated sequences, whereas H5 is required only for uniqueness of the weak cluster points of such sequences. We mention also that the factor $1/r$ in the definition of f_r is inessential for Corollary 6.2.5, whose results trivially hold for all positive multiples of $\|\cdot\|_X^r$.

We discuss next some properties of functions satisfying some of the assumptions above. A function $f : X \to \mathbb{R}$ is said to be *totally convex* if $\nu_f(x, t) > 0$ for all $t > 0$ and all $x \in X$.

Proposition 6.2.6. *Let X_1, X_2 be real Banach spaces, and $f : X_1 \to \mathbb{R} \cup \{\infty\}$ and $h : X_2 \to \mathbb{R} \cup \{\infty\}$ two proper convex functions whose domains have a nonempty interior. Define $F : X_1 \times X_2 \to \mathbb{R} \cup \{\infty\}$ as $F(x, y) = f(x) + h(y)$. Then*

(i) *For any $z = (x, y) \in (\mathrm{Dom}(F))^o$ the domain of $\nu_F(z, \cdot)$ is an interval $[0, \tau(z))$ or $[0, \tau(z)]$ with $0 < \tau(z) = \tau_f(x) + \tau_h(y) \leq \infty$, where $\tau_f(x) = \sup\{t \mid t \in \mathrm{Dom}(\nu_f(x, \cdot))\}$ and $\tau_h(y) = \sup\{t \mid t \in \mathrm{Dom}(\nu_h(y, \cdot))\}$. Moreover $\nu_F(z, t) = \inf\{\nu_f(x, s) + \nu_h(y, t - s) \mid s \in [0, t]\}$.*

(ii) *$\nu_F(z, t) \geq \min\{\nu_f(x, t/2), \nu_h(y, t/2)\}$ for all $z = (x, y) \in (\mathrm{Dom}(F))^o$ and all $t \in \mathrm{Dom}(\nu_F(z, \cdot))$; hence, if both f and h are totally convex then F is totally convex.*

(iii) *For $i = 1, \ldots, 5$, if both f and h satisfy Hi then F also satisfies Hi.*

Proof. Take $\bar{z} = (\bar{x}, \bar{y}) \in (\mathrm{Dom}(F))^o$ and $t \in \mathbb{R}_+$. By definition of ν_F and D_F ((6.11) and (6.12)),

$$\nu_F(\bar{z}, t) = \inf\{D_F(z, \bar{z}) \mid \|z - \bar{z}\|_{X_1 \times X_2} = t\}$$

$$= \inf\{D_f(x, \bar{x}) + D_h(y, \bar{y}) \mid \|x - \bar{x}\|_{X_1} + \|y - \bar{y}\|_{X_2} = t\}. \tag{6.31}$$

By Proposition 6.2.1(i), the domain of $\nu_f(x, \cdot)$ is an interval $[0, \tau_f(x))$ or $[0, \tau_f(x)]$ with $\tau_f(x) \in (0, +\infty]$, and the same holds for the domain of $\nu_h(y, \cdot)$, with some $\tau_h(y) \in (0, +\infty]$, which proves the first statement in (i). Because

$$\nu_f(\bar{x}, \|x - \bar{x}\|_{X_1}) \leq D_f(x, \bar{x})$$

$$\nu_h(\bar{y}, \|y - \bar{y}\|_{X_2}) \le D_h(y, \bar{y}),$$

we get from (6.31) that

$$\nu_F(\bar{z}, t) \le \inf\{\nu_f(\bar{x}, \|x - \bar{x}\|_{X_1}) + \nu_h(\bar{y}, \|y - \bar{y}\|_{X_2}) \mid \|x - \bar{x}\|_{X_1} + \|y - \bar{y}\|_{X_2} = t\}$$

$$= \inf\{\nu_f(\bar{x}, s) + \nu_h(\bar{y}, t - s) \mid s \in [0, t]\}. \tag{6.32}$$

In order to establish the opposite inequality, take $s \in \mathrm{Dom}(\nu_f(\bar{x}, \cdot))$ and t such that $t - s \in \mathrm{Dom}(\nu_h(\bar{y}, \cdot))$. By (6.31), we have,

$$\nu_F(\bar{z}, t) \le \inf\{D_f(x, \bar{x}) + D_h(y, \bar{y}) \mid \|x - \bar{x}\|_{X_1} = s, \|y - \bar{y}\|_{X_2} = t - s\}$$

$$= \inf\{D_f(x, \bar{x}) \mid \|x - \bar{x}\|_{X_1} = s\} + \inf\{D_h(y, \bar{y}) \mid \|y - \bar{y}\|_{X_2} = t - s\}$$

$$= \nu_f(\bar{x}, s) + \nu_h(\bar{y}, t - s).$$

Consequently,

$$\nu_F(\bar{z}, t) \le \inf\{\nu_f(\bar{x}, s) + \nu_h(\bar{y}, t - s) \mid s \in [0, t]\}. \tag{6.33}$$

Item (i) follows from (6.32) and (6.33).

For (ii) choose any $t_1 \in \mathrm{Dom}(\nu_f(\bar{x}, \cdot))$ and $t_2 \in \mathrm{Dom}(\nu_h(\bar{y}, \cdot))$ with $t_1 + t_2 = t$. Then $t_1 \ge t/2$ or $t_2 \ge t/2$. By Proposition 6.2.1(iii), we get $\nu_f(\bar{x}, t_1) \ge \nu_f(\bar{x}, t/2)$ or $\nu_h(\bar{y}, t_2) \ge \nu_h(\bar{y}, t/2)$. Because ν_f and ν_h are nonnegative, $\nu_f(\bar{x}, t_1) + \nu_h(\bar{y}, t_2) \ge \min\{\nu_f(\bar{x}, t/2), \nu_h(\bar{y}, t/2)\}$. Thus,

$$\inf\{\nu_f(\bar{x}, t_1) + \nu_h(\bar{y}, t_2) \mid t_1 + t_2 = t\} \ge \min\{\nu_f(\bar{x}, t/2), \nu_h(\bar{y}, t/2)\}. \tag{6.34}$$

Item (ii) follows from (i) and (6.34). The proof of (iii) is elementary, using (ii) for the case of H2. $\square$

We need the following consequence of H2 in our convergence analysis.

Proposition 6.2.7. *If f belongs to $\mathcal{F}$ and satisfies H2 then, for all $\{x^k\}$, $\{y^k\} \subset (\mathrm{Dom}(f))^\circ$ such that $\{x^k\}$ or $\{y^k\}$ is bounded and $\lim_{k \to \infty} D(y^k, x^k) = 0$, it holds that $\lim_{k \to \infty} \|x^k - y^k\| = 0$.*

Proof. If $\{x^k\}$ is bounded, then the result follows from Proposition 6.2.2. Assume now that $\{y^k\}$ is bounded. We claim that $\{x^k\}$ is also bounded. Otherwise, given $\theta > 0$, there exist subsequences $\{x^{i_k}\}$, $\{y^{i_k}\}$ of $\{x^k\}$, $\{y^k\}$, respectively, such that $\|x^{i_k} - y^{i_k}\| \ge \theta$ for all k. Define

$$\tilde{x}^k = y^{i_k} + \frac{\theta}{\|x^{i_k} - y^{i_k}\|}(x^{i_k} - y^{i_k}).$$

Note that $\|\tilde{x}^k - y^{i_k}\| = \theta$ for all k. Because $\{y^k\}$ is bounded, we conclude that $\{\tilde{x}^k\}$ is also bounded. Take a bounded $C \subset X$ such that $\tilde{x}^k \in C$ for all k. Then

$$D_f(\tilde{x}^k, y^{i_k}) \ge \nu_f(\tilde{x}^k, \|\tilde{x}^k - y^{i_k}\|) = \nu_f(\tilde{x}^k, \theta) \ge \inf_{x \in C} \nu_f(x, \theta) > 0, \tag{6.35}$$

where the first inequality follows from the definition of ν_f and the last one from H2. Note that $D_f(\cdot, y)$ is convex for all $y \in (\mathrm{Dom}(f))^o$. Because $\tilde{x}^k$ is in the segment between x^{i_k} and y^{i_k}, we have

$$0 \le D_f\left(\tilde{x}^k, y^{i_k}\right) \le \max\left\{D_f\left(x^{i_k}, y^{i_k}\right), D_f\left(y^{i_k}, y^{i_k}\right)\right\}$$

$$= \max\left\{D_f\left(x^{i_k}, y^{i_k}\right), 0\right\} = D_f\left(x^{i_k}, y^{i_k}\right). \tag{6.36}$$

Because $\lim_{k\to\infty} D_f(x^k, y^k) = 0$, (6.36) implies that $\lim_{k\to\infty} D_f\left(\tilde{x}^k, y^{i_k}\right) = 0$, which contradicts (6.35). This contradiction implies that $\{x^k\}$ is bounded, in which case we are back to the previous case; that is, the result follows from Proposition 6.2.2. $\square$

6.3 An inexact proximal point method in Banach spaces

We proceed to extend Algorithm (6.7)–(6.10) to the problem of finding zeroes of a maximal monotone operator $T : X \rightrightarrows X^*$, where X is a reflexive Banach space. Our algorithm requires the following exogenous data: a relative error tolerance $\sigma \in [0,1)$, a bounded sequence of regularization parameters $\{\lambda_k\} \subset \mathbb{R}_{++}$, and an auxiliary regularizing function $f \in \mathcal{F}$ such that $D(T) \subset (\mathrm{Dom}(f))^o$. It is defined as follows.

Algorithm IPPM: Inexact Proximal Point Method

1. Choose $x^0 \in (\mathrm{Dom}(f))^o$.

2. Given x^k, find $y^k \in X$, $v^k, e^k \in X^*$, and $\varepsilon_k \in \mathbb{R}_+$ such that

$$e^k = v^k + \lambda_k[f'(y^k) - f'(x^k)], \tag{6.37}$$

$$v^k \in T^e(\varepsilon_k, y^k), \tag{6.38}$$

$$D_f\left(y^k, (f')^{-1}[f'(x^k) - \lambda_k^{-1}v^k]\right) + \lambda_k^{-1}\varepsilon_k \le \sigma D_f(y^k, x^k), \tag{6.39}$$

where T^e is the enlargement of T defined in Section 5.2 and D_f is the Bregman distance induced by f, as defined in (6.12).

3. If $y^k = x^k$, then stop. Otherwise,

$$x^{k+1} = (f')^{-1}\left[f'(x^k) - \lambda_k^{-1}v^k\right]. \tag{6.40}$$

An algorithm similar to IPPM, but without errors in the inclusion (i.e., with $\varepsilon_k = 0$ for all k) was studied in [102]. In this section, we follow basically this reference. A related method is analyzed in [202].

Note that when X is a Hilbert space and $f(x) = 1/2\|x\|^2$, then $f' : H \to H$ is the identity and $D_f(x, x') = \|x - x'\|^2$ for all $x, x' \in X$. It is easy to check that in such a case IPPM reduces to the algorithm in [210]; that is, to (6.7)–(6.10).

Before proceeding to the convergence analysis, we make some general remarks. First, the assumption that $D(T) \subset (\text{Dom}(f))^o$, also used in [49], means basically that we are not attempting any penalization effect of the proximal iteration. Second, we remark that assumptions H1–H5 on f are used in the convergence analysis of our algorithm; we do not require them in the statement of the method in order to isolate which of them are required for each specific result. In this sense, for instance, existence of $(f')^{-1}$, as required in (6.39), (6.40), is a consequence of H4 for any $f \in \mathcal{F}$. We mention that for the case of strictly convex and smooth X and $f(x) = \|x\|_X^r$, we have an explicit formula for $(f')^{-1}$ in terms of g', where $g(\cdot) = (1/s)\|\cdot\|_{X*}^s$ with $(1/s) + (1/r) = 1$, namely $(f')^{-1} = r^{1-s}g'$. In fact, H4 is sufficient for existence of the iterates, as the following results show.

Proposition 6.3.1. *Take $f \in \mathcal{F}$ such that $D(T) \subset (\text{Dom}(f))^o$. If f satisfies H4, then for all $\lambda > 0$ and all $z \in (\text{Dom}(f))^o$ there exists a unique $x \in X$ such that $\lambda f'(z) \in T(x) + \lambda f'(x)$.*

Proof. It is immediate that if f satisfies H4 then λf also satisfies H4, in which case T and λf satisfy the assumptions of Theorem 4.4.12, and hence $T + \lambda f'$ is onto; that is, there exists $x \in X$ such that $\lambda f'(z) \in T(x) + \lambda f'(x)$. Because f belongs to $\mathcal{F}$, it is strictly convex, so that, $\lambda f'$ is strictly monotone by Corollary 4.7.2(i), and, because T is monotone, the same holds for $T + \lambda f'$, implying uniqueness of x, in view of Corollary 4.7.2(ii).   □

Corollary 6.3.2. *Take $f \in \mathcal{F}$ such that $D(T) \subset (\text{Dom}(f))^o$. If f satisfies H4, then, for all k and all $\sigma \in [0,1]$, the inexact proximal subproblems of Algorithm IPPM have solutions.*

Proof. Choose $\sigma = 0$, so that the right-hand side of (6.10) vanishes. Note that strict convexity of f implies that $D_f(x,x') \geq 0$ for all x, x', and $D_f(x,x') = 0$ if and only if $x = x'$. It follows from (6.39) that ε_k must be zero and that $y^k = (f')^{-1}[f'(y^k) - \lambda_k^{-1}v^k]$, in which case, by virtue of (6.37), $e^k = 0$, and the proximal subproblem takes the form: find $y^k \in X$ such that $\lambda_k[f'(x^k) - f'(y^k)] \in T(y^k)$, which always has a unique solution by Proposition 6.3.1. Of course, the pair (y^k, v^k) with $v^k = \lambda_k[f'(x^k) - f'(y^k)]$ found in this way satisfies (6.37)–(6.40) for any $\sigma \geq 0$, and so the result holds.   □

Corollary 6.3.2 ensures that exact solutions of (6.3) satisfy the error criteria in Algorithm IPPM with $e^k = 0$, so our subproblems will certainly have solutions without any assumption on the operator T, but obviously this is not good enough for an inexact algorithm: it is desirable that the error criterion be such that points close to the exact solutions are accepted as inexact solutions. Next we show that this is the case when T is single-valued and continuous: there exists a ball around the exact solution of the subproblem such that any point in its intersection with

$D(T)$ can be taken as an inexact solution. As a consequence, if we solve the equation

$$T(x) + \lambda_k f'(x) = \lambda_k f'(x^k) \tag{6.41}$$

with any feasible algorithm that generates a sequence strongly convergent to the unique solution of (6.41), a finite number of iterations of such an algorithm will provide an appropriate next iterate for Algorithm IPPM.

Proposition 6.3.3. *Assume that $f \in \mathcal{F}$, H4 holds, $D(T) \subset (\mathrm{Dom}(f))^o$, and T is single-valued and continuous. Assume additionally that $(f')^{-1} : X^* \to X$ is continuous. Let $\{x^k\}$ be the sequence generated by Algorithm IPPM. If x^k is not a solution of (6.3), and $\sigma > 0$, then there exists a ball U_k around the exact solution of (6.41) such that any $y \in U_k \cap D(T)$ solves (6.37)–(6.39).*

Proof. Let $\bar{x}^k$ be the unique solution x of (6.41), so that

$$\bar{x}^k = (f')^{-1}\big[f'(x^k) - \lambda_k^{-1} T(\bar{x}^k)\big]. \tag{6.42}$$

Inasmuch as $x^k \neq \bar{x}^k$, because otherwise $0 \in T(x^k)$, we have $\sigma D_f(\bar{x}^k, x^k) > 0$ by total convexity of f. Let us consider, now, the function $\psi_k : D(T) \to \mathbb{R}$ defined as

$$\psi_k(y) = D_f\big(y, (f')^{-1}\big[f'(x^k) - \lambda_k^{-1} T(y)\big]\big) - \sigma D_f(y, x^k).$$

Observe that ψ_k is continuous by continuity of T and $(f')^{-1}$, and that, in view of (6.42), $\psi_k(\bar{x}^k) = -\sigma D_f(\bar{x}^k, x^k) < 0$, so that there exist $\delta_k > 0$ such that $\psi_k(y) \leq 0$ for all $y \in D(T)$ with $\|y - \bar{x}^k\| < \delta_k$. Let $U_k := \{y \in X \; : \; \|y - \bar{x}^k\| < \delta_k\}$. Then

$$D_f\big(y, (f')^{-1}\big[f'(x^k) - \lambda_k^{-1} T(y)\big]\big) \leq \sigma D_f(y, x^k)$$

for all $y \in U_k \cap D(T)$. Thus, any pair $(y, T(y))$ with $y \in U_k \cap D(T)$ satisfies (6.37)–(6.39) with $\varepsilon_k = 0$. $\square$

In connection with the hypotheses of Proposition 6.3.3, we mention that $(f')^{-1}$ is continuous (e.g., for $f(x) = \|x\|_p^r$ in ℓ_p or $\mathcal{L}^p(\Omega)$) for any $p, r > 1$. Also, continuity of T is understood in the norm topology of both X and X^*, and so it does not follow from its single-valuedness (cf. Theorem 4.2.4).

6.4   Convergence analysis of Algorithm IPPM

We proceed now to the convergence analysis of IPPM. We settle first the issue of finite termination.

Proposition 6.4.1. *If Algorithm IPPM stops after k steps, then y^k is a zero of T.*

Proof. Finite termination in Algorithm IPPM occurs only if $y^k = x^k$, in which case $D_f(y^k, x^k) = 0$, and therefore, by (6.39), with the same argument as in the proof of Corollary 6.3.2, we get $e^k = 0$ and $\varepsilon_k = 0$, which in turn implies, by (6.37), (6.38), $0 \in T(y^k)$. $\square$

From now on, we assume that the sequence $\{x^k\}$ generated by Algorithm IPPM is infinite. We need first a basic property of D_f, which is already part of the folklore of Bregman distances (see, e.g., Corollary 2.6 in [211]).

Lemma 6.4.2 (Four-point lemma). *Take $f \in \mathcal{F}$. For all $x, z \in (\mathrm{Dom}(f))^\circ$ and $y, w \in \mathrm{Dom}(f)$, it holds that*

$$D_f(w, z) = D_f(w, x) + D_f(y, z) - D_f(y, x) + \langle f'(x) - f'(z), w - y \rangle.$$

Proof. It suffices to replace each occurrence of D_f in the previous equation by the definition in terms of (6.12), and proceed to some elementary algebra. $\square$

We gather in the next proposition two formulas, obtained by rewriting the iterative steps of Algorithm IPPM, which are used in the sequel.

Proposition 6.4.3. *With the notation of Algorithm IPPM, it holds that*

(i)
$$D_f(y^k, x^{k+1}) + \lambda_k^{-1} \varepsilon_k \le \sigma D_f(y^k, x^k), \tag{6.43}$$

(ii)
$$\lambda_k^{-1} v^k = f'(x^k) - f'(x^{k+1}), \tag{6.44}$$

for all k.

Proof. We get (6.43) by substituting (6.40) in (6.39), and (6.44) by rewriting (6.40) in an implicit way. $\square$

The next result, which applies Lemma 6.4.2 to the sequence generated by IPPM, establishes the basic convergence property of the sequence $\{x^k\}$ generated by IPPM, namely its Fejér convergence with respect to D_f to the set of zeroes of T (cf. Definition 5.5.1), or in other words, the fact that the Bregman distance with respect to f from the iterates of IPPM to any zero of T is nonincreasing.

Lemma 6.4.4. *Take $f \in \mathcal{F}$ satisfying H4, such that $\mathrm{Dom}(T) \subset (\mathrm{Dom}(f))^\circ$, $\bar{x} \in T^{-1}(0)$, and x^k, y^k, v^k, λ_k, and σ as in Algorithm IPPM. Then*

$$D_f(\bar{x}, x^{k+1}) \le D_f(\bar{x}, x^k) - (1 - \sigma) D_f(y^k, x^k)$$
$$- \lambda_k^{-1}(\langle v^k, y^k - \bar{x} \rangle + \varepsilon_k) \le D_f(\bar{x}, x^k). \tag{6.45}$$

Proof. From Lemma 6.4.2, we get

$$D_f(\bar{x}, x^{k+1}) = D_f(\bar{x}, x^k) + D_f(y^k, x^{k+1}) - D_f(y^k, x^k)$$
$$+ \langle f'(x^k) - f'(x^{k+1}), \bar{x} - y^k \rangle = D_f(\bar{x}, x^k) + D_f(y^k, x^{k+1}) - D_f(y^k, x^k)$$

$$+ \langle \lambda_k^{-1} v^k, \bar{x} - y^k \rangle \le D_f(\bar{x}, x^k) - (1 - \sigma) D_f(y^k, x^k) - \lambda_k^{-1}(\langle v^k, y^k - \bar{x} \rangle + \varepsilon_k),$$

where the second equality follows from Proposition 6.4.3(ii) and the inequality from Proposition 6.4.3(i). We have established the leftmost inequality of (6.45). Because D_f is nonnegative and $\sigma < 1$, the term $(1 - \sigma) D_f(y^k, x^k)$ is nonnegative. In order to establish the rightmost inequality in (6.45), it suffices thus to show that $\lambda_k^{-1}(\langle v^k, y^k - \bar{x} \rangle + \varepsilon_k) \ge 0$. We proceed to do so. By assumption, 0 belongs to $T(\bar{x})$, and by (6.38), v^k belongs to $T^e(\varepsilon_k, y^k)$. In view of the definition of the enlargement T^e (see (5.3) in Section 5.2), we have that $\langle v^k, y^k - \bar{x} \rangle + \varepsilon_k \ge 0$. Because $\lambda_k > 0$, we are done. $\square$

Proposition 6.4.5. *Consider the sequences generated by Algorithm IPPM, and assume that f belongs to $\mathcal{F}$, satisfies H1 and H4, and is such that $D(T) \subset (\text{Dom}(f))^\circ$. If T has zeroes, then*

(i) *$\{D_f(\bar{x}, x^k)\}$ is convergent and nonincreasing, for all $\bar{x} \in T^{-1}(0)$.*

(ii) *The sequence $\{x^k\}$ is bounded.*

(iii) *$\sum_{k=0}^{\infty} \lambda_k^{-1}(\langle v^k, y^k - \bar{x} \rangle + \varepsilon_k) < \infty$.*

(iv) *$\sum_{k=0}^{\infty} D_f(y^k, x^k) < \infty$.*

(v) *$\sum_{k=0}^{\infty} D_f(y^k, x^{k+1}) < \infty$.*

(vi) *If f satisfies H2, then*

   (a) *$\lim_{k \to \infty} \|y^k - x^k\| = 0$, and consequently $\{y^k\}$ is bounded.*

   (b) *$\lim_{k \to \infty} \|x^{k+1} - x^k\| = 0$.*

Proof. Take $\bar{x} \in T^{-1}(0)$, assumed to be nonempty. Then, by Lemma 6.4.4, $\{D_f(\bar{x}, x^k)\}$ is a nonnegative and nonincreasing sequence, henceforth convergent. Moreover, $\{x^k\}$ is contained in a level set of $D_f(\bar{x}, \cdot)$, which is bounded by H1. We have established (i) and (ii).

Using again Lemma 6.4.4, we get

$$\lambda_k^{-1}(\langle v^k, y^k - \bar{x} \rangle + \varepsilon_k) + (1 - \sigma) D_f(y^k, x^k) \le D_f(\bar{x}, x^k) - D_f(\bar{x}, x^{k+1}).$$

Summing the inequality above with k between 0 and an arbitrary M, it follows easily that the sums of the series in (iii) and (iv) are bounded by $D_f(\bar{x}, x^0)$, and, because all terms in both series are nonnegative, we conclude that both (iii) and (iv) hold. Item (v) follows from (iv) and Proposition 6.4.3(i). For (vi), observe that (iv) implies that $\lim_{k \to \infty} D_f(y^k, x^k) = 0$. Because $\{x^k\}$ is bounded by (i), we can apply H2 and Proposition 6.2.7 in order to conclude that $\lim_{k \to \infty} \|y^k - x^k\| = 0$, establishing (a). In the same way, using now (v), we get that $\lim_{k \to \infty} \|y^k - x^{k+1}\| = 0$, implying that $\lim_{k \to \infty} \|x^k - x^{k+1}\| = 0$, which gives (b). $\square$

Proposition 6.4.5 ensures existence of weak cluster points of the sequence $\{x^k\}$ and, also, that they coincide with those of the sequence $\{y^k\}$. The last step in the convergence analysis consists of proving that such weak cluster points are indeed zeroes of T, and that, under H5, the whole sequence $\{x^k\}$ is weakly convergent. These results are established in the following theorem.

Theorem 6.4.6. *Take $f \in \mathcal{F}$, satisfying H1, H2, H3, and H4, and such that $D(T) \subset (\mathrm{Dom}(f))^o$. If T has zeroes, then*

(i) *Any sequence $\{x^k\}$ generated by Algorithm IPPM has weak cluster points, and all of them are zeroes of T.*

(ii) *If f also satisfies H5, then any such sequence is weakly convergent to a zero of T.*

Proof. By H3, Proposition 6.4.5(ii), and 6.4.5(vi-b), $\lim_{k\to\infty} \|f'(x^{k+1}) - f'(x^k)\| = 0$. In view of Proposition 6.4.3(ii), $\lim_{k\to\infty} \lambda_k^{-1}\|v^k\| = 0$. Because $\{\lambda_k\}$ is bounded, we conclude that $\lim_{k\to\infty} \|v^k\| = 0$. Using Proposition 6.4.5(ii) and (vi-a), we conclude that the sequence $\{y^k\}$ is bounded, and hence it has a weakly convergent subsequence, say $\{y^{i_k}\}$, with weak limit $\hat{y}$. By 6.4.5(iii), $\lim_{k\to\infty} \varepsilon_{i_k} = 0$. Because $v^{i_k} \in T^e(\varepsilon_{i_k}, y^{i_k})$ by (6.38), we invoke Theorem 5.2.3(i) for concluding that $0 \in T^e(0, \hat{y}) = T(\hat{y})$ and hence $\hat{y}$ is a zero of T.

In order to prove (ii), let $\bar{x}_1$ and $\bar{x}_2$ be two weak cluster points of $\{x^k\}$, so that there exist subsequences $\{x^{j_k}\}$ and $\{x^{i_k}\}$ of $\{x^k\}$ such that $x^{j_k} \xrightarrow[k\to\infty]{w} \bar{x}_1$ and $x^{i_k} \xrightarrow[k\to\infty]{w} \bar{x}_2$. By (i), $\bar{x}_1$ and $\bar{x}_2$ are zeroes of T. Thus, Proposition 6.4.5(i) guarantees the existence of $\xi_1, \xi_2 \in \mathbb{R}_+$ such that

$$\lim_{k\to\infty} D_f(\bar{x}_1, x^k) = \xi_1, \qquad \lim_{k\to\infty} D_f(\bar{x}_2, x^k) = \xi_2. \tag{6.46}$$

Now, using Lemma 6.4.2, we get

$$\left|\langle f'(x^{i_k}) - f'(x^{j_k}), \bar{x}_1 - \bar{x}_2\rangle\right|$$

$$= \left|D_f(\bar{x}_1, x^{i_k}) - D_f(\bar{x}_1, x^{j_k}) - [D_f(\bar{x}_2, x^{i_k}) - D_f(\bar{x}_2, x^{j_k})]\right|$$

$$\leq \left|D_f(\bar{x}_1, x^{i_k}) - D_f(\bar{x}_1, x^{j_k})\right| + \left|D_f(\bar{x}_2, x^{i_k}) - D_f(\bar{x}_2, x^{j_k})\right|. \tag{6.47}$$

Taking limits as k goes to ∞ in the extreme expressions of (6.47) and using (6.46), we get that $\lim_{k\to\infty} |\langle f'(x^{i_k}) - f'(x^{j_k}), \bar{x}_1 - \bar{x}_2\rangle| = 0$, and then, because $\bar{x}_1$ and $\bar{x}_2$ belong to $D(T) \subset (\mathrm{Dom}(f))^o$, H5 and nonnegativity of D_f imply that

$$0 = \langle f'(\bar{x}_1) - f'(\bar{x}_2), \bar{x}_1 - \bar{x}_2\rangle = D_f(\bar{x}_1, \bar{x}_2) + D_f(\bar{x}_2, \bar{x}_1) \geq D_f(\bar{x}_1, \bar{x}_2). \tag{6.48}$$

It follows from (6.48) and H2 that $\bar{x}_1 = \bar{x}_2$, establishing the uniqueness of the weak cluster points of $\{x^k\}$. $\square$

As commented on in the introduction to this chapter, the proximal point algorithm is in a certain sense a conceptual scheme, rather than a numerical procedure for the solution of a given problem. As such, many interesting applications of the proximal theory are of a theoretical, rather than computational, nature. One of the more important applications of the former kind is discussed in the following three sections: the convergence analysis of the augmented Lagrangian methods, which are indeed implementable algorithms widely used in practical applications. They can be seen as specific instances of a proximal point method, applied to a certain maximal monotone operator. This insight brings significant progress in the convergence theory for these algorithms. We close this section with a cursory discussion of another application of this type, specific for the approximation scheme of IPPM. In this case, a method that cannot be reduced to an instance of the exact proximal point method, or even to instances of inexact methods prior to IPPM, turns out to be a special case of IPPM, which again brings improvements in the convergence theory, in this case related to the convergence rate, namely a linear rate when T^{-1} is Lipschitz continuous near 0.

The original method, presented in [61], deals with the following problem,

$$\min f_1(x_1) + f_2(x_2) \tag{6.49}$$

$$\text{s.t. } Ax_1 - Bx_2 = 0 \tag{6.50}$$

with both $f_1 : \mathbb{R}^{n_1} \to \mathbb{R}$ and $f_2 : \mathbb{R}^{n_2} \to \mathbb{R}$ convex, $A \in \mathbb{R}^{m \times n_1}$, and $B \in \mathbb{R}^{m \times n_2}$. The associated *Lagrangian operator* is $L : \mathbb{R}^{n_1} \times \mathbb{R}^{n_2} \times \mathbb{R}^m \to \mathbb{R}$, defined as

$$L(x_1, x_2, y) = f_1(x_1) + f_2(x_2) + y^t(Ax_1 - Bx_2),$$

and the *saddle point operator* is $T : \mathbb{R}^{n_1+n_2} \times \mathbb{R}^m \rightrightarrows \mathbb{R}^{n_1+n_2} \times \mathbb{R}^m$ defined as

$$T(x, y) = (\partial_x L(x, y), -\partial_y L(x, y)),$$

where $x = (x_1, x_2)$, $\partial_x L(x, y)$ is the subdifferential of the operator G defined as $G(\cdot) = L(\cdot, y)$, and $\partial_y L$ is the subdifferential of the operator H defined as $H(\cdot) = L(x, \cdot)$. As we show in the following sections, application of IPPM (or of its exact counterpart), to the problem of finding the zeroes of T, generates a sequence equivalent to the one produced by the augmented Lagrangian method. The method in [61] takes advantage of the separability of the objective $f_1 + f_2$, and proposes a two-step proximal approach: given (x^k, y^k), first a proximal step is taken in y, with fixed x^k, producing an auxiliary $\hat{y}^k$, and then a proximal step is taken in x, with fixed $\hat{y}^k$ (this step, by separability of $f_1 + f_2$, splits into two subproblems, one in x_1 and another one in x_2). Then the updated x-variables are used to generate y^{k+1}. Formally, the iterative step is given by

$$\hat{y}^k = y^k + \lambda_k^{-1} \left(Ax_1^k - Bx_2^k \right), \tag{6.51}$$

$$x_1^{k+1} = \operatorname{argmin}_{x_1 \in \mathbb{R}^{n_1}} \left\{ f_1(x_1) + (\hat{y}^k)^t Ax_1 + \lambda_k \|x_1 - x_1^k\|^2 \right\}, \tag{6.52}$$

$$x_2^{k+1} = \operatorname{argmin}_{x_2 \in \mathbb{R}^{n_2}} \left\{ f_2(x_2) - (\hat{y}^k)^t Bx_2 + \lambda_k \|x_2 - x_2^k\|^2 \right\}, \tag{6.53}$$

$$y_2^{k+1} = y^k + \lambda_k^{-1} \left(Ax_1^{k+1} - Bx_2^{k+1} \right). \tag{6.54}$$

The fact that the variables are updated consecutively and not simultaneously has as a consequence that this method is not a particular instance of the exact proximal point method. In [61], it is proved that the sequence generated by (6.51)–(6.54) converges to a zero of T, or equivalently to a solution of (6.49)–(6.50), with an ad hoc argument, whenever $\lambda_k < \max\{1, \|A\|, \|B\|\}$ for all k. On the other hand, it has been recently proved in [208] that (6.51)–(6.54) is indeed a particular instance of IPPM applied to the operator T (with $\varepsilon_k = 0$ for all k), taking as σ any value smaller than $(\max\{1, \|A\|, \|B\|\})^{-1}$, provided that the sequence of regularization parameters satisfies $\{\lambda_k\} \subset (0, \sigma^{-1})$. In fact, the method in [208] is much more general, encompassing also another two-step procedure for finding zeroes of maximal monotone operators with certain structure, introduced in [220]. Also, the method in [208] includes the option of inexact solution of the subproblems in (6.52) and (6.53), for example, with bundle methods (see [119]), in which case the method turns out to be equivalent to IPPM with inexact solutions of the inclusion; that is, with $\varepsilon_k > 0$.

6.5 Finite-dimensional augmented Lagrangians

In a finite-dimensional setting, the augmented Lagrangian method is devised for solving an inequality constrained optimization problem, denoted as (P), defined as

$$\min f_0(x) \tag{6.55}$$

$$\text{s.t.} \quad f_i(x) \leq 0 \quad (1 \leq i \leq m), \tag{6.56}$$

with $f_i : \mathbb{R}^n \to \mathbb{R}$ $(0 \leq i \leq m)$. The Lagrangian associated with (P) is $\mathcal{L} : \mathbb{R}^n \times \mathbb{R}^m \to \mathbb{R}$ defined as

$$\mathcal{L}(x, y) = f_0(x) + \sum_{i=1}^{m} y_i f_i(x). \tag{6.57}$$

Assuming differentiability of the f_i $(0 \leq i \leq n)$, the first-order optimality conditions for problem (P), called *Karush–Kuhn–Tucker conditions*, consist of the existence of a pair $(x, y) \in \mathbb{R}^n \times \mathbb{R}^m$ such that:

$$\nabla f_0(x) + \sum_{i=1}^{m} y_i \nabla f_i(x) = 0 \qquad \text{(Lagrangian condition)}, \tag{6.58}$$

$$y \geq 0 \qquad \text{(dual feasibility)}, \tag{6.59}$$

$$y_i f_i(x) = 0 \quad (1 \leq i \leq m) \qquad \text{(complementarity)}, \tag{6.60}$$

$$f_i(x) \leq 0 \quad (1 \leq i \leq m) \qquad \text{(primal feasibility)}. \tag{6.61}$$

Under some suitable constraint qualification, (e.g., linear independence of the set $\{\nabla f_i(x), 1 \leq i \leq m\}$), these conditions are necessary for x being a solution of (P).

The situation becomes considerably better for convex optimization problems; that is, when the f_is are convex $(0 \leq i \leq m)$, which is the case of interest in this section. In this case we can drop the differentiability assumption, substituting the subdifferentials of the f_is for their gradients in Equation (6.58), which becomes

$$0 \in \partial f_0(x) + \sum_{i=1}^{m} y_i \partial f_i(x) \quad \text{(Lagrangian condition)}. \tag{6.62}$$

We are assuming that the f_is are finite-valued on $\mathbb{R}^n$, hence by convexity they turn out to be indeed continuous on $\mathbb{R}^n$ (see Theorem 3.4.22). The constraint qualification required for necessity of (6.59)–(6.62) in the convex case is considerably weaker: it suffices any hypothesis on the f_is so that the subdifferential of the sum is equal to the sum of the subdifferentials (e.g., condition (CQ) in Section 3.5), in which case the right-hand side of (6.62) is the subdifferential of $L(\cdot, y)$ at x. This constraint qualification also ensures sufficiency of (6.59)–(6.62) for establishing that x solves (P).

Finally, in the convex case we have the whole machinery of convex duality at our disposal, beginning with the definition of the *dual problem* (D) (also a convex optimization problem), defined as

$$\min -\psi(y) \tag{6.63}$$

$$\text{s.t.} \quad y \geq 0, \tag{6.64}$$

where $\psi : \mathbb{R}^m \to \mathbb{R} \cup \{-\infty\}$, defined as

$$\psi(y) = \inf_{x \in \mathbb{R}^n} \mathcal{L}(x, y), \tag{6.65}$$

is the *dual objective* (in contrast with which f_0 is called the *primal objective*).

The solution sets of (P) and (D) are denoted S_P^* and S_D^*, respectively, and $S_P^* \times S_D^* \subset \mathbb{R}^n \times \mathbb{R}^m$ is called S^*. Under the constraint qualification commented upon, S^* consists of the set of pairs $(x, y) \in \mathbb{R}^n \times \mathbb{R}^m$ that satisfy the Karush–Kuhn–Tucker conditions for (P), namely (6.59)–(6.62) (see [94, Vol. I, p. 305]).

It is easy to check that S^* coincides also with the set of saddle points of the Lagrangian $\mathcal{L}$; that is, (x^*, y^*) belongs to S^* if and only if

$$\mathcal{L}(x^*, y) \leq \mathcal{L}(x^*, y^*) \leq \mathcal{L}(x, y^*), \tag{6.66}$$

for all $x \in \mathbb{R}^n$ and all $y \in \mathbb{R}_+^m$.

The augmented Lagrangian method generates a sequence $(x^k, y^k) \in \mathbb{R}^n \times \mathbb{R}^m$ expected to converge to a point $(x^*, y^*) \in S^*$. As is usually the case for problems consisting of finding points satisfying several conditions, the rationale behind the iterative procedure relies upon defining iterates that satisfy some of these conditions, in such a way that the remaining ones will hold for the limit of the sequence. In the case of the augmented Lagrangian method, it is intended that the Lagrangian condition (6.62) and the dual feasibility condition (6.59) hold for all k, whereas the primal feasibility condition (6.61) and the complementarity condition (6.60) are expected to hold only at the limit of the sequence. In this sense the method is dual feasible, but not primal feasible.

In this spirit, it seems reasonable to try a Lagrangian method consisting of, given a dual vector $y^k \geq 0$, determining x^{k+1} as a minimizer of $\mathcal{L}(\cdot, y^k)$ (which automatically ensures that (x^{k+1}, y^k) satisfy (6.62)), and then using somehow x^{k+1} in order to compute some $y^{k+1} \geq 0$ such that some progress is achieved in connection with problem D, for example, satisfying $\psi(y^{k+1}) > \psi(y^k)$. The problem lies in the nonnegativity constraints in (6.64). In other words, the Lagrangian, in order to be convex in x, must be defined as $-\infty$ when y is not nonnegative; that is, it is not smooth in y at the boundary of $\mathbb{R}_+^m$. As a consequence, a minimizer x^{k+1} of $\mathcal{L}(\cdot, y^k)$ does not provide, in an easy way, an increase direction for ψ. The solution consists of augmenting the Lagrangian, defining it over the whole $\mathbb{R}^m$, while keeping it convex in x. This augmentation, as we show, easily provides an increase direction for ψ, and even more, an appropriate closed formula for updating the sequence $\{y^k\}$. The price to be paid is the loss of second differentiability of the augmented Lagrangian in x, even when the problem data (i.e., the functions f_is $(0 \leq i \leq m)$) are as smooth as desired. Also, because x^{k+1} minimizes the augmented Lagrangian, rather than $L(\cdot, y^k)$, the updating of the ys is also needed to recover validity of the Lagrangian condition; in other words, (6.62) holds for the pair (x^{k+1}, y^{k+1}), rather than for the pair (x^{k+1}, y^k). This "shifting" of the desired properties of the sequence is indeed harmless for the convergence analysis. Also, it is fully harmless and indeed convenient, to add a positive parameter multiplying the summation in the definition of $\mathcal{L}$.

We define then the augmented Lagrangian $\overline{\mathcal{L}} : \mathbb{R}^n \times \mathbb{R}^m \times \mathbb{R}_{++} \to \mathbb{R}$ as

$$\overline{\mathcal{L}}(x, y, \rho) = f_0(x) + \rho \sum_{i=1}^m \left[\left(\max\{0, y_i + (2\rho)^{-1} f_i(x)\} \right)^2 - y_i^2 \right]. \tag{6.67}$$

Now we can formally introduce the *augmented Lagrangian method* (AL from now on) for problems (P) and (D).

AL generates a sequence $\{(x^k, y^k)\} \subset \mathbb{R}^n \times \mathbb{R}^m$, starting from any $y^0 \in \mathbb{R}^m$, through the following iterative formulas.

$$x^{k+1} \in \operatorname{argmin}_{x \in \mathbb{R}^n} \overline{\mathcal{L}}(x, y^k, \lambda_k), \tag{6.68}$$

$$y_i^{k+1} = \max\left\{0, y_i^k + (2\lambda_k)^{-1} f_i(x^{k+1})\right\} \quad (1 \leq i \leq m), \tag{6.69}$$

where $\{\lambda_k\}$ is an exogenous bounded sequence of positive real numbers.

Of course, one expects the sequence $\{(x^k, y^k)\}$ generated by (6.68)–(6.69) to converge to a point $(x^*, y^*) \in S^*$. There are some caveats, however. In the first place, the augmented Lagrangian $\overline{\mathcal{L}}$ is convex in x, but in principle $\overline{\mathcal{L}}(\cdot, y^k, \lambda^k)$ may fail to attain its minimum, in which case x^{k+1} is not defined. We remark that this situation may happen even if problems (P) and (D) have solutions. For instance, consider the problem of minimizing a constant function of one variable over the halfline $\{x \in \mathbb{R} : e^x \leq 1\}$. Clearly any nonpositive real is a primal solution, and it is immediate that any pair $(x, 0)$ with $x \leq 0$ satisfies (6.59)–(6.62), so that 0 is a dual solution, but taking $y = \rho = 1$ and the constant value of the function also equal to 1, we get $\overline{\mathcal{L}}(x, y, \rho) = 1 + \max\{0, 1 + e^x - 1\}^2 - 1 = e^{2x}$, which obviously does not attain its minimum. Even if $\overline{\mathcal{L}}(\cdot, y^k, \lambda_k)$ has minimizers, they might be

multiple, because $\overline{\mathcal{L}}$, although convex in x, is not strictly convex. Thus, there is
no way in principle to ensure that the sequence $\{x^k\}$ will be bounded. To give a
trivial example, if none of the f_is $(0 \le i \le m)$ depends upon x_n, we may choose the
last component of x^{k+1} arbitrarily (e.g., $x_n^{k+1} = k$), making $\{x^k\}$ unbounded. As
a consequence, all convergence results on the sequence $\{x^k\}$ will have to be stated
under the assumption that such a sequence exists and is bounded.

On the other hand, the convergence properties of the dual sequence $\{y^k\}$ are
much more solid, a fact that is a consequence of a remarkable result, proved for
the first time in [193], which establishes an unexpected link between the augmented
Lagrangian method and the proximal point method: starting from the same y^0, and
using the same exogenous sequence $\{\lambda_k\}$, the proximal sequence for problem (D)
(i.e., (6.2) with $\phi = -\psi$, $C = \mathbb{R}_+^n$) coincides with the sequence $\{y^k\}$ generated by
the augmented Lagrangian method (6.59)–(6.62) applied to problems (P)–(D).

As a consequence of the convergence results for IPPM, the whole sequence
$\{y^k\}$ converges to a point in S_D^* under the sole condition of existence of solutions of
(P)–(D). Of course, existence and boundedness of $\{x^k\}$ can be ensured by imposing
additional conditions on the problem data, such as, for instance, coerciveness of f_0
(meaning that its level sets are bounded) or of any of the constraint functions f_i
$(1 \le i \le m)$, in which case the feasible set for problem (P) is necessarily bounded.
However, the proximal theory provides also a very satisfactory alternative for the
case in which the data fail to satisfy these demanding properties. It consists of
adding a proximal regularization term to the primal subproblem (6.68), which leads
to the following *proximal augmented Lagrangian*, or *doubly augmented Lagrangian*.

$$x^{k+1} \in \mathrm{argmin}_{x \in \mathbb{R}^n} \left\{ \overline{\mathcal{L}}(x, y^k, \lambda_k) + \frac{\lambda_k}{2} \|x - x^k\|^2 \right\}, \qquad (6.70)$$

$$y_i^{k+1} = \max \left\{ 0, y_i^k + (2\lambda_k)^{-1} f_i(x^{k+1}) \right\} \quad (1 \le i \le m). \qquad (6.71)$$

Note that we have not modified the dual updating; that is, (6.71) is the same
as (6.69). This method, also introduced in [193], is also equivalent to a proximal
method, involving now both primal and dual variables; that is, applied to finding
zeroes of the saddle point operator $T_{\mathcal{L}} : \mathbb{R}^n \times \mathbb{R}^m \rightrightarrows \mathbb{R}^n \times \mathbb{R}^m$, defined as

$$T_{\mathcal{L}}(x, y) = (\partial_x \mathcal{L}, -\partial_y \mathcal{L}), \qquad (6.72)$$

where $\partial_x \mathcal{L}$, $\partial_y \mathcal{L}$ denote the subdifferential of $\mathcal{L}$ with respect to the x variables
and to the y variables, respectively. We commented above that S^* coincides with
the saddle points of the Lagrangian $\mathcal{L}$, and it is easy to check that these saddle
points are precisely the zeroes of $T_{\mathcal{L}}$. Another remarkable result in [193] establishes
that, starting from the same y^0 and using the same regularization parameters λ_k,
the proximal augmented Lagrangian sequence $\{(x^k, y^k)\}$ generated by (6.70)–(6.71)
coincides with the sequence generated by IPPM applied to the problem of finding
the zeroes of $T_{\mathcal{L}}$; that is, the sequence resulting from applying (6.1) with $T = T_{\mathcal{L}}$.
Now, the results on IPPM ensure convergence of both $\{x^k\}$ and $\{y^k\}$ to a pair
$(x^*, y^*) \in S^*$ under the only assumptions of nonemptiness of S^*; that is, existence
of solutions of problems (P)–(D).

The following section establishes this result in a rather more general situation: the primal objective is defined on a reflexive Banach space X, and the constraints are inequalities indexed by the elements of a measure space Ω, so that there is a possibly infinite number of them. Also, we consider inexact solutions of the primal optimization subproblem (6.70), in the spirit of the constant relative error tolerances of Section 6.4. Finally we present, but without detailed analysis, a further generalization, namely to optimization problems in a Banach space X_1, where the constraints demand that the image, in another Banach space X_2, of feasible points through the constraint function belongs to a fixed closed and convex cone $K \subset X_2$; that is, to the case of *cone-constrained* optimization problems in Banach spaces.

6.6 Augmented Lagrangians for $\mathcal{L}^p$-constrained problems

In this section we present an application of Algorithm IPPM: namely an augmented Lagrangian method for convex optimization problems in infinite-dimensional spaces with a possibly infinite number of constraints, introduced in Section 3.4 of [54]. We start with some material leading to the definition of the problem of interest.

Let X be a reflexive Banach space. Consider a measure space $(\Omega, \mathcal{A}, \mu)$ and functions $g : X \to \mathbb{R}$, $G : \Omega \times X \to \mathbb{R}$, such that:

(i) g and $G(\omega, \cdot)$ are convex and continuously Fréchet differentiable for all $\omega \in \Omega$.

(ii) For all $x \in X$, both $G(\cdot, x) : \Omega \to \mathbb{R}$ and $G'_x(\cdot, x) : \Omega \to X^*$ are p-integrable for some $p \in (1, \infty)$, where $G'_x(\omega, \cdot)$ denotes the Fréchet derivative of $G(\omega, \cdot)$ (in the case of G'_x, integrability is understood in the sense of the Bochner integral, see, e.g., [147]).

The primal problem (P) is defined as

$$(P) \quad \begin{cases} \min g(x) \\ \text{s.t.} \quad G(\omega, x) \leq 0 \ \text{a.e. in } \Omega, \end{cases}$$

and its dual (D), as

$$(D) \quad \begin{cases} \max \Phi(y) \\ \text{s.t.} \ y \in \mathcal{L}^q(\Omega), \quad y(\omega) \geq 0 \ \text{a.e. in } \Omega, \end{cases}$$

where $q = p/(p-1)$, the dual objective $\Phi : \mathcal{L}^q(\Omega) \to \mathbb{R} \cup \{-\infty\}$ is defined as $\Phi(y) = \inf_{x \in X} L(x, y)$, and the *Lagrangian* $L : X \times \mathcal{L}^q(\Omega) \to \mathbb{R}$, is given by

$$L(x, y) = g(x) + \int_\Omega G(\omega, x) y(\omega) d\mu(\omega). \tag{6.73}$$

A pair $(x, y) \in X \times \mathcal{L}^q(\Omega)$ is called *optimal* if x is an optimal solution of problem (P) and y an optimal solution of problem (D). A pair $(x, y) \in X \times \mathcal{L}^q(\Omega)$ is called a *KKT-pair* if:

$$0 = L'_x(x, y) = g'(x) + \int_\Omega G'_x(\omega, x) y(\omega) d\mu(\omega) \quad \text{(Lagrangian condition)}, \tag{6.74}$$

$$y(\omega) \geq 0 \text{ a.e. in } \Omega \qquad\qquad \text{(dual feasibility)}, \qquad (6.75)$$

$$\int_\Omega G(\omega, x) y(\omega) d\mu(\omega) = 0 \qquad\qquad \text{(complementarity)}, \qquad (6.76)$$

$$G(\omega, x) \leq 0 \text{ a.e. in } \Omega \qquad\qquad \text{(primal feasibility)}. \qquad (6.77)$$

See ([54, 3.3]) for this formulation of the Karush–Kuhn–Tucker conditions. Note also, that $L(x, \cdot) : \mathcal{L}^q(\Omega) \to \mathbb{R}$ is Fréchet differentiable for all $x \in X$. In fact,

$$\frac{L(x, y + t\tilde{y}) - L(x, y)}{t} = \int_\Omega G(\omega, x) \tilde{y}(\omega) d\mu(\omega),$$

for all $\tilde{y} \in \mathcal{L}^q(\Omega)$, so that $L_y'(x, y) = G(\cdot, x)$ for all $y \in \mathcal{L}^q(\Omega)$. Next we define the *saddle-point operator* $T_L : X \times \mathcal{L}^q(\Omega) \rightrightarrows X^* \times \mathcal{L}^p(\Omega)$ as

$$T_L(x, y) = \left(L_x'(x, y), -L_y'(x, y) + N_{\mathcal{L}_+^q(\Omega)}(y) \right) =$$

$$\left(g'(x) + \int_\Omega G_x'(\omega, x) y(\omega) d\mu(\omega), -G(\cdot, x) + N_{\mathcal{L}_+^q(\Omega)}(y) \right), \qquad (6.78)$$

where $\mathcal{L}_+^q(\Omega) = \{ y \in \mathcal{L}^q(\Omega) | \ y(\omega) \geq 0 \text{ a.e.}\}$, and $N_{\mathcal{L}_+^q(\Omega)} : \mathcal{L}^q(\Omega) \rightrightarrows \mathcal{L}^p(\Omega)$ denotes the normality operator of the cone $\mathcal{L}_+^q(\Omega)$, given by

$$N_{\mathcal{L}_+^q(\Omega)}(y)$$

$$= \begin{cases} \{v \in \mathcal{L}^p(\Omega) | \ \int_\Omega v(\omega)[y(\omega) - z(\omega)] d\mu(\omega) \geq 0, \ \forall z \in \mathcal{L}_+^q(\Omega)\} & \text{if } y \in \mathcal{L}_+^q(\Omega) \\ \emptyset & \text{otherwise.} \end{cases}$$
$$(6.79)$$

The following proposition presents some elementary properties of the saddle-point operator.

Proposition 6.6.1. *The operator T_L, defined in (6.78), satisfies*

(i) $0 \in T_L(x, y)$ if and only if (x, y) is a KKT-pair.

(ii) If $0 \in T_L(x, y)$, then (x, y) is an optimal pair.

(iii) T_L is a maximal monotone operator.

Proof.

(i) Note that $0 \in T_L(x, y)$ if and only if $(x, y) \in X \times \mathcal{L}^q(\Omega)$ is such that:

$$\begin{aligned} 0 &= g'(x) + \int_\Omega G_x'(\omega, x) y(\omega) d\mu(\omega) \\ 0 &\in -G(\cdot, x) + N_{\mathcal{L}_+^q(\Omega)}(y), \end{aligned}$$

or equivalently

$$0 = g'(x) + \int_\Omega G_x'(\omega, x) y(\omega) d\mu(\omega), \qquad (6.80)$$

$$0 \geq \int_\Omega G(\omega, x)[z(\omega) - y(\omega)]d\mu(\omega) \quad \text{for all } z \in \mathcal{L}^q_+(\Omega), \tag{6.81}$$

$$y \in \mathcal{L}^q_+(\Omega). \tag{6.82}$$

For the "if" statement of (i), observe that these conditions are immediately implied by the definition of KKT-pair. For the "only if" statement of (i), note first that (6.80) is just (6.74) and (6.82) is (6.75). If (6.77) does not hold, then we can find a subset $W \subset \Omega$ with finite nonzero measure such that $G(\omega, x) > 0$, for all $\omega \in W$. Take $z \in \mathcal{L}^q(\Omega)$ defined as $z(\omega) = y(\omega) + \mathcal{X}_W(\omega)$, where $\mathcal{X}_W$ is the characteristic function of the set W; that is, $\mathcal{X}_W(w) = 1$ if $w \in W$, $\mathcal{X}_W(w) = 0$ otherwise. Then $z \in \mathcal{L}^q_+(\Omega)$, and

$$0 < \int_W G(\omega, x)d\mu(\omega) = \int_\Omega G(w, x)\mathcal{X}_W(w)d\mu(\omega) =$$

$$\int_\Omega G(w, x)[z(\omega) - y(\omega)]d\mu(\omega),$$

contradicting (6.81), (6.82), so that (6.77) holds, which, in turn, implies

$$\int_\Omega G(\omega, x)y(\omega)d\mu(\omega) \leq 0. \tag{6.83}$$

Take $z \in \mathcal{L}^q_+(\Omega)$ defined as $z(\omega) = 0$ for all $\omega \in \Omega$, and get from (6.81),

$$-\int_\Omega G(\omega, x)y(\omega)d\mu(\omega) \leq 0. \tag{6.84}$$

In view of (6.83) and (6.84), (6.76) holds and the proof of item (i) is complete.

(ii) In view of (i), it suffices to verify that KKT-pairs are optimal. Let (x, y) be a KKT-pair. By (6.75), (6.77), we have $y(\omega)G(\omega, y) \leq 0$, a.e. Therefore,

$$\Phi(y) = \inf_{x \in X} L(x, y) \leq L(x, y) = g(x) + \int_\Omega y(\omega)G(\omega, x)d\mu(\omega) \leq g(x). \tag{6.85}$$

It follows immediately from (6.85) that the feasible pair (x, y) is optimal when $\Phi(y) = g(x)$.

By convexity of $G(\omega, \cdot)$, the function $c(t) = t^{-1}[G(\omega, x + td) - G(\omega, x)]$ is nondecreasing and bounded from above on $(0, 1]$ for all $x, d \in X$ and all $\omega \in \Omega$. Applying the monotone convergence theorem to c (see, e.g., page 35 in [177]), it follows that the convex function $L(\cdot, y)$ is continuously differentiable and that its Gâteaux derivative, $L'_x(\cdot, y)$, is given by

$$L'_x(\cdot, y) = g'(\cdot) + \int_\Omega y(\omega)G'_x(\omega, \cdot)d\mu(\omega). \tag{6.86}$$

By (6.74) and (6.86), $L'_x(x, y) = 0$, establishing that x is a minimizer of $L(\cdot, y)$, so that $\Phi(y) = L(x, y)$. Using now (6.76) and (6.85), we get $L(x, y) = g(x)$, implying $\Phi(y) = g(x)$, and hence optimality of (x, y).

(iii) The result follows from Theorem 4.7.5.   □

As is the case with the finite-dimensional algorithm (6.68)–(6.69), it is convenient to define an augmented Lagrangian algorithm with a proximal regularization
in the primal optimization subproblems, in order to ensure existence and boundedness of the primal sequence without rather demanding conditions, such as coercivity,
imposed on the data functions. We use for this purpose the regularizing function
$h : \mathcal{L}^q(\Omega) \to \mathbb{R}$ defined as $h(y) = (1/q)\|y\|_q^q$ (our full analysis still holds if we
use $(1/r)\| \cdot \|_q^r$ for any $r \in (1,\infty)$ instead of h; the cost of this generalization is
paid in the form of more involved formulas, see [102]). The *augmented Lagrangian*
$\bar{L} : X \times \mathcal{L}^q(\Omega) \times \mathbb{R}_{++} \to \mathbb{R}$ is defined as

$$\bar{L}(x,y,\rho) = g(x) + \frac{\rho}{p} \int_{\Omega} \left[\max\{0, h'(y)(\omega) + \rho^{-1}G(x,\omega)\}\right]^p d\mu(\omega), \qquad (6.87)$$

where $p = q/(q-1)$. The *proximal augmented Lagrangian* $\hat{L} : X \times X \times \mathcal{L}^q(\Omega) \times$
$\mathbb{R}_{++} \to \mathbb{R}$ is defined as

$$\hat{L}(x,z,y,\rho) = \bar{L}(x,y,\rho) + \rho D_f(x,z), \qquad (6.88)$$

where $f \in \mathcal{F}$ is such that dom $f = X$ and D_f is the Bregman distance defined in
(6.11). Our exact method with primal regularization requires a bounded sequence
of positive regularization parameters $\{\lambda_k\}$, and is defined as

Exact doubly augmented Lagrangian method (EDAL)

(i) Choose $(x^0, y^0) \in X \times \mathcal{L}_+^q(\Omega)$.

(ii) Given (x^k, y^k), define x^{k+1} as

$$x^{k+1} = \operatorname{argmin}_{x \in X} \hat{L}(x, x^k, y^k, \lambda_k) = \operatorname{argmin}_{x \in X} \left[\bar{L}(x, y^k, \lambda_k) + \lambda_k D_f(x, x^k)\right]. \tag{6.89}$$

(iii) Define y^{k+1} as

$$y^{k+1}(\omega) = \left[\max\left\{0, h'(y^k)(\omega) + \lambda_k^{-1}G(x^{k+1},\omega)\right\}\right]^{p-1}. \tag{6.90}$$

Although this version still requires exact solution of the subproblems, it solves
the issue of the existence and uniqueness of the primal iterates. In fact, we have
the following result.

Proposition 6.6.2. *Let $f \in \mathcal{F}$ such that dom $f = X$. If f satisfies H4, then there
exists a unique solution of each primal subproblem* (6.89).

Proof. Note that $T = \bar{L}_x'(\cdot, y^k, \lambda_k)$ is maximal monotone by convexity of $\bar{L}(\cdot, y^k, \lambda_k)$,
and use Proposition 6.3.1.   □

Next we translate the error criteria studied in Section 6.3 to this setting, allowing inexact solution of the primal subproblems (6.89). These criteria admit two different errors: the error in the proximal equation, given by e^k in (6.37), and the error in the inclusion, controlled by ε_k in (6.38). However, in this application the operator of interest, namely the saddle point operator T_L given by (6.78), is such that no error is expected in its evaluation (the situation would be different without assuming differentiability of g and G, but in such a case the augmented Lagrangian method loses one of its most attractive features, namely the closed formulas for the dual updates). This section is intended mainly as an illustration of the applications of the method, thus we have eliminated, for the sake of a simpler analysis, the error in the inclusion, meaning that our algorithm is equivalent to IPPM with $\varepsilon_k = 0$ for all k.

We consider the regularizing function $F : X \times \mathcal{L}^q(\Omega) \to \mathbb{R}$ defined as $F(x, y) = f(x) + h(y)$, with h as above; that is, $h(y) = (1/q)\|y\|_q^q$, and $f \in \mathcal{F}$ such that dom $f = X$. As in the case of IPPM, we have as exogenous parameters a bounded sequence of positive regularization parameters $\{\lambda_k\}$ and a relative error tolerance $\sigma \in [0, 1)$. Our inexact doubly augmented Lagrangian method (IDAL) is as follows.

Algorithm IDAL.

1. Choose $(x^0, y^0) \in X \times \mathcal{L}_+^q(\Omega)$.

2. Given $z^k = (x^k, y^k)$, find $\tilde{x}^k \in X$ such that

$$D_f\left(\tilde{x}^k, (f')^{-1}\left[f'(x^k) - \lambda_k^{-1}\bar{L}_x'(\tilde{x}^k, y^k, \lambda_k)\right]\right) \le \sigma\left[D_f(\tilde{x}^k, x^k) + D_h\left(u^k, y^k\right)\right], \tag{6.91}$$

where $u^k \in \mathcal{L}^q(\Omega)$ is defined as

$$u^k(\omega) := \left[\max\left\{0, h'(y^k)(\omega) + \lambda_k^{-1}G(\tilde{x}^k, \omega)\right\}\right]^{p-1}. \tag{6.92}$$

3. If $(\tilde{x}^k, u^k) = (x^k, y^k)$, then stop. Otherwise, define

$$y^{k+1}(\omega) := u^k(\omega) = \left[\max\left\{0, h'(y^k)(\omega) + \lambda_k^{-1}G(\tilde{x}^k, \omega)\right\}\right]^{p-1}, \tag{6.93}$$

$$x^{k+1} := (f')^{-1}\left[f'(x^k) - \lambda_k^{-1}L_x'(\tilde{x}^k, y^{k+1})\right]. \tag{6.94}$$

The following lemma is necessary for establishing the connection between IDAL and IPPM. Let

$$M^+(x, y, \rho)(\omega) := \max\{0, h'(y)(\omega) + \rho^{-1}G(x, \omega)\}, \tag{6.95}$$

$$M^-(x, y, \rho)(\omega) := \min\{0, h'(y)(\omega) + \rho^{-1}G(x, \omega)\}, \tag{6.96}$$

$$Q(x, y, \rho)(\omega) := \left[\max\left\{0, h'(y)(\omega) + \rho^{-1}G(x, \omega)\right\}\right]^{p-1}. \tag{6.97}$$

Lemma 6.6.3. *For all* $(x, y, \rho) \in X \times \mathcal{L}^q(\Omega) \times \mathbb{R}_{++}$ *it holds that*

(a) $h'(Q(x, y, \rho)) = M^+(x, y, \rho)$.

(b) $h'(Q(x, y, \rho)) - h'(y) = -\frac{1}{\rho}\left[-G(\cdot, x) + \rho M^-(x, y, \rho)\right]$.

(c) $M^-(x, y, \rho) \in N_{\mathcal{L}^q_+(\Omega)}(Q(x, y, \rho))$.

(d) $\bar{L}'_x(x, y, \rho) = L'_x(x, Q(x, y, \rho))$, with L as in (6.73).

Proof. It is well-known that $h'(z) = |z|^{q-1}\mathrm{sign}(z)$ for all $z \in \mathcal{L}^q(\Omega)$ (see, e.g., Propositions 2.1 and 4.9 of [65]). Replacing $z := Q(x, y, \rho)$ in this formula and using the fact that $q = p/(p - 1)$, we get

$$h'(Q(x, y, \rho)) = \left[\left(M^+(x, y, \rho)\right)^{p-1}\right]^{q-1} = M^+(x, y, \rho),$$

establishing (a). Item (b) follows from (a) and the definition of M^-. We proceed to prove (c). Take any $z \in \mathcal{L}^q_+(\Omega)$ and note that

$$\int_\Omega M^-(x, y, \rho)(\omega)[z(\omega) - Q(x, y, \rho)(\omega)]d\mu(\omega) =$$

$$\int_\Omega M^-(x, y, \rho)(\omega)z(\omega)d\mu(\omega) - \int_\Omega M^-(x, y, \rho)(\omega)\left[M^+(x, y, \rho)(\omega)\right]^{p-1}d\mu(\omega)$$

$$= \int_\Omega M^-(x, y, \rho)(\omega)z(\omega)d\mu(\omega) \le 0.$$

Then the result follows from the definition of $N_{\mathcal{L}^q_+(\Omega)}$.

Finally, we prove (d). We begin with the following elementary observation: if $\psi : X \to \mathbb{R}$ is a continuously differentiable convex function, then the function $\tilde{\psi} : X \to \mathbb{R}$, defined by $\tilde{\psi}(x) = [\max\{0, \psi(x)\}]^p$, is convex, continuously differentiable, and its derivative is given by

$$\tilde{\psi}'(x) = p\left[\max\{0, \psi(x)\}\right]^{p-1}\psi'(x). \tag{6.98}$$

Fix $y \in \mathcal{L}^q(\Omega)$ and $\rho > 0$, and let $\psi_\omega(x) := M^+(x, y, \rho)(\omega)$. Thus, $\tilde{\psi}_\omega(x) = [M^+(x, y, \rho)]^p(\omega)$. Note that $(\psi_\omega)'_x(x) = \rho^{-1}G'_x(\omega, x)$, so that, by (6.98), the derivative of $[M^+(\cdot, y, \rho)(\omega)]^p$ exists at any $x \in X$, and it is given by

$$\left\{[M^+(x, y, \rho)(\omega)]^p\right\}'_x = \frac{p}{\rho}\left[M^+(x, y, \rho)(\omega)\right]^{p-1}G'_x(\omega, x) \tag{6.99}$$

and, therefore, it is continuous.

By convexity of $\tilde{\psi}_\omega$, given $u \in X$ and $\{t_k\} \subset (0, 1]$ converging decreasingly to 0, the sequence $\left\{t_k^{-1}\left[\tilde{\psi}_\omega(x + t_k u) - \tilde{\psi}_\omega(x)\right]\right\}$ converges nonincreasingly to $\langle(\tilde{\psi}_\omega)'_x(x), u\rangle$, and is bounded above by $\tilde{\psi}_\omega(x + u) - \tilde{\psi}_\omega(x)$. Consequently, by the commutativity of the Bochner integral with continuous linear operators (see

Corollary 2, page 134 in [231]), and the monotone convergence theorem (see, e.g., page 35 in [177]), we get

$$\left\langle \int_\Omega (\tilde{\psi}_\omega)'_x(x)d\mu(\omega), u \right\rangle \int_\Omega \left\langle (\tilde{\psi}_\omega)'_x(x), u \right\rangle d\mu(\omega)$$

$$= \int_\Omega \left[\lim_{k\to\infty} \frac{\tilde{\psi}_\omega(x + t_k u) - \tilde{\psi}_\omega(x)}{t_k} \right] d\mu(\omega)$$

$$= \lim_{k\to\infty} \int_\Omega \frac{\tilde{\psi}_\omega(x + t_k u) - \tilde{\psi}_\omega(x)}{t_k} d\mu(\omega) = \left\langle \left[\int_\Omega \tilde{\psi}_\omega(x)d\mu(\omega) \right]'_x, u \right\rangle. \qquad (6.100)$$

Combining (6.99) and (6.100), we get

$$\left\{ \int_\Omega [M^+(x, y, \rho)]^p (\omega)d\mu(\omega) \right\}'_x = \int_\Omega (\tilde{\psi}_\omega)'_x(x)d\mu(\omega)$$

$$= \frac{p}{\rho} \int_\Omega [M^+(x, y, \rho)(\omega)]^{p-1} G'_x(\omega, x)d\mu(\omega). \qquad (6.101)$$

Using (6.101) and (6.87), we conclude that the function $\bar{L}(\cdot, y, \rho)$ is differentiable, and that its derivative is given by

$$\bar{L}'_x(x, y, \rho) = g'(x) + \frac{p}{\rho} \int_\Omega [M^+(x, y, \rho)(\omega)]^{p-1} G'_x(\omega, x)d\mu(\omega). \qquad (6.102)$$

The result follows from (6.102), (6.95), (6.97), and (6.86).   □

We point out that the exact solution of (6.89) obviously satisfies the error criterion of Algorithm IDAL; that is, the inequality in (6.91). We also mention that when $\sigma = 0$ Algorithm IDAL reduces to Algorithm EDAL, because in such a case $\tilde{x}^k$ coincides with the exact solution of (6.89) and (6.94) becomes redundant. Thus, our following convergence results for Algorithm IDAL also hold for Algorithm EDAL.

6.7   Convergence analysis of Algorithm IDAL

We prove next that Algorithm IDAL is a particular instance of Algorithm IPPM applied to the saddle-point operator T_L, with $\varepsilon_k = 0$ for all k.

Proposition 6.7.1. *Let* $\{x^k\}$, $\{y^k\}$, $\{\tilde{x}^k\}$, $\{\lambda_k\}$, σ, *and* f *be as in Algorithm IDAL. Define* $F : X \times \mathcal{L}^q(\Omega) \to \mathbb{R}$ *as* $F(x, y) = f(x) + h(y)$, $e^k = \left(\hat{L}'_x(\tilde{x}^k, x^k, y^k, \lambda_k), 0 \right)$, $\tilde{z}^k = (\tilde{x}^k, u^k)$, *and* $z^k = (x^k, y^k)$, *with* u^k *as in (6.92). Then*

(i) $e^k + \lambda_k[F'(z^k) - F'(\tilde{z}^k)] \in T_L(\tilde{z}^k)$.

(ii) $D_F\left(\tilde{z}^k, (F')^{-1}\left[F'(\tilde{z}^k) - \lambda_k^{-1}e^k \right] \right) \leq \sigma D_F(\tilde{z}^k, z^k)$.

(iii) $z^{k+1} = (F')^{-1} \left[F'(\tilde{z}^k) - \lambda_k^{-1} e^k \right].$

Proof. Using the definition of $\hat{L}$ in (6.88) and Lemma 6.6.3(b)–(d), we have

$$e^k + \lambda_k [F'(z^k) - F'(\tilde{z}^k)]$$

$$= \left(\hat{L}'_x(\tilde{x}^k, x^k, y^k, \lambda_k) + \lambda_k [f'(x^k) - f'(\tilde{x}^k)], \lambda_k [h'(y^k) - h'(u^k)] \right)$$

$$= \left(\bar{L}'_x(\tilde{x}^k, y^k, \lambda_k), -G(\cdot, \tilde{x}^k) + \lambda_k M^-(\tilde{x}^k, y^k, \lambda_k) \right)$$

$$= \left(L'_x(\tilde{x}^k, u^k), -G(\cdot, \tilde{x}^k) + \lambda_k M^-(\tilde{x}^k, y^k, \lambda_k) \right) \in T_L(\tilde{z}^k),$$

establishing (i), with M^- as in (6.96). Also,

$$D_F \left(\tilde{z}^k, (F')^{-1} \left[F'(\tilde{z}^k) - \lambda_k^{-1} e^k \right] \right)$$

$$= D_F \left((\tilde{x}^k, u^k), \left((f')^{-1} \left[f'(\tilde{x}^k) - \lambda_k^{-1} \hat{L}'_x(\tilde{x}^k, x^k, y^k, \lambda_k) \right], u^k \right) \right)$$

$$= D_f \left(\tilde{x}^k, (f')^{-1} \left[f'(x^k) - \lambda_k^{-1} \bar{L}'_x(\tilde{x}^k, y^k, \lambda_k) \right] \right) + D_h \left(u^k, u^k \right)$$

$$= D_f \left(\tilde{x}^k, (f')^{-1} \left[f'(x^k) - \lambda_k^{-1} \bar{L}'_x(\tilde{x}^k, y^k, \lambda_k) \right] \right)$$

$$\leq \sigma \left[D_f(\tilde{x}^k, x^k) + D_h \left(u^k, y^k \right) \right] = \sigma D_F(\tilde{z}^k, z^k),$$

using the error criterion given by (6.91) in the inequality, and the separability of F, establishing (ii). Finally, observe that

$$z^{k+1} = (F')^{-1} \left[F'(\tilde{z}^k) - \lambda_k^{-1} e^k \right]$$

if and only if

$$y^{k+1} = (h')^{-1} \left[h'(u^k) \right],$$

$$x^{k+1} = (f')^{-1} \left[f'(x^k) - \lambda_k^{-1} \bar{L}'_x(\tilde{x}^k, y^k, \lambda_k) \right]$$

if and only if

$$y^{k+1} = u^k$$

$$x^{k+1} = (f')^{-1} \left[f'(x^k) - \lambda_k^{-1} L'_x(\tilde{x}^k, u^k) \right],$$

which hold by (6.93) and (6.94). We have proved (iii). $\quad\square$

In view of Proposition 6.7.1, the convergence properties of IDAL will follow from those of IPPM. We start with the case of finite termination.

Proposition 6.7.2. *If Algorithm IDAL stops after k steps, then $\tilde{x}^k$ is an optimal solution of problem (P) and u^k is an optimal solution of problem (D), with $\tilde{x}^k$ and u^k as given by (6.91) and (6.92), respectively.*

Proof. Immediate from Propositions 6.6.1(ii), 6.7.1, and 6.4.1. $\quad\square$

Next we identify the cases in which the error criterion of Algorithm IDAL accepts as an approximate solution of the subproblem any point close enough to the exact solution.

Proposition 6.7.3. *Let $\{z^k\}$ be the sequence generated by Algorithm IDAL. Assume that f satisfies H4. If z^k is not a KKT-pair for (P)–(D), $(f')^{-1}$ is continuous, and $\sigma > 0$, then there exists an open subset $U_k \subset X$ such that any $x \in U_k$ solves (6.91).*

Proof. Let $\bar{x}^k$ denote the exact solution of (6.91); that is, the point in X satisfying

$$f'(\bar{x}^k) = f'(x^k) - \lambda_k^{-1}\bar{L}'_x(\bar{x}^k, y^k, \lambda_k), \tag{6.103}$$

whose existence is ensured by Proposition 6.6.2, and $\bar{z}^k = (\bar{x}^k, \bar{u}^k)$, where $\bar{u}^k$ is defined as

$$\bar{u}^k(\omega) = \left[\max\left\{0, h'(y^k)(\omega) + \lambda_k^{-1}G(\bar{x}^k, \omega)\right\}\right]^{p-1}.$$

Then $\bar{z}^k \neq z^k$, because otherwise, the kth iteration would leave z^k unchanged, in which case, in view of Proposition 6.4.1, z^k would be a solution of the problem (i.e., a KKT-pair), contradicting the assumption. Hence, $D_F(\bar{z}^k, z^k) > 0$, with $F(x, y) = f(x) + h(y)$, and by strict convexity of F, resulting from strict convexity of f and h, we get

$$\theta_k := \sigma D_F(\bar{z}^k, z^k) = \sigma\left[D_f(\bar{x}^k, x^k) + D_h(\bar{u}^k, y^k)\right] > 0. \tag{6.104}$$

Observe that the assumptions on the data functions of problem (P) and continuity of $(f')^{-1}$ ensure continuity of the function $\psi_k : B \to \mathbb{R}$ defined as

$$\psi_k(x) = D_f\left(x, (f')^{-1}\left[f'(x^k) - \lambda_k^{-1}\bar{L}'_x(x, u^k(x))\right]\right) -$$

$$\sigma\left[D_f(x, x^k) + D_h(Q(x, y^k, \lambda_k), y^k)\right],$$

with

$$u^k(x)(\omega) = \left[\max\left\{0, h'(y^k)(\omega) + \lambda_k^{-1}G(x, \omega)\right\}\right]^{p-1}.$$

Also, $\psi_k(\bar{x}^k) = -\theta_k < 0$, and consequently there exists $\delta_k > 0$ such that $\psi_k(x) \leq 0$ for all $x \in U_k := \{x \in X | \|x - \bar{x}^k\| < \delta_k\}$. The result follows then from the previous inequality. $\square$

As mentioned after Proposition 6.3.3, $f(x) = \|x\|_p^r$ in ℓ_p or $\mathcal{L}^p(\Omega)$ $(p, r > 1)$, satisfies the hypothesis of continuity of $(f')^{-1}$.

Finally, we establish the convergence properties of the sequence generated by Algorithm IDAL.

Theorem 6.7.4. *Let X be a uniformly smooth and uniformly convex Banach space. Take $f \in \mathcal{F}$ satisfying H1–H4 and such that $\mathrm{dom}\, f = X$, and assume that $\{\lambda_k\}$ is bounded. Let $\{z^k\} = \{(x^k, y^k)\}$ be the sequence generated by Algorithm IDAL. If there exist KKT-pairs for problems (P) and (D), then*

(i) The sequence $\{z^k\}$ is bounded and all its weak cluster points are optimal pairs for problems (P) and (D).

(ii) If f also satisfies H5 and either Ω is countable (i.e., the dual variables belong to ℓ_q) or $p = q = 2$ (i.e., the dual variables belong to $\mathcal{L}^2(\Omega)$), then the whole sequence $\{z^k\}$ converges weakly to an optimal pair for problems (P)-(D).

Proof. By Proposition 6.7.1 the sequence $\{z^k\}$ is a particular instance of the sequences generated by Algorithm IPPM for finding zeroes of the operator T_L, with regularizing function $F : X \times \mathcal{L}^q(\Omega) \to \mathbb{R}$ given by $F(x,y) = f(x) + \frac{1}{q}\|y\|_q^q$. By Corollary 6.2.5(i) and Proposition 6.2.6(iii), F satisfies H1–H4. By Corollary 6.2.5(ii), it also satisfies H5 under the assumptions of item (ii). By our assumption on existence of KKT-pairs, and items (i) and (iii) of Proposition 6.6.1, T_L is a maximal monotone operator with zeroes. The result follows then from Theorem 6.4.6 and Proposition 6.6.1(ii). $\square$

6.8   Augmented Lagrangians for cone constrained problems

Note that the constraints in the problem studied in Section 6.6 can be thought of as given by a $\widehat{G} : X \to \mathcal{L}^q(\Omega)$, taking $\widehat{G}(x) = G(\cdot, x)$. In this section we consider the more general case in which an arbitrary reflexive Banach space plays the role of $\mathcal{L}^q(\Omega)$. In this case the nonnegativity constraints have to be replaced by more general constraints, namely *cone constraints*.

Let X_1 and X_2 be real reflexive Banach spaces, $K \subset X_2$ a closed and convex cone, and consider the partial order "$\precsim$" induced by K in X_2, namely $z \precsim z'$ if and only if $z' - z \in K$. K also induces a natural extension of convexity, namely K-convexity, given in Definition 4.7.3.

The problem of interest, denoted as (P), is

$$\min g(x) \tag{6.105}$$

$$\text{s.t. } G(x) \precsim 0, \tag{6.106}$$

with $g : X_1 \to \mathbb{R}$, $G : X_1 \to X_2$, satisfying

(A1) g is convex and G is K-convex.

(A2) g and G are Fréchet differentiable functions with Gâteaux derivatives denoted g' and G', respectively.

We define the Lagrangian $L : X_1 \times X_2^* \to \mathbb{R}$, as

$$L(x,y) = g(x) + \langle y, G(x)\rangle, \tag{6.107}$$

where $\langle \cdot, \cdot \rangle$ denotes the duality pairing in $X_2^* \times X_2$, and the dual objective $\Phi : X_2^* \to \mathbb{R} \cup \{-\infty\}$ as $\Phi(y) = \inf_{x \in X_1} L(x,y)$.

We also need the positive polar cone K^* of K (see Definition 3.7.7), the *metric projection* $P_{-K} : X_2 \to -K$, defined as $P_{-K}(y) = \mathrm{argmin}_{v \in -K} \|v - y\|$, and the weighted duality mapping $J_s : X_2 \to X_2^*$, uniquely defined, when X_2 is smooth, by the following relations,

$$\langle J_s(x), x \rangle = \|J_s(x)\| \|x\|, \quad \|J_s(x)\| = \|x\|^{s-1}.$$

In the notation of Section 6.2, $J_s = J_\varphi$, with $\varphi(t) = t^{s-1}$. Let $\succsim_*$ be the partial order induced by K^* in X_2^*.

With this notation, the dual problem (D) can be stated as

$$(D) \quad \begin{cases} \max \ \Phi(y) \\ \text{s.t. } y \succsim_* 0. \end{cases}$$

In order to define an augmented Lagrangian method for this problem, we consider regularizing functions $f : X_1 \to \mathbb{R}$, $h : X_2^* \to \mathbb{R}$. As in the previous section, we fix the dual regularization function as a power of the norm, namely $h(y) = (1/r)\|y\|^r$ for some $r \in (1, \infty)$, where $\| \cdot \|$ is the norm in X_2^*. For $z \in X_2$ and $C \subset X_2$, $d(z, C)$ denotes the metric distance from z to C, namely $\inf_{v \in C} \|z - v\|$. With this notation, the *doubly augmented Lagrangian* $\bar{L} : X_1 \times X_1 \times X_2^* \times \mathbb{R}_{++} \to \mathbb{R}$ is defined as

$$\hat{L}(x, z, y, \rho) = g(x) + \frac{\rho}{s} d(h'(y) + \rho^{-1} G(x), -K)^s + D_f(x, z), \tag{6.108}$$

with $s = r/(r-1)$.

It is easy to check that when $X_2 = \mathcal{L}^q(\Omega)$, the doubly augmented Lagrangian given by (6.108) reduces to the one in Section 6.7, namely the one defined by (6.88).

The exact version of the doubly augmented Lagrangian method for problems (P)–(D), is given by:

1. Choose $(x^0, y^0) \in X_1 \times K^*$.

2. Given (x^k, y^k), take

$$x^{k+1} = \mathrm{argmin}_{x \in X_1} \hat{L}(x, x^k, y^k, \lambda_k). \tag{6.109}$$

$$y^{k+1} = J_s\left(h'(y^k) + \lambda_k^{-1} G(x^{k+1}) - P_{-K}\left[h'(y^k) + \lambda_k^{-1} G(x^{k+1})\right]\right). \tag{6.110}$$

The inexact version incorporates the possibility of error in the solution of the minimization problem in (6.109). It is given by:

1. Choose $(x^0, y^0) \in X_1 \times K^*$.

2. Given $z^k = (x^k, y^k)$, find $\tilde{x}^k \in X_1$ such that

$$D_f\left(\tilde{x}^k, (f')^{-1}\left[f'(x^k) - \lambda_k^{-1} \bar{L}'_x(\tilde{x}^k, y^k, \lambda_k)\right]\right)$$

$$\leq \sigma\left[D_f(\tilde{x}^k, x^k) + D_h(u^k, y^k)\right] \tag{6.111}$$

with u^k given by

$$u^k = J_s\left(h'(y^k) + \lambda_k^{-1} G(\tilde{x}^k) - P_{-K}\left[h'(y^k) + \lambda_k^{-1} G(\tilde{x}^k)\right]\right).$$

3. If $(\tilde{x}^k, u^k) = (x^k, y^k)$, then stop. Otherwise, define

$$y^{k+1} = u^k = J_s\left(h'(y^k) + \lambda_k^{-1}G(\tilde{x}^k) - P_{-K}\left[h'(y^k) + \lambda_k^{-1}G(\tilde{x}^k)\right]\right), \quad (6.112)$$

$$x^{k+1} = (f')^{-1}\left[f'(x^k) - \lambda_k^{-1}\bar{L}_x'(\tilde{x}^k, y^k, \lambda_k)\right]. \quad (6.113)$$

The convergence analysis of these two algorithms proceeds in the same way as was done for EDAL and IDAL in Section 6.7; namely both generate, with the same initial iterate and same regularization parameters, sequences identical to those of IPPM applied to the problem of finding zeroes of the saddle-point operator $T_L : X_1 \times X_2^* \rightrightarrows X_1^* \times X_2$ defined as

$$T_L(x, y) = (g'(x) + [G'(x)]^*(y), -G(x) + N_{K^*}(y)), \quad (6.114)$$

where $[G'(x)]^* : X_2^* \to X_1^*$ is the adjoint of the linear operator $G'(x) : X_1 \to X_2$, and $N_{K^*} : X_2^* \rightrightarrows X_2$ denotes the normality operator of the cone K^*, as defined in Definition 3.6.1.

More specifically, Algorithm (6.109)–(6.110) is equivalent to IPPM with $\sigma = 0$ (i.e., both $e^k = 0$ and $\varepsilon_k = 0$ for all k), applied to the operator T_L defined by (6.114), and Algorithm (6.111)–(6.113) is equivalent to IPPM, with $\varepsilon_k = 0$ for all k, applied to the same operator. The zeroes of T_L are precisely the saddle points of the Lagrangian L defined by (6.107), and also the optimal pairs for problems (P) and (D), therefore the result of Theorem 6.7.4(i) holds verbatim in this case. The result of Theorem 6.7.4(ii) also holds if both f and h satisfy H5. The detailed proofs can be found in [103].

We remark that for $X_2 = \mathcal{L}^p(\Omega)$, $K = \mathcal{L}_+^p(\Omega)$, and $s = p$, the algorithms given by (6.109)–(6.110) and (6.111)–(6.113) reduce to EDAL and IDAL of Section 6.6, respectively. We have chosen to present the full convergence analysis only for the particular case of nonnegativity constraints in $\mathcal{L}^p(\Omega)$, because the introduction of an arbitrary cone K in a general space X_2 makes the proofs considerably more technical, and somewhat obscure.

It is worthwhile to mention that the inexact proximal method in [211] has also been extended to Banach spaces, together with a resulting augmented Lagrangian method for $\mathcal{L}_+^p$ constrained problems, in [102], and furthermore, in [103], to an augmented Lagrangian method for cone constrained optimization in Banach spaces. The main difference between such a proximal method and IPPM lies in the nonproximal step: instead of (6.40), which gives an explicit expression for x^{k+1} (at least when an explicit formula for $(f')^{-1}$ is available, which is the case when f is a power of the norm), the nonproximal step in [211] requires the *Bregman projection* of x^k onto a hyperplane E separating x^k from $T^{-1}(0)$; that is, solving a problem of the type $x^{k+1} = \operatorname{argmin}_{x \in E} D_f(x, x^k)$. In non-Hilbertian spaces, no explicit formulas exist for Bregman projections onto hyperplanes (not even in ℓ_p or $\mathcal{L}^p(\Omega)$ with $f(x) = \|x\|_p^p$). A similar comparison applies to the augmented Lagrangian methods derived from [211] with respect to those presented here, namely IDAL and (6.111)–(6.113). In this sense, the methods analyzed in this chapter seem preferable, and for this reason we chose not to develop here the algorithms derived from the procedure in [211].

6.9   Nonmonotone proximal point methods

Another issue of interest in the proximal theory is the validity of the convergence results when the operator whose zeroes are looked for is not monotone. A very interesting approach to the subject was recently presented in [169], which deals with a class of nonmonotone operators in Hilbert spaces that, when restricted to a neighborhood of the solution set, are not far from being monotone. More precisely, it is assumed that, for some $\rho > 0$, the mapping $T^{-1} + \rho I$ is monotone when its graph is restricted to a neighborhood of $\hat{S}^* \times \{0\}$, where $\hat{S}^*$ is a connected component of the solution set $S^* = T^{-1}(0)$ (remember that the set of zeroes of nonmonotone operators may fail to be convex). When this happens, the main convergence result of [169] states that a "localized" version of (6.1) generates a sequence that converges to a point in $\hat{S}^*$, provided that x^0 is close enough to $\hat{S}^*$ and $\sup \lambda_k < (2\rho)^{-1}$. We present next an inexact version of the method in [169], developed in [105], which incorporates the error criteria introduced in [209] and [210]. The convergence analysis in the nonmonotone case is rather delicate, and the case of Banach spaces has not yet been adequately addressed. Thus, in this section we confine our attention to operators $T : H \rightrightarrows H$, where H is a Hilbert space. The issue of convergence of the proximal point method applied to nonmonotone operators in Banach spaces remains as an important open question.

We define next the class of operators to which our results apply. From now on we identify, in a set-theoretic fashion, a point-to-set operator $T : H \rightrightarrows H$ with its graph; that is, with $\{(x, v) \in H \times H : v \in T(x)\}$. Thus, $(x, v) \in T$ has the same meaning as $v \in T(x)$. We emphasize that (x, v) is seen here as an ordered pair: $(x, v) \in T$ (or equivalently to $(v, x) \in T^{-1}$) is not the same as $(v, x) \in T$.

Definition 6.9.1. *Given a positive $\rho \in \mathbb{R}$ and a subset W of $H \times H$, an operator $T : H \rightrightarrows H$ is said to be*

(a) *ρ-hypomonotone if and only if $\langle x-y, u-v \rangle \geq -\rho \|x-y\|^2$ for all $(x, u), (y, v) \in T$.*

(b) *Maximal ρ-hypomonotone if and only if T is ρ-hypomonotone and in addition $T = T'$ whenever $T' \subset H \times H$ is ρ-hypomonotone and $T \subset T'$.*

(c) *ρ-hypomonotone in W if and only if $T \cap W$ is ρ-hypomonotone.*

(d) *Maximal ρ-hypomonotone in W if and only if T is ρ-hypomonotone in W and in addition $T \cap W = T' \cap W$ whenever $T' \in H \times H$ is ρ-hypomonotone and $T \cap W \subset T' \cap W$.*

The notion of hypomonotonicity was introduced in [194]. For practical reasons, it is convenient in this section to rewrite the basic proximal iteration (6.1) using regularization parameters $\gamma_k = \lambda_k^{-1}$, so that (6.1) becomes $x^k - x^{k+1} \in \gamma_k T(x^{k+1})$, or, separating the equation and the inclusion, as in (6.5) and (6.6), the exact iteration is:

$$\gamma_k v^k + x^{k+1} - x^k = 0, \tag{6.115}$$

$$v^k \in T(x^{k+1}).$$ (6.116)

We proceed to introduce an error term in (6.115). We refrain from using the enlargement T^e for allowing errors in the inclusion (6.116), as done in (6.8), because the properties of this enlargement have not yet been studied for the case of a nonmonotone T. This is also a relevant open issue. In this section we consider the following inexact procedure for finding zeroes of an operator $T : H \rightrightarrows H$, whose inverse is maximal ρ-hypomonotone on a set $U \times V \subset H \times H$.

Algorithms IPPH1 and IPPH2

Given $x^k \in H$, find $(y^k, v^k) \in U \times V$ such that

$$v^k \in T(y^k),$$ (6.117)

$$\gamma_k v^k + y^k - x^k = e^k,$$ (6.118)

where the error term e^k satisfies either

$$\|e^k\| \le \sigma \left(\frac{\hat{\gamma}}{2} - \rho \right) \|v^k\|,$$ (6.119)

or

$$\|e^k\| \le \nu \|y^k - x^k\|,$$ (6.120)

with

$$\nu = \frac{\sqrt{\sigma + (1 - \sigma)\left(2\rho/\hat{\gamma}\right)^2} - 2\rho/\hat{\gamma}}{1 + 2\rho/\hat{\gamma}},$$ (6.121)

where $\sigma \in [0, 1)$, $\{\gamma_k\} \subset \mathbb{R}_{++}$ is an exogenous sequence bounded away from 0, $\hat{\gamma} = \inf\{\gamma_k\}$, and ρ is the hypomonotonicity constant of T^{-1}. Then, under any of our two error criteria, the next iterate x^{k+1} is given by

$$x^{k+1} = x^k - \gamma_k v^k.$$ (6.122)

From now on, Algorithm IPPH1 refers to the algorithm given by (6.117)–(6.119) and (6.122), and Algorithm IPPH2 to the one given by (6.117), (6.118) and (6.120)–(6.122) (the H in IPPH stands for "hypomonotone").

We prove that, when ρ-hypomonotonicity of T^{-1} holds on the whole space (i.e., $U = V = H$), both Algorithm IPPH1 and Algorithm IPPH2 generate sequences that are weakly convergent to a zero of T, starting from any $x^0 \in H$, under the assumption of existence of zeroes of T, whenever $2\rho < \hat{\gamma} = \inf\{\gamma_k\}$. Note that this is the same as demanding that $\lambda_k \in \left(0, (2\rho)^{-1}\right]$ for all k. This condition illustrates the trade-off imposed by the admission of nonmonotone operators: if ρ is big, meaning that T^{-1} (and therefore T) is far from being monotone, then the regularization parameters λ_k must be close to zero, so that the regularized operator $T + \lambda_k I$ is close to T, hence ill-conditioned when T is ill-conditioned. At the same time, this "negligible" regularization of T is expected to have its zero close to the set S^* of zeroes of T, so that the lack of monotonicity of T is compensated by a choice of the regularization parameters that prevents the generated sequence from moving very far away from S^*.

For the case in which the set $U \times V$ where T^{-1} is ρ-hypomonotone in an appropriate neighborhood of $\hat{S}^* \times \{0\} \subset H \times H$, where $\hat{S}^*$ is a connected component of S^*, we still get a local convergence result, meaning weak convergence of $\{x^k\}$ to a zero of T, but requiring additionally that x^0 be sufficiently close to $\hat{S}^* \cap U$, in a sense that is presented in a precise way in Theorem 6.10.4.

We remark that when the tolerance σ vanishes, we get $e^k = 0$ from either (6.119) or (6.120)–(6.121), and then $x^k - y^k = \gamma_k v^k$ from (6.118), so that $x^{k+1} = y^k$ from (6.122). Thus, with $\sigma = 0$ our algorithm reduces to the classical exact proximal point algorithm, whose convergence properties, when applied to operators whose inverse is ρ-hypomonotone, have been studied in [169].

It follows from 13.33 and 13.36 of [194] that, in a finite-dimensional setting, if a function $f : H \to \mathbb{R} \cup \{+\infty\}$ can be written as $g - h$ in a neighborhood of a point $x \in H$, where g is finite and h is C^2, then the subdifferential ∂f of f, suitably defined, is ρ-hypomonotone for some $\rho > 0$ in a neighborhood of any point $(x, v) \in H \times H$ with $v \in \partial f(x)$. It is also easy to check that a locally Lipschitz continuous mapping is hypomonotone for every ρ greater than the Lipschitz constant. In particular, if H is finite-dimensional and $T : H \rightrightarrows H$ is such that T^{-1} is point-to-point and differentiable in a neighborhood of some $v \in H$, then T is ρ-hypomonotone in a neighborhood of (x, v) for any x such that $v \in T(x)$, and for any ρ larger than the absolute value of the most negative eigenvalue of $J + J^t$, where J is the Jacobian matrix of T^{-1} at v. In other words, local ρ-hypomonotonicity for some $\rho > 0$ is to be expected of any T that is not too badly behaved.

We establish next several properties of hypomonotone operators, needed for the convergence analysis of Algorithms IPPH1 and IPPH2. Note that a mapping T is ρ-hypomonotone (maximal ρ-hypomonotone) if and only if $T + \rho I$ is monotone (maximal monotone). We also have the following result.

Proposition 6.9.2. *If $T : H \rightrightarrows H$ is ρ-hypomonotone, then there exists a maximal ρ-hypomonotone $\hat{T} : H \rightrightarrows H$ such that $T \subset \hat{T}$.*

Proof. The proof is a routine application of Zorn's lemma, with exactly the same argument as in the proof of Proposition 4.1.3. □

Next we introduce in a slightly different way the Yosida regularization of an operator. For $\rho \geq 0$, define $Y_\rho : H \times H \to H \times H$ (Y for Yosida) as

$$Y_\rho(x, v) = (x + \rho v, v). \tag{6.123}$$

Observe that Y_ρ is a bijection, and $(Y_\rho)^{-1}(y, u) = (y - \rho u, u)$. Note also that, with the identification mentioned above,

$$Y_\rho(T) = (T^{-1} + \rho I)^{-1}, \tag{6.124}$$

because $(v, x + \rho v) \in T^{-1} + \rho I$.

Proposition 6.9.3. *Take $\rho \geq 0$, $T : H \rightrightarrows H$, and Y_ρ as in (6.123). Then*

(i) T^{-1} is ρ-hypomonotone if and only if $Y_\rho(T)$ is monotone.

(ii) T^{-1} is maximal ρ-hypomonotone if and only if $Y_\rho(T)$ is maximal monotone.

Proof.

(i) Monotonicity of the Yosida regularization means that $(T^{-1} + \rho I)^{-1}$ is monotone, which is equivalent to monotonicity of $T^{-1} + \rho I$.

(ii) Assume that T^{-1} is maximal ρ-hypomonotone. We prove maximal monotonicity of $Y_\rho(T)$. The monotonicity follows from item (a). Assume that $Y_\rho(T) \subset Q$ for some monotone $Q \subset H \times H$. Note that $Q = Y_\rho(T_Q)$ for some T_Q because Y_ρ is a bijection. It follows, in view of (i) and the monotonicity of Q, that T_Q^{-1} is ρ-hypomonotone, and therefore, using again the bijectivity of Y_ρ, we have $T^{-1} \subset T_Q^{-1}$. Because T^{-1} is maximal ρ-hypomonotone, we conclude that $T^{-1} = T_Q^{-1}$; that is, $T = T_Q$, so that $Q = Y_\rho(T') = Y_\rho(T)$, proving that $Y_\rho(T)$ is maximal monotone. The converse statement is proved with a similar argument.

□

We continue with an elementary result on the Yosida regularization $Y_\rho(T)$.

Proposition 6.9.4. *For all $T : H \rightrightarrows H$ and all $\rho \geq 0$, $0 \in T(x)$ if and only if $0 \in [Y_\rho(T)](x)$.*

Proof. The result follows immediately from (6.124).   □

Assume that T is a maximal monotone operator defined in a reflexive Banach space. By Corollary 4.2.8 the set of zeroes of a maximal monotone operator is a convex set. In view of Propositions 6.9.3 and 6.9.4, the same holds for mappings whose inverses are ρ-hypomonotone. Thus, although reasonably well-behaved operators can be expected to be locally ρ-hypomonotone for some $\rho > 0$, as discussed above, global ρ-hypomonotonicity is not at all generic; looking, for instance, at point-to-point operators in $\mathbb{R}$, we observe that polynomials with more than one real root, or analytic functions such as $T(x) = \sin x$, are not ρ-hypomonotone for any $\rho > 0$.

We prove now that ρ-hypomonotone operators are (locally) sequentially (sw*)-osc, with a proof which mirrors the one on sequential outer-semicontinuity of maximal monotone operators (cf. Proposition 4.2.1).

Proposition 6.9.5. *Assume that $T^{-1} : H \rightrightarrows H$ is maximal ρ-hypomonotone in W^{-1} for some $W \subset H \times H$, and consider a sequence $\{(x^k, v^k)\} \subset T \cap W$. If $\{v^k\}$ is strongly convergent to $\bar{v}$, $\{x^k\}$ is weakly convergent to $\bar{x}$, and $(\bar{x}, \bar{v}) \in W$, then $\bar{v} \in T(\bar{x})$.*

Proof. Define $T' : H \rightrightarrows H$ as $T' = T \cup \{(\bar{x}, \bar{v})\}$. We claim that $(T')^{-1}$ is ρ-hypomonotone in W^{-1}. Because T^{-1} is ρ-hypomonotone in W^{-1}, clearly it suffices to prove that

$$-\rho\|\bar{v} - v\|^2 \leq \langle \bar{x} - x, \bar{v} - v \rangle \tag{6.125}$$

for all $(x, v) \in T \cap W$. Observe that, for all $(x, v) \in T \cap W$,

$$-\rho\|v^k - v\|^2 \leq \langle x^k - x, v^k - v \rangle. \tag{6.126}$$

Because $\{v^k\}$ is strongly convergent to $\bar{v}$ and $\{x^k\}$ is weakly convergent to $\bar{x}$, taking limits in (6.126) as $k \to \infty$ we obtain (6.125), and the claim is established. Because $T \subset T'$, $(T')^{-1}$ is ρ-hypomonotone in W^{-1} and T^{-1} is maximal ρ-hypomonotone in W^{-1}, we have that $T \cap W = T' \cap W$, by Definition 6.9.1(d). Because $\bar{v} \in T'(\bar{x})$ and $(\bar{x}, \bar{v}) \in W$, we conclude that $\bar{v} \in T(\bar{x})$. □

We close this section with a result on convexity and weak closedness of some sets related to the set of zeroes of operators whose inverses are ρ-hypomonotone.

Proposition 6.9.6. *Assume that $T^{-1} : H \rightrightarrows H$ is maximal ρ-hypomonotone in a subset $V \times U \subset H \times H$, where U is convex and $0 \in V$. Let $S^* \subset H$ be the set of zeroes of T. Then*

(i) $S^* \cap U$ is convex.

(ii) If $S^* \cap U$ is closed, then $(S^* \cap U) + B(0, \delta)$ is weakly closed for all $\delta \geq 0$.

Proof.
    (i) By Proposition 6.9.2, $T \cap (U \times V) \subset \hat{T}$ for some $\hat{T} : H \rightrightarrows H$ such that $\hat{T}^{-1}$ is maximal ρ-hypomonotone . Let $\hat{S}^*$ be the set of zeroes of $\hat{T}$. By Proposition 6.9.4, $\hat{S}^*$ is also the set of zeroes of $Y_\rho(\hat{T})$, which is maximal monotone by Proposition 6.9.3(ii). The set of zeroes of a maximal monotone operator is convex by Corollary 4.2.6, thus we conclude that $\hat{S}^*$ is convex, and therefore, $\hat{S}^* \cap U$ is convex, because U is convex. Because T^{-1} is maximal ρ-hypomonotone in $V \times U$, $\hat{T}^{-1}$ is ρ-hypomonotone and $T \subset \hat{T}$, we have that $\hat{T} \cap (U \times V) = T \cap (U \times V)$, and then, because $0 \in V$, it follows easily that $\hat{S}^* \cap U = S^* \cap U$. The result follows.
    (ii) Because H is a Hilbert space, $B(0, \delta)$ is weakly compact by Theorem 2.5.16(ii), and $S^* \cap U$, being closed by assumption and convex by item (i), is weakly closed. Thus $(S^* \cap U) + B(0, \delta)$ is weakly closed, being the sum of a closed and a compact set, both with respect to the weak topology. □

6.10   Convergence analysis of IPPH1 and IPPH2

We start with the issue of existence of iterates, which is rather delicate, forcing us to go through some technicalities, where the notion of ρ-hypomonotonicity becomes crucial. These technicalities are encapsulated in the following lemma.

Lemma 6.10.1. *Let $T : H \rightrightarrows H$ be an operator such that T^{-1} is maximal ρ-hypomonotone in a subset $V \times U$ of $H \times H$. Assume that T has a nonempty set of zeroes S^*, that U is convex and that*

(i) $S^* \cap U$ is nonempty and closed.

(ii) There exists $\beta > 0$ such that $B(0, \beta) \subset V$.

(iii) There exists $\delta > 0$ such that $(S^* \cap U) + B(0, \delta) \subset U$.

Take any $\gamma > 2\rho$ and define $\varepsilon = \min\{\delta, \beta\gamma/2\}$. If $x \in H$ is such that $d(x, S^* \cap U) \le \varepsilon$, then there exists $y \in H$ such that $\gamma^{-1}(x - y) \in T(y)$ and $d(y, S^* \cap U) \le \varepsilon$.

Proof. By Definition 6.9.1(c) and (d), $T^{-1} \cap (V \times U)$ is ρ-hypomonotone. By Proposition 6.9.2, there exists a maximal ρ-hypomonotone $\hat{T}^{-1} \subset H \times H$ such that

$$[T^{-1} \cap (V \times U)] \subset \hat{T}^{-1}. \tag{6.127}$$

By Proposition 6.9.3(ii), $Y_\rho(\hat{T})$ is maximal monotone, with Y_ρ as defined in (6.123). Let $\hat{\gamma} = \gamma - \rho$. Because $\hat{\gamma} > 0$ by assumption, it follows from Theorem 4.4.12 that the operator $I + \hat{\gamma}Y_\rho(\hat{T})$ is onto; that is, there exists $z \in H$ such that $x \in \left[I + \hat{\gamma}Y_\rho(\hat{T})\right](z)$, or equivalently

$$\hat{\gamma}^{-1}(x - z) \in \left[Y_\rho(\hat{T})\right](z). \tag{6.128}$$

Letting

$$v := \hat{\gamma}^{-1}(x - z), \tag{6.129}$$

we can rewrite (6.128) as $(z, v) \in Y_\rho(\hat{T})$, which is equivalent, in view of (6.123), to

$$(z - \rho v, v) \in \hat{T}. \tag{6.130}$$

Let now $y = z - \rho v$. In view of (6.129) and the definition of $\hat{\gamma}$, (6.130) is in turn equivalent to

$$(y, \hat{\gamma}^{-1}(x - z)) \in \hat{T}. \tag{6.131}$$

It follows easily from (6.129) and the definitions of y and $\hat{\gamma}$ that

$$\hat{\gamma}^{-1}(x - z) = (\gamma - \rho)^{-1}(x - z) = \gamma^{-1}(x - y) = \rho^{-1}(z - y). \tag{6.132}$$

We conclude from (6.131) and (6.132) that

$$\gamma^{-1}(x - y) \in \hat{T}(y). \tag{6.133}$$

Note that (6.133) looks pretty much like the statement of the lemma, except that we have $\hat{T}$ instead of T. The operators T and $\hat{T}$ do coincide on $U \times V$, as we show, but in order to use this fact we must first establish that $(y, \gamma^{-1}(x - y))$ indeed belongs to $U \times V$, which will result from the assumption on $d(x, S^* \cap U)$.

Take any $\bar{x} \in S^* \cap U$, nonempty by condition (i), and z as in (6.128). Note that $\bar{x}$ is a zero of $T \cap (U \times V)$, because it belongs to $S^* \cap U$ and $0 \in V$ by condition (ii). Thus $\bar{x}$ is a zero of $\hat{T}$, which contains $T \cap (U \times V)$. By Proposition 6.9.4, $\bar{x}$ is a zero of $Y_\rho(\hat{T})$. Then

$$\|x - \bar{x}\|^2 = \|x - z\|^2 + \|z - \bar{x}\|^2 + 2\langle x - z, z - \bar{x}\rangle =$$

$$\|x - z\|^2 + \|z - \bar{x}\|^2 + 2\hat{\gamma}\langle\hat{\gamma}^{-1}(x - z) - 0, z - \bar{x}\rangle \geq \|x - z\|^2 + \|z - \bar{x}\|^2, \quad (6.134)$$

using (6.128), monotonicity of $Y_\rho(\hat{T})$, nonnegativity of $\hat{\gamma}$, and the fact that $\bar{x}$ is a zero of $Y_\rho(\hat{T})$ in the inequality. Take now y as defined after (6.130). Then

$$\|y - \bar{x}\|^2 = \|y - z\|^2 + \|z - \bar{x}\|^2 - 2\langle y - z, \bar{x} - z\rangle$$

$$= \left(\frac{\rho}{\gamma - \rho}\right)^2 \|z - x\|^2 + \|z - \bar{x}\|^2 - \frac{2\rho}{\gamma - \rho}\langle z - x, \bar{x} - z\rangle$$

$$\leq \|x - \bar{x}\|^2 - \left[1 - \left(\frac{\rho}{\gamma - \rho}\right)^2\right]\|z - x\|^2 - 2\rho\langle 0 - (\gamma - \rho)^{-1}(x - z), \bar{x} - z\rangle$$

$$\leq \|x - \bar{x}\|^2 - \left[1 - \left(\frac{\rho}{\gamma - \rho}\right)^2\right]\|z - x\|^2$$

$$= \|x - \bar{x}\|^2 - \frac{\gamma(\gamma - 2\rho)}{(\gamma - \rho)^2}\|z - x\|^2 \leq \|x - \bar{x}\|^2, \quad (6.135)$$

using (6.132) in the first equality, (6.134) in the first inequality, (6.128) and monotonicity of $Y_\rho(\hat{T})$ in the second inequality, and the assumption that $\gamma > 2\rho$ in the third inequality. It follows from (6.135) that $\|y - \bar{x}\| \leq \|x - \bar{x}\|$ for all $\bar{x} \in S^* \cap U$, in particular when $\bar{x}$ is the orthogonal projection of x onto $S^* \cap U$, which exists because $S^* \cap U$ is closed by condition (i) and convex by Proposition 6.9.6(i). For this choice of $\bar{x}$ we have that

$$\|y - \bar{x}\| \leq \|x - \bar{x}\| = d(x, S^* \cap U) \leq \varepsilon = \min\{\delta, \beta\gamma/2\} \leq \delta, \quad (6.136)$$

where the second inequality holds by the assumption on x. Because $\bar{x}$ belongs to $S^* \cap U$, we get from (6.136) that

$$y \in (S^* \cap U) + B(0, \delta) \subset U, \quad (6.137)$$

using condition (iii) in the inclusion.

Observe now that, with the same choice of $\bar{x}$,

$$\gamma^{-1}\|x - y\| \leq \gamma^{-1}(\|x - \bar{x}\| + \|y - \bar{x}\|) \leq 2\varepsilon\gamma^{-1} \leq \beta, \quad (6.138)$$

using (6.136) and the assumption on x in the second inequality, and the fact that $\varepsilon = \min\{\delta, \beta\gamma/2\}$ in the third one. It follows from (6.137), (6.138), and condition (ii) that

$$(y, \gamma^{-1}(x - y)) \in U \times V. \quad (6.139)$$

Because T is maximal ρ-hypomonotone in $U \times V$ and $\hat{T}$ is ρ-hypomonotone, it follows from (6.127) and Definition 6.9.1(d) that $T \cap (U \times V) = \hat{T} \cap (U \times V)$. In view of (6.133), we conclude from (6.139) that $\gamma^{-1}(x - y) \in T(y)$. Finally, using (6.136), $d(y, S^* \cap U) \leq \|y - \bar{x}\| \leq \varepsilon$, completing the proof. $\quad\square$

Corollary 6.10.2. *Consider either Algorithm IPPH1 or Algorithm IPPH2 applied to an operator $T : H \rightrightarrows H$ such that T^{-1} is ρ-hypomonotone on a subset $U \times V$*

*of $H \times H$ satisfying conditions (i)–(iii) of Lemma 6.10.1. If $d(x^k, S^* \cap U) \leq \varepsilon$,
with ε as in the statement of Lemma 6.10.1, and $\gamma_k > 2\rho$, then there exists a pair
$(y^k, v^k) \in U \times V$ satisfying (6.117) and (6.118), and consequently a vector x^{k+1}
satisfying (6.122).*

Proof. Apply Lemma 6.10.1 with $x = x^k$, $\gamma = \gamma_k$. Take y^k as the vector y whose
existence is ensured by the lemma and $v^k = \gamma_k^{-1}(x^k - y^k)$. Then y^k and v^k satisfy
(6.117) and (6.118) with $e^k = 0$, so that (6.119) or (6.120)–(6.121) hold for any
$\sigma \geq 0$. Once a pair (y^k, v^k) exists, the conclusion about x^{k+1} is obvious, inasmuch
as (6.122) raises no existence issues.   ☐

In order to ensure existence of the iterates, we still have to prove, in view of
Corollary 6.10.2, that the whole sequence $\{x^k\}$ is contained in $B(\bar{x}, \varepsilon)$, where $\bar{x}$ is
the orthogonal projection of x^0 onto $S^* \cap U$ and ε is as in Lemma 6.10.1. This
will be a consequence of the Fejér monotonicity properties of $\{x^k\}$ (see Definition
5.5.1), which we establish next. We have not yet proved yet existence of $\{x^k\}$, but
the lemma is phrased so as to circumvent the existential issue, for the time being.

Lemma 6.10.3. *Let $\{x^k\} \subset H$ be a sequence generated by either Algorithm IPPH1
or Algorithm IPPH2 applied to an operator $T : H \rightrightarrows H$ such that T^{-1} is ρ-
hypomonotone in a subset W^{-1} of $H \times H$, and take x^* in the set S^* of zeroes
of T. If $2\rho < \hat{\gamma} = \inf\{\gamma_k\}$ and both $(x^*, 0)$ and (y^k, v^k) belong to W, then*

(i)

$$\|x^{k+1} - x^*\|^2 \leq \|x^k - x^*\|^2 - (1 - \sigma)\gamma_k(\hat{\gamma} - 2\rho)\|v^k\|^2$$

for Algorithm IPPH1, and

(ii)

$$\|x^* - x^{k+1}\|^2 \leq \|x^* - x^k\|^2 - (1 - \sigma)\left(1 - \frac{2\rho}{\hat{\gamma}}\right)\|y^k - x^k\|^2,$$

for Algorithm IPPH2.

Proof. We start with the following elementary algebraic equality,

$$\|x^* - x^k\|^2 - \|x^* - x^{k+1}\|^2 - \|y^k - x^k\|^2 + \|y^k - x^{k+1}\|^2$$

$$= 2\langle x^* - y^k, x^{k+1} - x^k\rangle. \tag{6.140}$$

Using first (6.122) in the right-hand side of (6.140), and then ρ-hypomonotonicity of
T^{-1} in W^{-1}, together with the fact that both $(x^*, 0)$ and (y^k, v^k) belong to $T \cap W$,
by (6.117) and the assumptions of the lemma, we get

$$\|x^* - x^k\|^2 - \|x^* - x^{k+1}\|^2 - \|y^k - x^k\|^2 + \|y^k - x^{k+1}\|^2$$

$$= 2\gamma_k\langle x^* - y^k, 0 - v^k\rangle \geq -2\rho\gamma_k\|v^k\|^2. \tag{6.141}$$

From this point the computation differs according to the error criterion. We start with the one given by (6.119). It follows from (6.118) and (6.122) that $y^k - x^k = e^k - \gamma_k v^k$ and $y^k - x^{k+1} = e^k$. Substituting these two equalities in (6.141) we get

$$\|x^* - x^k\|^2 - \|x^* - x^{k+1}\|^2 \geq \gamma_k^2 \|v^k\|^2 - 2\gamma_k \langle v^k, e^k \rangle - 2\rho\gamma_k \|v^k\|^2$$

$$\geq \gamma_k^2 \|v^k\|^2 - 2\gamma_k \|v^k\| \|e^k\| - 2\rho\gamma_k \|v^k\|^2$$

$$= \gamma_k \|v^k\| \left[(\gamma_k - 2\rho)\|v^k\| - 2\|e^k\| \right]$$

$$\geq \gamma_k \|v^k\| \left[(\gamma_k - 2\rho)\|v^k\| - \sigma(\hat{\gamma} - 2\rho)\|v^k\| \right]$$

$$\geq \gamma_k \|v^k\| \left[(1-\sigma)(\hat{\gamma} - 2\rho)\|v^k\| \right] = (1-\sigma)\gamma_k(\hat{\gamma} - 2\rho)\|v^k\|^2, \tag{6.142}$$

using (6.119) in the third inequality and the definition of $\hat{\gamma}$ in the last inequality. The result follows immediately from (6.142).

Now we look at the error criterion given by (6.120)–(6.121). Using again (6.118) and (6.122), we can replace $y^k - x^{k+1}$ by e^k and $-v^k$ by $\gamma_k^{-1}(y^k - x^k - e^k)$ in (6.141), obtaining

$$\|x^* - x^k\|^2 - \|x^* - x^{k+1}\|^2$$

$$\geq \|y^k - x^k\|^2 - \left(\|e^k\|^2 + 2\rho\gamma_k^{-1}\|y^k - x^k - e^k\|^2 \right)$$

$$\geq \|y^k - x^k\|^2 - \left[\|e^k\|^2 + 2\rho\gamma_k^{-1} \left(\|y^k - x^k\| + \|e^k\| \right)^2 \right]. \tag{6.143}$$

Using now (6.120) in (6.143) we get

$$\|x^* - x^k\|^2 - \|x^* - x^{k+1}\|^2 \geq \left[1 - \nu^2 - \frac{2\rho}{\gamma_k}(1+\nu)^2 \right] \|y^k - x^k\|^2$$

$$\geq \left[1 - \nu^2 - \frac{2\rho}{\hat{\gamma}}(1+\nu)^2 \right] \|y^k - x^k\|^2. \tag{6.144}$$

It follows from (6.121), after some elementary algebra, that

$$\left[1 - \nu^2 - \frac{2\rho}{\hat{\gamma}}(1+\nu)^2 \right] = (1-\sigma)\left(1 - \frac{2\rho}{\hat{\gamma}} \right). \tag{6.145}$$

Replacing (6.145) in (6.144), we obtain

$$\|x^* - x^k\|^2 - \|x^* - x^{k+1}\|^2 \geq (1-\sigma)\left(1 - \frac{2\rho}{\hat{\gamma}} \right) \|y^k - x^k\|^2, \tag{6.146}$$

and the result follows immediately from (6.146) □

Next we combine the results of Lemmas 6.10.1 and 6.10.3 in order to obtain our convergence theorem.

Theorem 6.10.4. *Let* $T : H \rightrightarrows H$ *so that* T^{-1} *is maximal* ρ*-hypomonotone in a subset* $V \times U$ *of* $H \times H$ *satisfying*

(i) $S^* \cap U$ is nonempty and closed,

(ii) There exists $\beta > 0$ such that $B(0, \beta) \subset V$,

(iii) There exists $\delta > 0$ such that $(S^* \cap U) + B(0, \delta) \subset U$,

(iv) U is convex,

where S^* is the set of zeroes of T. Take a sequence $\{\gamma_k\}$ of positive real numbers such that $2\rho < \hat{\gamma} = \inf\{\gamma_k\}$. Define $\varepsilon = \min\{\delta, \beta\hat{\gamma}/2\}$. If $d(x^0, S^* \cap U) \leq \varepsilon$, then, both for Algorithm IPPH1 and Algorithm IPPH2,

(a) For all k there exist $y^k, v^k, e^k, x^{k+1} \in H$ satisfying (6.117)–(6.119) and (6.122), in the case of Algorithm IPPH1, and (6.117)–(6.118) and (6.120)–(6.122), in the case of Algorithm IPPH2, and such that $(y^k, v^k) \in U \times V$, $d(x^{k+1}, S^* \cap U) \leq \varepsilon$.

(b) For any sequence as in (a), we have that $\{x^k\}$ converges weakly to a point in $S^* \cap U$.

Proof.

(a) We proceed by induction. Take any $k \geq 0$. We have that

$$d(x^k, S^* \cap U) \leq \varepsilon, \qquad\qquad (6.147)$$

by inductive hypothesis, if $k \geq 1$, and by assumption, if $k = 0$. We are within the hypotheses of Corollary 6.10.2. Applying this corollary we conclude that the desired vectors exist and that $(y^k, v^k) \in U \times V$. It remains to establish that $d(x^{k+1}, S^* \cap U) \leq \varepsilon$. Let $\bar{x}$ be the orthogonal projection of x^k onto $S^* \cap U$, which exists by conditions (i) and (iv) and Proposition 6.9.6. Note that both $(\bar{x}, 0)$ and (y^k, v^k) belong to $U \times V$. Thus we are within the hypotheses of Lemma 6.10.3, with $W = U \times V$, and both for Algorithm IPPH1 and Algorithm IPPH2 we get from either Lemma 6.10.3(i) or Lemma 6.10.3(ii) that

$$\|x^* - x^{k+1}\| \leq \|x^* - x^k\| \qquad\qquad (6.148)$$

for all $x^* \in S^* \cap U$. By (6.148) with $\bar{x}$ instead of x^*,

$$d(x^{k+1}, S^* \cap U) \leq \|\bar{x} - x^{k+1}\| \leq \|\bar{x} - x^k\| = d(x^k, S^* \cap U) \leq \varepsilon,$$

using (6.147) in the last inequality.

(b) We follow here with minor variations the standard convergence proof for the proximal point algorithm in Hilbert spaces; see, for example, [192]. In view of (6.148), for all $x^* \in S^* \cap U$ the sequence $\{\|x^k - x^*\|\}$ is nonincreasing, and certainly nonnegative, hence convergent. Also, because $\|x^k - x^*\| \leq \|x^0 - x^*\|$ for all k, we get that $\{x^k\}$ is bounded.

Now we consider separately both algorithms. In the case of Algorithm IPPH1, we get from Lemma 6.10.3(i),

$$(1 - \sigma)(\hat{\gamma} - 2\rho)\gamma_k\|v^k\|^2 \leq \|x^k - x^*\|^2 - \|x^{k+1} - x^*\|^2, \qquad\qquad (6.149)$$

Because the right-hand side of (6.149) converges to 0, we have $\lim_{k\to\infty} \gamma_k \|v^k\|$ = 0, and therefore, because $\gamma_k \geq \hat{\gamma} > 0$ for all k

$$\lim_{k\to\infty} v^k = 0, \tag{6.150}$$

which implies, in view of (6.119), that $\lim_{k\to\infty} e^k = 0$, and therefore, by (6.118),

$$\lim_{k\to\infty} (y^k - x^k) = 0. \tag{6.151}$$

In the case of Algorithm IPPH2, we get from Lemma 6.10.3(ii),

$$(1 - \sigma)\left(1 - \frac{2\rho}{\hat{\gamma}}\right) \|y^k - x^k\|^2 \leq \|x^* - x^k\|^2 - \|x^* - x^{k+1}\|^2. \tag{6.152}$$

Again, the right-hand side of (6.152) converges to 0, and thus (6.151) also holds in this case, so that, in view of (6.120), $\lim_{k\to\infty} e^k = 0$, which gives, in view of (6.151) and (6.118), $\lim_{k\to\infty} \gamma_k v^k = 0$, so that in this case we also have (6.150). We have proved that (6.150) and (6.151) hold both for Algorithm IPPH1 and Algorithm IPPH2, and we proceed from now on with an argument that holds for both algorithms.

Because $\{x^k\}$ is bounded, it has weak cluster points. Let $\tilde{x}$ be any weak cluster point of $\{x^k\}$; that is, $\tilde{x}$ is the weak limit of a subsequence $\{x^{i_k}\}$ of $\{x^k\}$. By (6.151), $\tilde{x}$ is also the weak limit of $\{y^{i_k}\}$. We claim that $(\tilde{x}, 0)$ belongs to $U \times V$. In view of condition (ii), it suffices to check that $\tilde{x} \in U$. Note that $\{x^k\} \subset (S^* \cap U) + B(0, \varepsilon)$ by item (a). Because U is convex by condition (iv) and $S^* \cap U$ is closed by condition (i), we can apply Proposition 6.9.6(ii) to conclude that $(S^* \cap U) + B(0, \varepsilon)$ is weakly closed. Thus, the weak limit $\tilde{x}$ of $\{x^{i_k}\}$ belongs to $(S^* \cap U) + B(0, \varepsilon)$, and henceforth to U, in view of condition (iii) and the fact that $\varepsilon \leq \delta$. The claim holds, and we are within the hypotheses of Proposition 6.9.5: $\{v^{i_k}\}$ is strongly convergent to 0 by (6.150), $\{x^{i_k}\}$ is weakly convergent to $\tilde{x}$, and $(0, \tilde{x})$ belongs to $V \times U$, where T^{-1} is maximal ρ-hypomonotone. Then $0 \in T(\tilde{x})$; that is, $\tilde{x} \in S^* \cap U$.

Finally we establish uniqueness of the weak cluster point of $\{x^k\}$, with the standard argument (e.g., [192]) which we include just for the sake of completeness (cf. also the proof of Theorem 6.4.6(ii) in Section 6.4). Let $\tilde{x}$, $\hat{x}$ be two weak cluster points of $\{x^k\}$, say the weak limits of $\{x^{i_k}\}$ and $\{x^{j_k}\}$, respectively. We have just proved that both $\tilde{x}$ and $\hat{x}$ belong to $S^* \times U$, and thus, by (6.148), both $\{\|\hat{x} - x^k\|\}$ and $\{\|\tilde{x} - x^k\|\}$ are nonincreasing, hence convergent, say to $\hat{\alpha} \geq 0$ and to $\tilde{\alpha} \geq 0$, respectively. Now,

$$\|\hat{x} - x^k\|^2 = \|\hat{x} - \tilde{x}\|^2 + \|\tilde{x} - x^k\|^2 + 2\langle \hat{x} - \tilde{x}, \tilde{x} - x^k \rangle. \tag{6.153}$$

Taking limits in (6.153) as $k \to \infty$ along the subsequence $\{x^{i_k}\}$, we get

$$\|\hat{x} - \tilde{x}\|^2 = \hat{\alpha}^2 - \tilde{\alpha}^2. \tag{6.154}$$

Reversing now the roles of $\tilde{x}$, $\hat{x}$ in (6.153), and taking limits along the subsequence $\{x^{j_k}\}$, we get

$$\|\hat{x} - \tilde{x}\|^2 = \tilde{\alpha}^2 - \hat{\alpha}^2. \tag{6.155}$$

It follows from (6.154) and (6.155) that $\tilde{x} = \hat{x}$, and thus the whole sequence $\{x^k\}$ has a weak limit which is a zero of T and belongs to U. $\square$

The next corollary states the global result for the case in which T^{-1} is ρ-hypomonotone in the whole $H \times H$.

Corollary 6.10.5. *Assume that $T : H \rightrightarrows H$ has a nonempty set of zeroes S^* and that T^{-1} is maximal ρ-hypomonotone. Take a sequence $\{\gamma_k\}$ of positive real numbers such that $2\rho < \hat{\gamma} = \inf\{\gamma_k\}$. Then, both for Algorithm IPPH1 and Algorithm IPPH2, given any $x^0 \in H$,*

(a) *For all k there exist $y^k, v^k, e^k, x^{k+1} \in H$ satisfying (6.117)–(6.119) and (6.122), in the case of Algorithm IPPH1, and (6.117)–(6.118) and (6.120)–(6.122), in the case of Algorithm IPPH2.*

(b) *Any sequence generated by Algorithm IPPH1 or by Algorithm IPPH2 is weakly convergent to a point in S^*.*

Proof. This is just Theorem 6.10.4 for the case of $U = V = H$. In this case all the assumptions above hold trivially. Regarding condition (i), note that S^* is closed because, by Proposition 6.9.4, it is also the set of zeroes of the maximal monotone operator $Y_\rho(T)$, which is closed (by Corollary 4.2.6). Conditions (ii) and (iii) hold for any $\beta, \delta > 0$, so that the result will hold for any $\varepsilon > 0$, in particular for $\varepsilon > d(x^0, S^*)$. $\square$

Exercises

6.1. Prove that the set S^* of primal–dual optimal pairs for problems (P) and (D), defined by (6.55)–(6.56) and (6.63)–(6.64), respectively, coincides with the set of saddle points of the Lagrangian given by (6.57); that is, the points that satisfy (6.66).

6.2. Prove that the set S^* of Exercise 1 coincides with the set of zeroes of the operator $T_{\mathcal{L}}$ defined in (6.72).

6.11   Historical notes

It is not easy to determine the exact origin of the proximal point method. Possibly, it started with works by members of Tykhonov's school dealing with regularization methods for solving ill-posed problems, and one of the earliest relevant references is [126]. It received its current name in the 1960s, through the works of Moreau, Yosida, and Martinet, among others (see [160, 142, 143]) and attained the form of (6.1) in Rockafellar's works in the 1970s ([193, 192]). A good survey on the proximal point method, including its development up to 1989, can be found in [129].

We mention next some previous references on the proximal point method in Banach spaces. A scheme similar to (6.3) was considered in [68, 69], but requesting a condition on f much stronger than H1–H4, namely strong convexity. It was proved in [4] that if there exists $f : X \to \mathbb{R}$ which is strongly convex and twice continuously differentiable at least at one point, then X is Hilbertian, so that the analysis in these references holds only for Hilbert spaces. With $f(x) = \|x\|_X^2$, the scheme given by (6.3) was analyzed in [114], where only partial convergence results are given (excluding, e.g., boundedness of $\{x^k\}$). For the optimization case (i.e., when T is the subdifferential of a convex function $\phi : X \to \mathbb{R}$, so that the zeroes of T are the minimizers of ϕ), the scheme given by (6.3) was analyzed in [2] for the case of $f(x) = \|x\|_X^2$ and in [53] for the case of a general f.

In finite-dimensional or Hilbert spaces, where the square of the norm leads always to simpler computations, the scheme given by (6.3) with a nonquadratic f has been proposed mainly with penalization purposes: if the problem includes a feasible set C (e.g., constrained convex optimization or monotone variational inequalities) then the use of an f whose domain is C and whose gradient diverges at the boundary of C, forces the sequence $\{x^k\}$ to stay in the interior of C and makes the subproblems unconstrained. Divergence of the gradient precludes validity of H3, and so several additional hypothesis on f are required. These methods, usually known as proximal point methods with Bregman distances, have been studied, for example, in [59, 60, 99, 121] for the optimization case and in [44, 75, 100] for finding zeroes of monotone operators. The notion of Bregman distance appeared for the first time in [36]. Other variants of the proximal point method in finite-dimensional spaces, also with nonquadratic regularization terms but with iteration formulas different from (6.3), can be found in [20, 21, 51, 77, 108] and [109].

The analysis of inexact variants of the method in Hilbert spaces, with error criteria demanding summability conditions on the error, started in [192]. A similar error criterion for the case of optimization problems in Banach spaces, using a quadratic regularization function, was proposed in [114]. Related error criteria, including summability conditions and using nonquadratic regularization functions, appear in [75, 121] for optimization, and in [46] for variational inequalities.

Error schemes for the proximal point method, accepting constant relative errors, have been developed in [209, 210] with a quadratic regularization function in Hilbert spaces and in [211, 51] with a nonquadratic regularization function in finite-dimensional spaces.

Algorithm IPPM, without allowing for errors in the inclusion (6.6) (i.e., with $\varepsilon_k = 0$) has been studied in [102], which is the basis for Sections 6.2–6.7 in this chapter. A related algorithm, allowing for errors in the inclusion, has been studied in [202]. A proximal point method for cone constrained optimization in a Hilbert space H was introduced in [228]. Convergence results in this reference require either that $\lim_{k\to\infty} \lambda_k = 0$ or at least that λ_k be small enough for large k.

The augmented Lagrangian method for equality constrained optimization problems (nonconvex, in general) was introduced in [92, 173]. Its extension to inequality constrained problems started with [56] and was continued in [27, 124, 191]. Its connections with the proximal point method are treated in [193, 21, 28, 75] and [108]. A method for the finite-dimensional case, with $f(x) = \|x\|^2$ and inexact solution of the subproblems in the spirit of [209, 211], appears in [97].

There are few references for augmented Lagrangian methods in infinite-dimensional spaces. Algorithm IDAL was presented in [54]. The method presented in Section 6.8 was analyzed in [103].

In connection with the proximal point method applied to nonmonotone operators, they have been implicitly treated in several references dealing with augmented Lagrangian methods for minimization of nonconvex functions (e.g., [28, 76] and [213]). A survey of results on the convergence of the proximal point algorithm without monotonicity up to 1997 can be found in [113]. The notion of hypomonotone operator was introduced in [194]. The exact version of Algorithms IPPH1 and IPPH2 was presented in [169], and the inexact version in [105]. These two references also introduce an exact and an inexact version, respectively, of a multiplier method for rather general variational problems with hypomonotone operators, which can be seen as a generalization of the augmented Lagrangian method.

Bibliography

[1] Alber, Ya.I. Recurrence relations and variational inequalities. *Soviet Mathematics, Doklady* **27** (1983) 511–517.

[2] Alber, Ya.I, Burachik, R.S., Iusem, A.N. A proximal point method for nonsmooth convex optimization problems in Banach spaces. *Abstract and Applied Analysis* **2** (1997) 97–120.

[3] Alber, Ya.I., Iusem, A.N., Solodov, M.V. On the projected subgradient method for nonsmooth convex optimization in a Hilbert space. *Mathematical Programming* **81** (1998) 23–37.

[4] Araujo, A. The non-existence of smooth demands in general Banach spaces. *Journal of Mathematical Economics* **17** (1988) 309–319.

[5] Armijo, L. Minimization of functions having continuous partial derivatives. *Pacific Journal of Mathematics* **16** (1966) 1–3.

[6] Arrow, K.J., Debreu, G. Existence of an equilibrium for a competitive economy. *Econometrica* **22** (1954) 597–607.

[7] Arrow, K.J., Hahn, F.H. *General Competitive Analysis.* Holden-Day, Edinburgh (1971).

[8] Asplund, E. Averaged norms. *Israel Journal of Mathematics* **5** (1967) 227–233.

[9] Asplund, E. Fréchet differentiability of convex functions. *Acta Mathematica* **121** (1968) 31–47.

[10] Attouch, H., Baillon, J.-B., Théra, M. Variational sum of monotone operators. *Journal of Convex Analysis* **1** (1994) 1–29.

[11] Attouch, H., Riahi, H., Théra, M. Somme ponctuelle d'operateurs maximaux monotones. *Serdica Mathematical Journal* **22** (1996) 267–292.

[12] Attouch, H.,Théra, M. A general duality principle for the sum of two operators. *Journal of Convex Analysis* **3** (1996) 1–24.

[13] Aubin, J.-P. *Mathematical Methods of Game and Economic Theory. Studies in Mathematics and Its Applications* **7**. North Holland, Amsterdam (1979).

[14] Aubin, J.-P. Contingent derivatives of set valued maps and existence of solutions to nonlinear inclusions and differential inclusions. In *Advances in Mathematics: Supplementary Studies 7A* (L. Nachbin, editor). Academic Press, New York (1981) 160–232.

[15] Aubin, J.-P. *Viability Theory*. Birkhäuser, Boston (1991).

[16] Aubin, J.-P. *Optima and Equilibria, an Introduction to Nonlinear Analysis*. Springer, Berlin (1993).

[17] Aubin, J.-P., Ekeland, I. *Applied Nonlinear Analysis*. John Wiley, New York (1984).

[18] Aubin, J.-P., Frankowska, H. *Set-Valued Analysis*. Birkhäuser, Boston (1990).

[19] Auchmuty, G. Duality algorithms for nonconvex variational principles. *Numerical Functional Analysis and Optimization* **10** (1989) 211–264.

[20] Auslender, A., Teboulle, M., Ben-Tiba, S., A logarithmic-quadratic proximal method for variational inequalities. *Computational Optimization and Applications* **12** (1999) 31–40.

[21] Auslender, A., Teboulle, M., Ben-Tiba, S. Interior proximal and multiplier methods based on second order homogeneous kernels. *Mathematics of Operations Research* **24** (1999) 645–668.

[22] Bachman, G., Narici, L. *Functional Analysis*. Academic Press, New York (1966).

[23] Beckenbach, E.F. Convex functions. *Bulletin of the American Mathematical Society* **54** (1948) 439–460.

[24] Beer, G. *Topology of Closed Convex Sets*. Kluwer, Dordrecht (1993).

[25] Berestycki, H., Brezis, H. Sur certains problèmes de frontière libre. *Comptes Rendues de l'Académie des Sciences de Paris* 283 (1976) 1091–1094.

[26] Berge, C. *Espaces Topologiques et Fonctions Multivoques*. Dunod, Paris (1959).

[27] Bertsekas, D.P. On penalty and multiplier methods for constrained optimization problems. *SIAM Journal on Control and Optimization* **14** (1976) 216–235.

[28] Bertsekas, D.P. *Constrained Optimization and Lagrange Multipliers*. Academic Press, New York (1982).

[29] Bertsekas, D.P. *Nonlinear Programming*. Athena, Belmont (1995).

[30] Bishop, E., Phelps, R.R. The support functionals of a convex set. *Proceedings of Symposia in Pure Mathematics, American Mathematical Society* **7** (1963) 27–35.

[31] Bonnesen, T., Fenchel, W. *Theorie der Convexen Körper.* Springer, Berlin (1934).

[32] Borwein, J.M., Fitzpatrick, S. Local boundedness of monotone operators under minimal hypotheses. *Bulletin of the Australian Mathematical Society* **39** (1989) 439–441.

[33] Borwein, J.M., Fitzpatrick, S., Girgensohn, R. Subdifferentials whose graphs are not norm-weak* closed. *Canadian Mathematical Bulletin* **46** (2003) 538–545.

[34] Bouligand, G. Sur les surfaces dépourvues de points hyperlimites. *Annales de la Société Polonaise de Mathématique* **9** (1930) 32–41.

[35] Bouligand, G. Sur la semi-continuité d'inclusions et quelques sujets connexes. *Enseignement Mathématique* **31** (1932) 14–22.

[36] Bregman, L.M. The relaxation method of finding the common points of convex sets and its application to the solution of problems in convex programming. *USSR Computational Mathematics and Mathematical Physics* **7** (1967) 200–217.

[37] Brezis, H. *Opérateurs maximaux monotones et semi-groups de contractions dans les espaces de Hilbert.* North Holland, Amsterdam (1973).

[38] Brezis, H. *Analyse Fonctionelle, Théorie et Applications.* Masson, Paris (1983).

[39] Brezis, H. Periodic solutions of nonlinear vibrating strings and duality principles. *Bulletin of the American Mathematical Society* **4** (1983) 411–425.

[40] Brøndsted, A. Conjugate convex functions in topological vector spaces. *Matematiskfysiske Meddelelser udgivet af det Kongelige Danske Videnskabernes Selskab* **34** (1964) 1–26.

[41] Brøndsted, A., Rockafellar, R.T. On the subdifferentiability of convex functions. *Proceedings of the American Mathematical Society* **16** (1965) 605–611.

[42] Brouwer, L.E.J. Über Abbildung von Mannigfaltigkeiten. *Mathematische Annalen* **71** (1911) 97–115.

[43] Browder, F.E. Nonlinear operators and nonlinear equations of evolution in Banach spaces. *Proceedings of Symposia in Pure Mathematics, American Mathematical Society* **18** (1976).

[44] Burachik, R.S., Iusem, A.N. A generalized proximal point algorithm for the variational inequality problem in a Hilbert space. *SIAM Journal on Optimization* **8** (1998) 197–216.

[45] Burachik, R.S., Iusem, A.N. On enlargeable and non-enlargeable maximal monotone operators. *Journal of Convex Analysis* **13** (2006) 603–622.

[46] Burachik, R.S., Iusem, A.N., Svaiter, B.F. Enlargements of maximal monotone operators with application to variational inequalities. *Set Valued Analysis* **5** (1997) 159–180.

[47] Burachik, R.S., Sagastizábal, C.A., Svaiter, B.F. ε-enlargement of maximal monotone operators with application to variational inequalities. In *Reformulation - Nonsmooth, Piecewise Smooth, Semismooth and Smoothing Methods* (M. Fukushima, and L. Qi, editors). Kluwer, Dordrecht (1997) 25–43.

[48] Burachik, R.S., Sagastizábal, C.A., Svaiter, B.F. Bundle methods for maximal monotone operators. In *Ill–posed Variational Problems and Regularization Techniques* (R. Tichatschke and M. Théra, editors). Springer, Berlin (1999) 49–64.

[49] Burachik, R.S., Scheimberg, S. A proximal point algorithm for the variational inequality problem in Banach spaces. *SIAM Journal on Control and Optimization* **39** (2001) 1633–1649.

[50] Burachik, R.S., Svaiter, B.F. ε-enlargements of maximal monotone operators in Banach spaces. *Set Valued Analysis* **7** (1999) 117–132.

[51] Burachik, R.S., Svaiter, B.F. A relative error tolerance for a family of generalized proximal point methods. *Mathematics of Operations Research* **26** (2001) 816–831.

[52] Burachik, R.S., Svaiter, B.F. Maximal monotone operators, convex functions and a special family of enlargements. *Set Valued Analysis* **10** (2002) 297–316.

[53] Butnariu, D., Iusem, A.N. On a proximal point method for convex optimization in Banach spaces. *Numerical Functional Analysis and Optimization* **18** (1997) 723–744.

[54] Butnariu, D., Iusem, A.N. *Totally Convex Functions for Fixed Points Computation and Infinite Dimensional Optimization*. Kluwer, Dordrecht (2000).

[55] Butnariu, D., Iusem, A.N., Resmerita, E. Total convexity of the powers of the norm in uniformly convex Banach spaces. *Journal of Convex Analysis* **7** (2000) 319–334.

[56] Buys, J.D. *Dual Algorithms for Constrained Optimization Problems*. PhD. Thesis. University of Leiden (1972).

[57] Cambini, R. Some new classes of generalized concave vector valued functions. *Optimization* **36** (1996) 11–24.

[58] Caristi, J. Fixed point theorems for mappings satisfying inwardness conditions. *Transactions of the American Mathematical Society* **215** (1976) 241–251.

[59] Censor, Y., Zenios, S.A. The proximal minimization algorithm with D-functions. *Journal of Optimization Theory and Applications* **73** (1992) 451–464.

[60] Chen, G., Teboulle, M. Convergence analysis of a proximal-like optimization algorithm using Bregman functions. *SIAM Journal on Optimization* **3** (1993) 538–543.

[61] Chen, G., Teboulle, M. A proximal based decomposition method for convex optimization problems. *Mathematical Programming* **64** (1994) 81–101.

[62] Choquet, G. Convergences. *Annales de l'Université de Grenoble* **23** (1947) 55–112.

[63] Christenson, C.O., Voxman, W.L. *Aspects of Topology*. Marcel Dekker, New York (1977).

[64] Chu, L.J. On the sum of monotone operators. *Michigan Mathematical Journal* **43** (1996) 273-289.

[65] Ciorănescu, I. *Geometry of Banach Spaces, Duality Mappings and Nonlinear Problems*. Kluwer, Dordrecht (1990).

[66] Clarke, F.H. *Optimization and Nonsmooth Analysis*. SIAM, New York (1990).

[67] Clarke, F.H., Ekeland, I. Hamiltonian trajectories with prescribed minimal period. *Communications in Pure and Applied Mathematics* **33** (1980) 103–116.

[68] Cohen, G. Auxiliary problem principle and decomposition in optimization problems. *Journal of Optimization Theory and Applications* **32** (1980) 277–305.

[69] Cohen, G. Auxiliary problem principle extended to variational inequalities. *Journal of Optimization Theory and Applications* **59** (1988) 325–333.

[70] Cotlar, M., Cignoli, R. *Nociones de Espacios Normados*. Editorial Universitaria de Buenos Aires, Buenos Aires (1967).

[71] Debreu, G. A social equilibrium existence theorem. *Proceedings of the National Academy of Sciences USA* **38** (1952) 886–893.

[72] Debrunner, H., Flor, P. Ein Erweiterungssatz für monotone Mengen. *Archiv der Mathematik* **15** (1964) 445–447.

[73] Dugundji, J. *Topology*. Allyn and Bacon, Boston, Mass. (1966).

[74] Dunford, N., Schwartz, J.T. *Linear Operators*. Interscience, New York (1958).

[75] Eckstein, J. Nonlinear proximal point algorithms using Bregman functions, with applications to convex programming. *Mathematics of Operations Research* **18** (1993) 202–226.

[76] Eckstein, J., Ferris, M. Smooth methods of multipliers for complementarity problems. *Mathematical Programming* **86** (1999) 65–90.

[77] Eckstein, J., Humes Jr, C., Silva, P.J.S. Rescaling and stepsize selection in proximal methods using separable generalized distances. *SIAM Journal on Optimization* **12** (2001) 238–261.

[78] Ekeland, I. Remarques sur les problèmes variationels I. *Comptes Rendues de l'Académie des Sciences de Paris* **275** (1972) 1057–1059.

[79] Ekeland, I. Remarques sur les problèmes variationels II. *Comptes Rendues de l'Académie des Sciences de Paris* **276** (1973) 1347–1348.

[80] Ekeland, I. On the variational principle. *Journal of Mathematical Analysis and Applications* **47** (1974) 324–353.

[81] Ekeland, I., Teman, R. *Convex Analysis and Variational Problems*. North Holland, Amsterdam (1976).

[82] Fan, K. A generalization of Tychonoff's fixed point theorem. *Mathematische Annalen* **142** (1961) 305–310.

[83] Fan, K. A minimax inequality and applications. In *Inequality III* (O. Shisha, editor). Academic Press, New York (1972) 103–113.

[84] Fenchel, W. On conjugate convex functions. *Canadian Journal of Mathematics* **1** (1949) 73–77.

[85] Fenchel, W. *Convex Cones, Sets and Functions*. Mimeographed lecture notes, Princeton University Press, NJ (1951).

[86] Gárciga Otero, R., Iusem, A.N., Svaiter, B.F. On the need for hybrid steps in hybrid proximal point methods. *Operations Research Letters* **29** (2001) 217–220.

[87] Giles, J.R. *Convex Analysis with Application to the Differentiation of Convex Functions. Pitman Research Notes in Mathematics* **58**. Pitman, Boston (1982).

[88] Golumb, M. Zur Theorie der nichtlinearen Integralgleichungen, Integralgleichungs Systeme und allgemeinen Functionalgleichungen. *Mathematische Zeitschrift* **39** (1935) 45–75.

[89] Harker, P.T., Pang, J.S. Finite dimensional variational inequalities and nonlinear complementarity problems: A survey of theory, algorithms and applications. *Mathematical Programming* **48** (1990) 161-220.

[90] Hausdorff, F. *Mengenlehre*. Walter de Gruyter, Berlin (1927).

[91] Hestenes, M.R. *Calculus of Variations and Optimal Control Theory*. John Wiley, New York (1966).

[92] Hestenes, M.R. Multiplier and gradient methods. *Journal of Optimization Theory and Applications* **4** (1969) 303–320.

[93] Hiriart-Urruty, J.-B. Lipschitz *r*-continuity of the approximate subdifferential of a convex function. *Mathematica Scandinavica* **47** (1980) 123–134.

[94] Hiriart-Urruty, J.-B., Lemaréchal, C. *Convex Analysis and Minimization Algorithms*. Springer, Berlin (1993).

[95] Hiriart-Urruty J.-B., Phelps, R.R. Subdifferential calculus using ε-subdifferentials. *Journal of Functional Analysis* **118** (1993) 154–166.

[96] Hofbauer, J., Sigmund, K. *The Theory of Evolution and Dynamical Systems*. Cambridge University Press, London (1988).

[97] Humes Jr, C., Silva, P.J.S., Svaiter, B.F. Some inexact hybrid proximal augmented Lagrangian methods. *Numerical Algorithms* **35** (2004) 175–184.

[98] Iusem, A.N. An iterative algorithm for the variational inequality problem. *Computational and Applied Mathematics* **13** (1994) 103–114.

[99] Iusem, A.N. On some properties of generalized proximal point methods for quadratic and linear programming. *Journal of Optimization Theory and Applications* **85** (1995) 593–612.

[100] Iusem, A.N. On some properties of generalized proximal point methods for variational inequalities. *Journal of Optimization Theory and Applications* **96** (1998) 337–362.

[101] Iusem, A.N. On the convergence properties of the projected gradient method for convex optimization. *Computational and Applied Mathematics* **22** (2003) 37–52.

[102] Iusem, A.N., Gárciga Otero, R. Inexact versions of proximal point and augmented Lagrangian algorithms in Banach spaces. *Numerical Functional Analysis and Optimization* **2** (2001) 609–640.

[103] Iusem, A.N., Gárciga Otero, R. Augmented Lagrangian methods for cone-constrained convex optimization in Banach spaces. *Journal of Nonlinear and Convex Analysis* **2** (2002) 155–176.

[104] Iusem, A.N., Lucambio Pérez, L.R. An extragradient-type algorithm for nonsmooth variational inequalities. *Optimization* **48** (2000) 309–332.

[105] Iusem, A.N., Pennanen, T., Svaiter, B.F. Inexact variants of the proximal point method without monotonicity. *SIAM Journal on Optimization* **13** (2003) 1080–1097.

[106] Iusem, A.N., Sosa, W. New existence results for equilibrium problems. *Nonlinear Analysis* **52** (2003) 621–635.

[107] Iusem, A.N., Svaiter, B.F. A variant of Korpelevich's method for variational inequalities with a new search strategy. *Optimization* **42** (1997) 309–321.

[108] Iusem, A.N., Svaiter, B.F., Teboulle, M. Entropy-like proximal methods in convex programming. *Mathematics of Operations Research* **19** (1994) 790–814.

[109] Iusem, A.N., Teboulle, M. Convergence rate analysis of nonquadratic proximal and augmented Lagrangian methods for convex and linear programming. *Mathematics of Operations Research* **20** (1995) 657–677.

[110] Kachurovskii, R.I. On monotone operators and convex functionals. *Uspekhi Matematicheskikh Nauk* **15** (1960) 213–214.

[111] Kachurovskii, R.I. Nonlinear monotone operators in Banach spaces. *Uspekhi Matematicheskikh Nauk* **23** (1968) 121–168.

[112] Kakutani, S. A generalization of Brouwer's fixed point theorem. *Duke Mathematical Journal* **8** (1941) 457–489.

[113] Kaplan, A., Tichatschke, R. Proximal point methods and nonconvex optimization. *Journal of Global Optimization* **13** (1998) 389–406.

[114] Kassay, G. The proximal point algorithm for reflexive Banach spaces. *Studia Mathematica* **30** (1985) 9–17.

[115] Kelley, J.L. *General Topology*. Van Nostrand, New York (1955).

[116] Kelley, J.L., Namioka, I., and co-authors, *Linear Topological Spaces*, Van Nostrand, Princeton (1963).

[117] Khobotov, E.N. Modifications of the extragradient method for solving variational inequalities and certain optimization problems. *USSR Computational Mathematics and Mathematical Physics* **27** (1987) 120–127.

[118] Kinderlehrer, D., Stampacchia, G. *An Introduction to Variational Inequalities and Their Applications*. Academic Press, New York (1980).

[119] Kiwiel, K.C. *Methods of Descent for Nondifferentiable Optimization. Lecture Notes in Mathematics* **1133**. Springer, Berlin (1985).

[120] Kiwiel, K.C. Proximity control in bundle methods for convex nondifferentiable minimization. *Mathematical Programming* **46** (1990) 105–122.

[121] Kiwiel, K.C. Proximal minimization methods with generalized Bregman functions. *SIAM Journal on Control and Optimization* **35** (1997) 1142–1168.

[122] Knaster, B., Kuratowski, K., Mazurkiewicz, S. Ein Beweiss des Fixpunkt-satzes für n-dimensional Simplexe. *Fundamenta Mathematica* **14** (1929) 132–137.

[123] Korpelevich, G. The extragradient method for finding saddle points and other problems. *Ekonomika i Matematcheskie Metody* **12** (1976) 747–756.

[124] Kort, B.W., Bertsekas, D.P. Combined primal-dual and penalty methods for convex programming. *SIAM Journal on Control and Optimization* **14** (1976) 268–294.

[125] Köthe, G. *Topological Vector Spaces*. Springer, New York (1969).

[126] Krasnoselskii, M.A. Two observations about the method of succesive approximations. *Uspekhi Matematicheskikh Nauk* **10** (1955) 123–127.

[127] Kuratowski, K. Les fonctions semi-continues dans l'espace des ensembles fermés. *Fundamenta Mathematica* **18** (1932) 148–159.

[128] Kuratowski, K. *Topologie*. Panstowowe Wyd Nauk, Warsaw (1933).

[129] Lemaire, B. The proximal algorithm. In *International Series of Numerical Mathematics* (J.P. Penot, editor). Birkhauser, Basel **87** (1989) 73–87.

[130] Lemaire, B., Salem, O.A., Revalski, J. Well-posedness by perturbations of variational problems. *Journal of Optimization Theory and Applications* **115** (2002) 345–368.

[131] Lemaréchal, C. *Extensions Diverses des Méthodes de Gradient et Applications*. Thèse d'État, Université de Paris IX (1980).

[132] Lemaréchal, C., Sagastizábal, C. Variable metric bundle methods: from conceptual to implementable forms. *Mathematical Programming* **76** (1997) 393–410.

[133] Lemaréchal, C., Strodiot, J.-J., Bihain, A. On a bundle method for nonsmooth optimization. In *Nonlinear Programming 4* (O.L. Mangasarian, R.R. Meyer, and S.M. Robinson, editors). Academic Press, New York (1981) 245–282.

[134] Lima, E.L. *Espaços Métricos*. IMPA, Rio de Janeiro (1993).

[135] Limaye, B.V. *Functional Analysis*. New Age International, New Delhi (1996).

[136] Luc, T.D. *Theory of Vector Optimization. Lecture Notes in Economics and Mathematical Systems* **319**. Springer, Berlin (1989).

[137] Luchetti, R., Torre, A. Classical set convergences and topologies. *Set Valued Analysis* **36** (1994) 219–240.

[138] Makarov, V.L., Rubinov, A.M. *Mathematical Theory of Economic Dynamics and Equilibria*. Nauka, Moskow (1973).

[139] Mandelbrojt, S. Sur les fonctions convexes. *Comptes Rendues de l'Académie des Sciences de Paris* **209** (1939) 977–978.

[140] Marcotte, P. Application of Khobotov's algorithm to variational inequalities and network equilibrium problems. *Information Systems and Operational Research* **29** (1991) 258–270.

[141] Marcotte, P., Zhu, D.L. Co-coercivity and its role in the convergence of iterative schemes for solving variational inequalities. *SIAM Journal on Optimization* **6** (1996) 714–726.

[142] Martinet, B. Régularisation d'inéquations variationelles par approximations succesives. *Revue Française de Informatique et Recherche Opérationelle* **2** (1970) 154–159.

[143] Martinet, B. *Algorithmes pour la Résolution de Problèmes d'Optimisation et Minimax*. Thèse d'État, Université de Grenoble (1972).

[144] Martínez-Legaz, J.E., Théra, M. ε-subdifferentials in terms of subdifferentials. *Set Valued Analysis* **4** (1996) 327–332.

[145] Mazur, S. Über konvexe Mengen in linearen normieren Räumen. *Studia Mathematica* **4** (1933) 70–84.

[146] Mazur, S., Orlicz, W. Sur les espaces métriques linéaires II. *Studia Mathematica* **13** (1953) 137–179.

[147] Mikusinski, J. *The Bochner Integral*. Academic Press, New York (1978).

[148] Minkowski, H. Theorie der konvexen Körper, insbesondere Begründung ihres ober Flächenbegriffs. In *Gesammelte Abhandlungen II*. Teubner, Leipzig (1911).

[149] Minty, G. Monotone networks. *Proceedings of the Royal Society* **A 257** (1960) 194–212.

[150] Minty, G. On the maximal domain of a "monotone" function. *Michigan Mathematical Journal* **8** (1961) 135–137.

[151] Minty, G. Monotone (nonlinear) operators in Hilbert space. *Duke Mathematical Journal* **29** (1962) 341–346.

[152] Minty, G. On the monotonicity of the gradient of a convex function. *Pacific Journal of Mathematics* **14** (1964) 243–247.

[153] Minty, G. A theorem on monotone sets in Hilbert spaces. *Journal of Mathematical Analysis and Applications* **11** (1967) 434–439.

[154] Mordukhovich, B.S. Metric approximations and necessary optimality conditions for general classes of extremal problems. Soviet Mathematics Doklady **22** (1980) 526–530.

[155] Mordukhovich, B.S. *Approximation Methods in Problems of Optimization and Control*. Nauka, Moscow (1988).

[156] Mordukhovich, B.S. *Variational Analysis and Generalized Differentiation*. Springer, New York (2006).

[157] Moreau, J. Fonctions convexes duales et points proximaux dans un espace Hilbertien. *Comptes Rendues de l'Académie des Sciences de Paris* **255** (1962) 2897–2899.

[158] Moreau, J. Propriétés des applications "prox". *Comptes Rendues de l'Académie des Sciences de Paris* **256** (1963) 1069–1071.

[159] Moreau, J. Inf-convolution des fonctions numériques sur un espace vectoriel. *Comptes Rendues de l'Académie des Sciences de Paris* **256** (1963) 5047–5049.

[160] Moreau, J. Proximité et dualité dans un espace Hilbertien. *Bulletin de la Societé Mathématique de France* **93** (1965) 273–299.

[161] Moreau, J. *Fonctionelles Convexes*. Collège de France, Paris (1967).

[162] Mosco, U. Dual variational inequalities. *Journal of Mathematical Analysis and Applications* **40** (1972) 202–206.

[163] Mrowka, S. Some comments on the space of subsets, In *Set-Valued Mappings, Selection and Topological Properties of 2^X* (W. Fleishman, editor) *Lecture Notes in Mathematics* **171** (1970) 59–63.

[164] Nurminskii, E.A. The continuity of ε-subgradient mappings. *Kibernetika* **5** (1977) 148–149.

[165] Painlevé, P. Observations au sujet de la communication précédente. *Comptes Rendues de l'Académie des Sciences de Paris* **148** (1909) 156–1157.

[166] Pascali, D., Sburlan, S. *Nonlinear Mappings of Monotone Type*. Editura Academiei, Bucarest (1978).

[167] Pchnitchny, B.W. *Convex Analysis and Extremal Problems*. Nauka, Moscow (1980).

[168] Pennanen, T. Dualization of generalized equations of maximal monotone type. *SIAM Journal on Optimization* **10** (2000) 803–835.

[169] Pennanen, T. Local convergence of the proximal point algorithm and multiplier methods without monotonicity. *Mathematics of Operations Research* **27** (2002) 170–191.

[170] Penot, J.-P. Subdifferential calculus without qualification conditions. *Journal of Convex Analysis* **3** (1996) 1–13.

[171] Phelps, R.R. *Convex Functions, Monotone Operators and Differentiability*. *Lecture Notes in Mathematics* **1364**. Springer, Berlin (1993).

[172] Phelps, R.R. Lectures on maximal monotone operators. *Extracta Mathematica* **12** (1997) 193–230.

[173] Powell, M.J.D. A method for nonlinear constraints in minimization problems. In *Optimization* (R. Fletcher, editor). Academic Press, London (1969).

[174] Preiss, D., Zajicek, L. Stronger estimates of smallness of sets of Fréchet non-differentiability of convex functions. *Rendiconti del Circolo Matematico di Palermo* **II-3** (1984) 219–223.

[175] Revalski, J., Théra, M. Generalized sums of monotone operators. *Comptes Rendues de la Académie de Sciences de Paris* **329** (1999) 979–984.

[176] Revalski, J., Théra, M. Enlargements and sums of monotone operators. *Nonlinear Analysis* **48** (2002) 505–519.

[177] Riesz, F., Nagy, B. *Functional Analysis*. Frederick Ungar, New York (1955).

[178] Roberts, A.W., Varberg, D.E. Another proof that convex functions are locally Lipschitz. *American Mathematical Monthly* **81** (1974) 1014–1016.

[179] Robinson, S.M. Regularity and stability for convex multivalued functions. *Mathematics of Operations Research* **1** (1976) 130–143.

[180] Robinson, S.M. Normal maps induced by linear transformations. *Mathematics of Operations Research* **17** (1992) 691–714.

[181] Robinson, S.M. A reduction method for variational inequalities. *Mathematical Programming* **80** (1998) 161–169.

[182] Robinson, S.M. Composition duality and maximal monotonicity. *Mathematical Programming* **85** (1999) 1–13.

[183] Rockafellar, R.T. *Convex Functions and Dual Extremum Problems*. PhD Dissertation, Harvard University, Cambridge, MA (1963).

[184] Rockafellar, R.T. Characterization of the subdifferentials of convex functions. *Pacific Journal of Mathematics* **17** (1966) 497–510.

[185] Rockafellar, R.T. Monotone processes of convex and concave type. *Memoirs of the American Mathematical Society* **77** (1967).

[186] Rockafellar, R.T. Local boundedness of nonlinear monotone operators. *Michigan Mathematical Journal* **16** (1969) 397–407.

[187] Rockafellar, R.T. *Convex Analysis*. Princeton University Press, Princeton, NJ (1970).

[188] Rockafellar, R.T. Monotone operators associated with saddle functions and minimax problems. *Proceedings of Symposia in Pure Mathematics, American Mathematical Society* **18**, Part 1 (1970) 241–250.

[189] Rockafellar, R.T. On the maximal monotonicity of subdifferential mappings. *Pacific Journal of Mathematics* **33** (1970) 209–216.

[190] Rockafellar, R.T. On the maximality of sums of nonlinear monotone operators. *Transactions of the American Mathematical Society* **149** (1970) 75–88.

[191] Rockafellar, R.T. The multiplier method of Hestenes and Powell applied to convex programming. *Journal of Optimization Theory and Applications* **12** (1973) 555–562.

[192] Rockafellar, R.T. Monotone operators and the proximal point algorithm. *SIAM Journal on Control and Optimization* **14** (1976) 877–898.

[193] Rockafellar, R.T. Augmented Lagrangians and applications of the proximal point algorithm in convex programming. *Mathematics of Operations Research* **1** (1976) 97–116.

[194] Rockafellar, R.T., Wets, R.J-B. *Variational Analysis.* Springer, Berlin (1998).

[195] Rubinov, A.M. *Superlinear Multivalued Mappings and Their Applications to Problems of Mathematical Economics.* Nauka, Leningrad (1980).

[196] Rubinov, A.M. Upper semicontinuously directionally differentiable functions, In *Nondifferentiable Optimization: Motivations and Applications* (V.F. Demyanov and D. Pallaschke, editors) *Lecture Notes in Mathematical Economics* **225** (1985) 74–86.

[197] Rudin, W. *Functional Analysis.* McGraw-Hill, New York (1991).

[198] Scarf, H. *The Computation of Economic Equilibria.* Cowles Foundation Monographs **24**, Yale University Press, New Haven, CT (1973).

[199] Schauder, J. Der Fixpunktsatz in Funktionalräumen. *Studia Mathematica* **2** (1930) 171–180.

[200] Schramm, H., Zowe, J. A version of the bundle idea for minimizing a nonsmooth function: conceptual idea, convergence analysis, numerical results. *SIAM Journal on Optimization* **2** (1992) 121–152.

[201] Siegel, J. A new proof of Caristi's fixed point theorem. *Proceedings of the American Mathematical Society* **66** (1977) 54–56.

[202] Silva, G.J.P., Svaiter, B.F. An inexact generalized proximal point algorithm in Banach spaces (to appear).

[203] Simons, S. Cyclical coincidences of multivalued maps. *Journal of the Mathematical Society of Japan* **38** (1986) 515–525.

[204] Simons, S. The least slope of a convex function and the maximal monotonicity of its subdifferential. *Journal of Optimization Theory and Applications* **71** (1991) 127–136.

[205] Simons, S. The range of a maximal monotone operator. *Journal of Mathematical Analysis and Applications* **199** (1996) 176–201.

[206] Simons, S. *Minimax and Monotonicity. Lecture Notes in Mathematics* **1693**. Springer, Berlin (1998).

[207] Simons, S. Maximal monotone operators of Brøndsted-Rockefellar type. *Set Valued Analysis* **7** (1999) 255–294.

[208] Solodov, M.V. A class of decomposition methods for convex optimization and monotone variational inclusions via the hybrid inexact proximal point framework. *Optimization Methods and Software* **19** (2004) 557–575.

[209] Solodov, M.V., Svaiter, B.F. A hybrid projection–proximal point algorithm. *Journal of Convex Analysis* **6** (1999) 59–70.

[210] Solodov, M.V., Svaiter, B.F. A hybrid approximate extragradient-proximal point algorithm using the enlargement of a maximal monotone operator. *Set-Valued Analysis* **7** (1999) 323–345.

[211] Solodov, M.V., Svaiter, B.F. An inexact hybrid generalized proximal point algorithm and some new results on the theory of Bregman functions. *Mathematics of Operations Research* **25** (2000) 214–230.

[212] Sonntag, Y., Zălinescu, C. Set convergence, an attempt of classification. *Transactions of the American Mathematical Society* **340** (1993) 199–226.

[213] Spingarn, J.E. Submonotone mappings and the proximal point algorithm. *Numerical Functional Analysis and Optimization* **4** (1981) 123–150.

[214] Strodiot, J.-J., Nguyen, V.H. On the numerical treatment of the inclusion $0 \in \partial f(x)$. In *Topics in Nonsmooth Mechanics* (J.J. Moreau, P.D. Panagiotopulos, and G. Strang, editors). Birkhäuser, Basel (1988) 267–294.

[215] Sussmann, H.J. New theories of set-valued differentials and new versions of the maximum principle in optimal control theory. In *Nonlinear Control in the Year 2000. Lecture Notes in Control and Information Theory* **259**. Springer, London (2001) 487–526.

[216] Svaiter, B.F. A family of enlargements of maximal monotone operators. *Set Valued Analysis* **8** (2000) 311–328.

[217] Teman, R. Remarks on a free boundary value problem arising in plasma physics. *Communications in Partial Differential Equations* **2** (1977) 563–586.

[218] Toland, J. Duality in nonconvex optimization. *Journal of Mathematical Analysis and Applications* **66** (1978) 399–415.

[219] Torralba, D. *Convergence Épigraphique et Changements d'Échelle en Analyse Variationnelle et Optimisation*. Thèse de Doctorat, Université de Montpelier II (1996).

[220] Tseng, P. Alternating projection-proximal methods for convex programming and variational inequalities. *SIAM Journal on Optimization* **7** (1997) 951–965.

[221] Tykhonov, A.N. On the stability of the functional optimization problem. *USSR Journal of Computational Mathematics and Mathematical Physics* **6** (1966) 631–634.

[222] Ursescu, C. Multifunctions with closed convex graph. *Czechoslovak Mathematical Journal* **25** (1975) 438–441.

[223] van Tiel, J. *Convex Analysis, an Introductory Text.* John Wiley, New York (1984).

[224] Vasilesco, F. *Essai sur les Fonctions Multiformes de Variables Réels.* Thèse, Université de Paris (1925).

[225] Veselý, L. Local uniform boundedness principle for families of ε-monotone operators. *Nonlinear Analysis* **24** (1995) 1299–1304.

[226] Vorobiev, N.N. *Foundations of Game Theory.* Birkhäuser, Basel (1994).

[227] Wang, C., Xiu, N. Convergence of the gradient projection method for generalized convex minimization. *Computational Optimization and Applications* **16** (2000) 111–120.

[228] Wierzbicki, A.P., Kurcyusz, S. Projection on a cone, penalty functionals and duality theory for problems with inequality constraints in Hilbert space. *SIAM Journal on Control and Optimization* **15** (1977) 25–56.

[229] Xu, H., Rubinov, A.M., Glover, B.M. Continuous approximation to generalized Jacobians. *Optimization* **46** (1999) 221–246.

[230] Xu, Z.-B., Roach, G.F. Characteristic inequalities of uniformly convex and uniformly smooth Banach spaces. *Journal of Mathematical Analysis and Applications* **157** (1991) 189–210.

[231] Yosida, K. *Functional Analysis.* Springer, Berlin (1968).

[232] Young, W.H. On classes of summable functions and their Fourier series. *Proceedings of the Royal Society* **87** (1912) 225–229.

[233] Zarankiewicz, C. Sur les points de division des ensembles connexes. *Fundamenta Mathematica* **9** (1927) 124–171.

[234] Zarantonello, E.H. Solving functional equations by contractive averaging. U.S. Army Mathematics Research Center, Technical Report No. 160 (1960).

[235] Zeidler, E. *Nonlinear Functional Analysis and its Applications* II/B. Springer, New York (1990).

[236] Zhong, S.S. Semicontinuités génériques de multiapplications. *Comptes Rendues de l'Académie des Sciences de Paris* **293** (1981) 27–29.

[237] Zolezzi, T. Well-posed criteria in optimization with application to the calculus of variations. *Nonlinear Analysis* **25** (1995) 437–453.

[238] Zolezzi, T. Extended well-posedness of optimization problems. *Journal of Optimization Theory and Applications* **91** (1996) 257–268.

[239] Zoretti, L. Sur les fonctions analytiques uniformes qui possèdent un ensamble discontinu parfait de points singuliers. *Journal des Mathématiques Pures et Appliquées* **1** (1905) 1–51.

Notation

$\mathbb{R}$: the real numbers
$\mathbb{R}_+$: the nonnegative real numbers
$\mathbb{R}_{++}$: the positive real numbers
X^*: the topological dual of the normed space X
$\overline{\mathbb{R}}$: $\mathbb{R} \cup \{\infty\}$
$\mathbb{N}$: the natural numbers
$\mathcal{N}_\infty$: the neighborhoods of infinity in $\mathbb{N}$
$\mathcal{N}_\sharp$: the subsequences of $\mathbb{N}$
$x_n \to_J x$: the subsequence $\{x_n\}_{n \in J}$ converges to x, with $J \in \mathcal{N}_\sharp$
N_1: first countability axiom
N_2: second countability axiom
$T3$: third separability axiom
$\liminf C_n$: the internal limit of the sequence of sets $\{C_n\}$
$\limsup C_n$: the external limit of the sequence of sets $\{C_n\}$
$d(x, C)$: the distance of x to the set C
$B(x, r)$: the closed ball of center x and radius r
$\| \cdot \|$: the norm
$\| \cdot \|_{X^*}$: the norm in X^*
$\langle \cdot, \cdot \rangle$: the duality pairing
C^c: the complement of the set C
C^o: the topological interior of the set C
$(C)^o_A$: the topological interior of the set C relative to A
$\overline{C}$: the topological closure of C
$\overline{C}_A$: the topological closure of C relative to A
$A \setminus B$: the set $A \cap B^c$
$\mathrm{ri}(A)$: the relative interior of the set A
∂C: the topological boundary of C
$\mathcal{CP}$: cluster points of a sequence of sets
F^{-1}: the inverse of the point-to-set mapping F
$F^{-1}(A)$: the inverse image of the set A
$F^{+1}(A)$: the core of the set A
$\mathrm{Gph}(F)$: the graph of the point-to-set mapping F
$D(F)$: the domain of the point-to-set mapping F
$R(F)$: the range of the point-to-set mapping F
$x' \to_F x$: $x' \in D(F)$ converging to x
$\ell_+ F(x) = \limsup_{x' \to_F x} F(x')$: the outer limit of F at x
$\ell_- F(x) = \liminf_{x' \to_F x} F(x')$: the inner limit of F at x
$\mathrm{Dom}(f)$: the effective domain of the function f
$\mathrm{Epi}(f)$: the epigraph of the function f
∂f: the subdifferential of the function f
$\partial_\varepsilon f$: the ε-subdifferential of the function f
$f_{/K}$: the restriction of the function f to the set K
δ_A: the indicator function of the set A

$S_f(\lambda)$: the level mapping of f with value λ

E_f: the epigraphic profile of f

$\mathrm{diam}(A)$: the diameter of the set A

f^*: the convex conjugate of the function f

f^{**}: the convex bicongugate of the function f (i.e., the convex conjugate of f^*)

$\overline{A}^w$: the closure of the set A with respect to the weak topology

$\overline{A}^{w^*}$ the closure of the set A with respect to the weak* topology

$N_K(x)$: the normal cone of K at the point x

$T_K(x)$: the tangent cone of K at the point x

K^-: the polar cone of the set K

K^*: the positive polar cone of the set K

K°: the antipolar cone of the set K

$DF(x,y)$: the derivative of the mapping F at $(x,y) \in \mathrm{Gph}(F)$

$DF(x,y)^*$: the coderivative of the mapping F at $(x,y) \in \mathrm{Gph}(F)$

$\precsim$: the partial order induced by a cone

σ_A: the support function of the set A

∇f: the gradient of the function f

Δf: the Laplacian of the function f

P_C: the orthogonal projection onto the set C

L^*: the adjoint of the linear mapping L

$\delta T(x)$: the Gâteaux derivative of T at the point x

w-lim: weak limit

s-lim: strong limit

$\mathrm{co}\, A$: the convex hull of A

$\overline{\mathrm{co}}A$: the closed convex hull of A

J: the normalized duality mapping

J_φ: the duality mapping of weight φ

X/L: the quotient space of X modulus L

$\mathbb{E}(T)$: the family of enlargements of T

$\mathbb{E}_c(T)$: the family of closed enlargements of T

T^e: the largest element of $\mathbb{E}(T)$

T^{se}: the smallest element of $\mathbb{E}(T)$

$\breve{\partial}f$: the Brøndsted and Rockafellar enlargement of the subdifferential of f

$\mathrm{VIP}(T,C)$: the variational inequality problem associated with T and C

$T_1 +_{ext} T_2$: the extended sum of the operators T_1 and T_2

ν_f: the modulus of total convexity of the function f

ρ_X: the modulus of smoothness of the space X

θ_X: the modulus of uniform convexity of the space X

$D_f(x,y)$: the Bregman distance related to the function f

$\mathcal{F}$: the family of lsc, differentiable, strictly convex functions

f': the Gâteaux derivative of the function f

$\mathrm{argmin}_C f$: the set of minimizers of the function f on the set C

$Y_\rho(T)$: the Yosida regularization of the operator T with parameter ρ

Index